# Z80 PROGRAMMING
# FOR LOGIC DESIGN

By Adam Osborne
Jerry Kane
Russell Rector
Susanna Jacobson

Osborne & Associates, Inc.
Berkeley, California

Library of Congress 78-104532
ISBN 0-931988-11-X

1st printing March 1978
2nd printing July 1978

Copyright © 1978 by Adam Osborne & Associates, Incorporated.

All rights reserved. Printed in the United States of America. No part of this publication may be reproduced, stored in any retrieval system, or transmitted in any form or by any means, electronic, mechanical, photocopying, recording or otherwise, without the prior written permission of the publishers.

Published by Adam Osborne & Associates, Incorporated
P.O. Box 2036, Berkeley, California 94702

## DISTRIBUTORS OF OSBORNE & ASSOCIATES, INC. PUBLICATIONS

For information on translations and on book distributors outside of the United States of America, please call or write:

**Osborne & Associates, Inc.**
P.O. Box 2036
Berkeley, California 94702
United States of America
(415) 548-2805
TWX 910-366-7277
9:00 a.m. - 5:00 p.m. Pacific Standard Time

## ACKNOWLEDGMENTS

The authors wish to thank Mr. Sten Engström for his extensive work on Chapter 6 of this book.

# TABLE OF CONTENTS

| CHAPTER | | PAGE |
|---|---|---|
| 1 | INTRODUCTION | 1-1 |
| | WHAT THIS BOOK ASSUMES YOU KNOW | 1-1 |
| | UNDERSTANDING ASSEMBLY LANGUAGE | 1-1 |
| | HOW THIS BOOK HAS BEEN PRINTED | 1-2 |
| 2 | ASSEMBLY LANGUAGE AND DIGITAL LOGIC | 2-1 |
| | THE DESIGN CYCLE | 2-1 |
| | SIMULATING DIGITAL LOGIC | 2-4 |
| | MICROCOMPUTER SIMULATION OF A SIGNAL INVERTER | 2-5 |
| |   A MICROCOMPUTER EVENT SEQUENCE | 2-5 |
| |   IMPLEMENTING THE TRANSFER FUNCTION | 2-7 |
| |   DETERMINING DATA SOURCES AND DESTINATIONS | 2-7 |
| |   EVENT TIMING | 2-13 |
| | BUFFERS, AMPLIFIERS AND SIGNAL LOADS | 2-15 |
| | MICROCOMPUTER SIMULATION OF 7404/05/06 HEX INVERTERS | 2-22 |
| | MICROCOMPUTER SIMULATION OF 7408/09 QUADRUPLE, TWO-INPUT POSITIVE AND GATES | 2-23 |
| |   TWO-INPUT FUNCTIONS | 2-24 |
| | THE MICROCOMPUTER SIMULATION OF A 7411 TRIPLE, THREE-INPUT POSITIVE AND GATE | 2-25 |
| |   THREE-INPUT FUNCTIONS | 2-26 |
| |   MINIMIZING CPU REGISTER ACCESSES | 2-28 |
| |   COMPARING MEMORY UTILIZATION AND EXECUTION SPEED | 2-32 |
| | THE MICROCOMPUTER SIMULATION OF A 7474 DUAL, D-TYPE POSITIVE EDGE TRIGGERED FLIP-FLOP WITH PRESET AND CLEAR | 2-34 |
| |   A DIGITAL LOGIC DESCRIPTION OF FLIP-FLOPS | 2-34 |
| |   AN ASSEMBLY LANGUAGE SIMULATION OF FLIP-FLOPS | 2-37 |
| |   MICROCOMPUTER SIMULATION OF FLIP-FLOPS IN GENERAL | 2-38 |
| | THE MICROCOMPUTER SIMULATION OF REAL TIME DEVICES | 2-38 |
| |   THE 555 MONOSTABLE MULTIVIBRATOR | 2-39 |
| |   THE 74121 MONOSTABLE MULTIVIBRATOR | 2-40 |
| |   THE 74107 DUAL J-K MASTER-SLAVE FLIP-FLOP WITH CLEAR | 2-42 |
| |   MICROCOMPUTER SIMULATION OF REAL TIME | 2-43 |
| |   MICROCOMPUTER TIMING INSTRUCTION LOOPS | 2-44 |
| |   THE LIMITS OF DIGITAL LOGIC SIMULATION | 2-46 |
| |   INTERFACING WITH EXTERNAL ONE-SHOTS | 2-47 |
| |   TIME OUT AND INTERRUPTS | 2-49 |
| |   INTERFACING WITH PROGRAMMABLE TIMERS | 2-49 |
| 3 | A DIRECT DIGITAL LOGIC SIMULATION | 3-1 |
| | HOW THE QUME PRINTER WORKS | 3-2 |
| | INPUT AND OUTPUT SIGNALS | 3-9 |
| |   INPUT/OUTPUT DEVICES | 3-9 |
| |   THE Z80 PARALLEL I/O INTERFACE (PIO) | 3-9 |
| |   INPUT SIGNALS | 3-15 |

# TABLE OF CONTENTS (Continued)

| CHAPTER | | PAGE |
|---|---|---|
| | $\overline{\text{RETURN STROBE}}$ | 3-15 |
| | PFL REL | 3-16 |
| | RIB LIFT RDY | 3-17 |
| | $\overline{\text{PW STROBE}}$ | 3-17 |
| | $\overline{\text{FFA}}$ | 3-17 |
| | RESET | 3-18 |
| | PFR REL | 3-18 |
| | CA REL | 3-19 |
| | FFI | 3-19 |
| | $\overline{\text{EOR DET}}$ | 3-19 |
| | HAMMER ENABLE FF | 3-21 |
| | CLK | 3-21 |
| | H1 - H6 | 3-21 |
| | INPUT SIGNAL SUMMARY | 3-22 |
| | OUTPUT SIGNALS | 3-22 |
| | A DIGITAL-LOGIC ORIENTED SIMULATION | 3-23 |
| | A LOGIC OVERVIEW | 3-23 |
| | FLIP-FLOP $\text{FFA}_W$ | 3-25 |
| | SIMULATING FLIP-FLOP $\text{FFA}_W$ | 3-27 |
| | FLIP-FLOP $\text{FFB}_W$ | 3-34 |
| | SIMULATING FLIP-FLOP FFB | 3-36 |
| | FLIP-FLOP FFC | 3-41 |
| | SIMULATING FLIP-FLOP FFC | 3-43 |
| | START RIBBON MOTION PULSE SIMULATION | 3-44 |
| | FLIP-FLOP FFD | 3-46 |
| | SIMULATING FLIP-FLOP FFD | 3-46 |
| | FLIP-FLOP FFE | 3-48 |
| | PW SETTLING ONE-SHOT | 3-50 |
| | SIMULATING THE PW SETTLING ONE-SHOT | 3-51 |
| | FLIP-FLOP FFF | 3-51 |
| | SIMULATING FLIP-FLOP FFF | 3-52 |
| | THE 555 MULTIVIBRATOR | 3-55 |
| | SIMULATING THE MULTIVIBRATOR 555 | 3-56 |
| | THE PW RELEASE ENABLE FLIP-FLOP | 3-63 |
| | SIMULATING THE PW RELEASE ENABLE FLIP-FLOP | 3-63 |
| | SIMULATING THE PW READY ENABLE ONE-SHOT | 3-65 |
| | SIMULATION SUMMARY | 3-67 |
| 4 | A SIMPLE PROGRAM | 4-1 |
| | ASSEMBLY LANGUAGE TIMING VERSUS DIGITAL LOGIC TIMING | 4-1 |
| | INPUT AND OUTPUT SIGNALS | 4-1 |
| | MICROCOMPUTER DEVICE CONFIGURATION | 4-3 |
| | GENERAL DESIGN CONCEPTS | 4-3 |
| | Z80 PARALLEL INPUT/OUTPUT INTERFACE (PIO) | 4-5 |
| | ROM AND RAM MEMORY | 4-6 |
| | SYSTEM INITIALIZATION | 4-7 |
| | PROGRAM FLOWCHART | 4-9 |
| | PROGRAM LOGIC ERRORS | 4-27 |
| | RESET AND INITIALIZATION | 4-30 |
| | A PROGRAM SUMMARY | 4-31 |

# TABLE OF CONTENTS (Continued)

| CHAPTER | | PAGE |
|---|---|---|
| 5 | A PROGRAMMER'S PERSPECTIVE | 5-1 |
| | SIMPLE PROGRAMMING EFFICIENCY | 5-1 |
| |    EFFICIENT TABLE LOOKUPS | 5-1 |
| | SUBROUTINES | 5-7 |
| |    SUBROUTINE CALL | 5-9 |
| |    SUBROUTINE RETURN | 5-13 |
| |    WHEN TO USE SUBROUTINES | 5-14 |
| |    CONDITIONAL SUBROUTINE RETURNS | 5-15 |
| |    MULTIPLE SUBROUTINE RETURNS | 5-17 |
| |    CONDITIONAL SUBROUTINE CALLS | 5-20 |
| | MACROS | 5-21 |
| |    WHAT IS A MACRO? | 5-21 |
| |    MACROS WITH PARAMETERS | 5-23 |
| | INTERRUPTS | 5-24 |
| |    INTERRUPT HARDWARE CONSIDERATIONS | 5-24 |
| |    MULTIPLE INTERRUPTS | 5-35 |
| |    JUSTIFYING INTERRUPTS | 5-37 |
| 6 | THE Z80 INSTRUCTION SET | 6-1 |
| |    ABBREVIATIONS | 6-2 |
| |    STATUS | 6-3 |
| |    INSTRUCTION MNEMONICS | 6-4 |
| |    INSTRUCTION OBJECT CODES | 6-4 |
| |    INSTRUCTION EXECUTION TIMES AND CODES | 6-4 |
| |    ADC A,data — ADD IMMEDIATE WITH CARRY TO ACCUMULATOR | 6-25 |
| |    ADC A,reg — ADD REGISTER WITH CARRY TO ACCUMULATOR | 6-26 |
| |    ADC A,(HL) — ADD MEMORY AND CARRY TO ACCUMULATOR | 6-27 |
| |    ADC A,(IX+disp) | |
| |    ADC A,(IY+disp) | |
| |    ADC HL,rp — ADD REGISTER PAIR WITH CARRY TO H AND L | 6-28 |
| |    ADD A,data — ADD IMMEDIATE INTO ACCUMULATOR | 6-29 |
| |    ADD A,reg — ADD CONTENTS OF REGISTER TO ACCUMULATOR | 6-30 |
| |    ADD A,(HL) — ADD MEMORY TO ACCUMULATOR | 6-31 |
| |    ADD A,(IX+disp) | |
| |    ADD A,(IY+disp) | |
| |    ADD HL,rp — ADD REGISTER PAIR TO H AND L | 6-32 |
| |    ADD xy,rp — ADD REGISTER PAIR TO INDEX REGISTER | 6-33 |
| |    AND data — AND IMMEDIATE WITH ACCUMULATOR | 6-34 |
| |    AND reg — AND REGISTER WITH ACCUMULATOR | 6-35 |
| |    AND (HL) — AND MEMORY WITH ACCUMULATOR | 6-36 |
| |    AND (IX+disp) | |
| |    AND (IY+disp) | |
| |    BIT b,reg — TEST BIT b IN REGISTER reg | 6-37 |

# TABLE OF CONTENTS (Continued)

**CHAPTER**                                                                                          **PAGE**

BIT b,(HL) — TEST BIT b OF INDICATED MEMORY POSITION   6-38
BIT b,(IX+disp)
BIT b,(IY+disp)
CALL label — CALL THE SUBROUTINE IDENTIFIED IN THE OPERAND   6-39
CALL condition,label — CALL THE SUBROUTINE IDENTIFIED IN THE OPERAND IF CONDITION IS SATISFIED   6-40
CCF — COMPLEMENT CARRY FLAG   6-41
CP data — COMPARE IMMEDIATE DATA WITH ACCUMULATOR CONTENTS   6-42
CP reg — COMPARE REGISTER WITH ACCUMULATOR   6-43
CP (HL) — COMPARE MEMORY WITH ACCUMULATOR   6-44
CP (IX+disp)
CP (IY+disp)
CPD — COMPARE ACCUMULATOR WITH MEMORY. DECREMENT ADDRESS AND BYTE COUNTER   6-45
CPDR — COMPARE ACCUMULATOR WITH MEMORY. DECREMENT ADDRESS AND BYTE COUNTER. CONTINUE UNTIL MATCH IS FOUND OR BYTE COUNTER IS ZERO   6-46
CPI — COMPARE ACCUMULATOR WITH MEMORY. DECREMENT BYTE COUNTER. INCREMENT ADDRESS   6-47
CPIR — COMPARE ACCUMULATOR WITH MEMORY. DECREMENT BYTE COUNTER. INCREMENT ADDRESS. CONTINUE UNTIL MATCH IS FOUND OR BYTE COUNTER IS ZERO   6-48
CPL — COMPLEMENT THE ACCUMULATOR   6-49
DAA — DECIMAL ADJUST ACCUMULATOR   6-50
DEC reg — DECREMENT REGISTER CONTENTS   6-51
DEC rp — DECREMENT CONTENTS OF SPECIFIED REGISTER PAIR   6-52
DEC IX
DEC IY
DEC HL — DECREMENT MEMORY CONTENTS   6-53
DEC (IX+disp)
DEC (IY+disp)
DI — DISABLE INTERRUPTS   6-54
DJNZ disp — JUMP RELATIVE TO PRESENT CONTENTS OF PROGRAM COUNTER IF REG B IS NOT ZERO   6-55
EX AF,AF' — EXCHANGE PROGRAM STATUS AND ALTERNATE PROGRAM STATUS   6-57
EX DE,HL — EXCHANGE DE AND HL CONTENTS   6-58
EX (SP),HL — EXCHANGE CONTENTS OF REGISTER AND TOP OF STACK   6-59
EX (SP),IX
EX (SP),IY
EXX — EXCHANGE REGISTER PAIRS AND ALTERNATE REGISTER PAIRS   6-60

# TABLE OF CONTENTS (Continued)

| CHAPTER | | PAGE |
|---|---|---|
| | HALT | 6-61 |
| | IM 0 — INTERRUPT MODE 0 | 6-62 |
| | IM 1 — INTERRUPT MODE 1 | 6-62 |
| | IM 2 — INTERRUPT MODE 2 | 6-62 |
| | IN A,(port) — INPUT TO ACCUMULATOR | 6-63 |
| | INC reg — INCREMENT REGISTER CONTENTS | 6-64 |
| | INC rp — INCREMENT CONTENTS OF SPECIFIED REGISTER PAIR | 6-65 |
| | INC IX | |
| | INC IY | |
| | INC (HL) — INCREMENT MEMORY CONTENTS | 6-66 |
| | INC (IX+disp) | |
| | INC (IY+disp) | |
| | IND — INPUT TO MEMORY AND DECREMENT POINTER | 6-67 |
| | INDR — INPUT TO MEMORY AND DECREMENT POINTER UNTIL BYTE COUNTER IS ZERO | 6-67 |
| | INI — INPUT TO MEMORY AND INCREMENT POINTER | 6-68 |
| | INIR — INPUT TO MEMORY AND INCREMENT POINTER UNTIL BYTE COUNTER IS ZERO | 6-68 |
| | IN reg,(C) — INPUT TO REGISTER | 6-69 |
| | JP label — JUMP TO THE INSTRUCTION IDENTIFIED IN THE OPERAND | 6-70 |
| | JP condition,label — JUMP TO ADDRESS IDENTIFIED IN THE OPERAND IF CONDITION IS SATISFIED | 6-71 |
| | JP (HL) — JUMP TO ADDRESS SPECIFIED BY CONTENTS OF 16-BIT REGISTER | 6-72 |
| | JP (IX) | |
| | JP (IY) | |
| | JR C,disp — JUMP RELATIVE TO CONTENTS OF PROGRAM COUNTER IF CARRY IS SET | 6-73 |
| | JR disp — JUMP RELATIVE TO PRESENT CONTENTS OF PROGRAM COUNTER | 6-74 |
| | JR NC,disp — JUMP RELATIVE TO CONTENTS OF PROGRAM COUNTER IF CARRY FLAG IS RESET | 6-75 |
| | JR NZ,disp — JUMP RELATIVE TO CONTENTS OF PROGRAM COUNTER IF ZERO FLAG IS RESET | 6-75 |
| | JR Z,disp — JUMP RELATIVE TO CONTENTS OF PROGRAM COUNTER IF ZERO FLAG IS SET | 6-76 |
| | LD A,IV — MOVE CONTENTS OF INTERRUPT VECTOR OR REFRESH REGISTER TO ACCUMULATOR | 6-76 |
| | LD A,R | |
| | LD A,(addr) — LOAD ACCUMULATOR FROM MEMORY USING DIRECT ADDRESSING | 6-77 |
| | LD A,(rp) — LOAD ACCUMULATOR FROM MEMORY LOCATION ADDRESSED BY REGISTER PAIR | 6-78 |

# TABLE OF CONTENTS (Continued)

**CHAPTER**                                                            **PAGE**

| | |
|---|---|
| LD dst,src — MOVE CONTENTS OF SOURCE REGISTER TO DESTINATION REGISTER | 6-79 |
| LD HL,(addr) — LOAD REGISTER PAIR OR INDEX REGISTER FROM MEMORY USING DIRECT ADDRESSING | 6-80 |
| LD rp,(addr) | |
| LD IX,(addr) | |
| LD IY,(addr) | |
| LD IV,A — LOAD INTERRUPT VECTOR OR REFRESH REGISTER FROM ACCUMULATOR | 6-81 |
| LD R,A | |
| LD reg,data — LOAD IMMEDIATE INTO REGISTER | 6-82 |
| LD rp,data — LOAD 16 BITS OF DATA IMMEDIATE INTO REGISTER | 6-83 |
| LD IX,data | |
| LD IY,data | |
| LD reg,(HL) — LOAD REGISTER FROM MEMORY | 6-84 |
| LD reg,(IX+disp) | |
| LD reg,(IY+disp) | |
| LD SP,HL — MOVE CONTENTS OF HL OR INDEX REGISTER TO STACK POINTER | 6-85 |
| LD SP,IX | |
| LD SP,IY | |
| LD (addr),A — STORE ACCUMULATOR IN MEMORY USING DIRECT ADDRESSING | 6-86 |
| LD (addr),HL — STORE REGISTER PAIR OR INDEX REGISTER IN MEMORY USING DIRECT ADDRESSING | 6-87 |
| LD (addr),rp | |
| LD (addr),XY | |
| LD (HL),data — LOAD IMMEDIATE INTO MEMORY | 6-89 |
| LD (IX+disp),data | |
| LD (IY+disp),data | |
| LD (HL),reg — LOAD MEMORY FROM REGISTER | 6-90 |
| LD (IX+disp),reg | |
| LD (IY+disp),reg | |
| LD (rp),A — LOAD ACCUMULATOR INTO THE MEMORY LOCATION ADDRESSED BY REGISTER PAIR | 6-91 |
| LDD — TRANSFER DATA BETWEEN MEMORY LOCATIONS, DECREMENT DESTINATION AND SOURCE ADDRESSES | 6-92 |
| LDDR — TRANSFER DATA BETWEEN MEMORY LOCATIONS UNTIL BYTE COUNTER IS ZERO. DECREMENT DESTINATION AND SOURCE ADDRESSES | 6-93 |
| LDI — TRANSFER DATA BETWEEN MEMORY LOCATIONS. INCREMENT DESTINATION AND SOURCE ADDRESSES | 6-94 |
| LDIR — TRANSFER DATA BETWEEN MEMORY LOCATIONS UNTIL BYTE COUNTER IS ZERO. INCREMENT DESTINATION AND SOURCE ADDRESSES | 6-95 |

# TABLE OF CONTENTS (Continued)

**CHAPTER** **PAGE**

| | |
|---|---|
| NEG — NEGATE CONTENTS OF ACCUMULATOR | 6-95 |
| NOP — NO OPERATION | 6-96 |
| OR data — OR IMMEDIATE WITH ACCUMULATOR | 6-97 |
| OR reg — OR REGISTER WITH ACCUMULATOR | 6-98 |
| OR (HL) — OR MEMORY WITH ACCUMULATOR | 6-99 |
| OR (IX+disp) | |
| OR (IY+disp) | |
| OUT (C),reg — OUTPUT FROM REGISTER | 6-100 |
| OUTD — OUTPUT FROM MEMORY. DECREMENT ADDRESS | 6-101 |
| OTDR — OUTPUT FROM MEMORY. DECREMENT ADDRESS, CONTINUE UNTIL REGISTER B=0 | 6-101 |
| OUTI — OUTPUT FROM MEMORY. INCREMENT ADDRESS | 6-102 |
| OTIR — OUTPUT FROM MEMORY. INCREMENT ADDRESS, CONTINUE UNTIL REGISTER B=0 | 6-102 |
| OUT (port),A — OUTPUT FROM ACCUMULATOR | 6-103 |
| POP rp — READ FROM THE TOP OF THE STACK | 6-104 |
| POP IX | |
| POP IY | |
| PUSH rp — WRITE TO THE TOP OF THE STACK | 6-105 |
| PUSH IX | |
| PUSH IY | |
| RES b,reg — RESET INDICATED REGISTER BIT | 6-106 |
| RES b,(HL) — RESET BIT B OF INDICATED MEMORY POSITION | 6-107 |
| RES b,(IX+disp) | |
| RES b,(IY+disp) | |
| RET — RETURN FROM SUBROUTINE | 6-108 |
| RET cond — RETURN FROM SUBROUTINE IF CONDITION IS SATISFIED | 6-109 |
| RETI — RETURN FROM INTERRUPT | 6-110 |
| RETN — RETURN FROM NON-MASKABLE INTERRUPT | 6-111 |
| RL reg — ROTATE CONTENTS OF REGISTER LEFT THROUGH CARRY | 6-112 |
| RL (HL) — ROTATE CONTENTS OF MEMORY LOCATION LEFT THROUGH CARRY | 6-113 |
| RL (IX+disp) | |
| RL (IY+disp) | |
| RLA — ROTATE ACCUMULATOR LEFT THROUGH CARRY | 6-114 |
| RLC reg — ROTATE CONTENTS OF REGISTER LEFT CIRCULAR | 6-115 |
| RLC (HL) — ROTATE CONTENTS OF MEMORY LOCATION LEFT CIRCULAR | 6-116 |
| RLC (IX+disp) | |
| RLC (IY+disp) | |
| RLCA — ROTATE ACCUMULATOR LEFT CIRCULAR | 6-117 |
| RLD — ROTATE ONE BCD DIGIT LEFT BETWEEN THE ACCUMLATOR AND MEMORY LOCATION | 6-118 |
| RR reg — ROTATE CONTENTS OF REGISTER RIGHT THROUGH CARRY | 6-119 |
| RR (HL) — ROTATE CONTENTS OF MEMORY LOCATION RIGHT THROUGH CARRY | 6-120 |

# TABLE OF CONTENTS (Continued)

**CHAPTER**                                                                 **PAGE**

| | |
|---|---|
| RR (IX+disp) | |
| RR (IY+disp) | |
| RRA — ROTATE ACCUMULATOR RIGHT THROUGH CARRY | 6-121 |
| RRC reg — ROTATE CONTENTS OF REGISTER RIGHT CIRCULAR | 6-122 |
| RRC (HL) — ROTATE CONTENTS OF MEMORY LOCATION RIGHT CIRCULAR | 6-123 |
| RRC (IX+disp) | |
| RRC (IY+disp) | |
| RRCA — ROTATE ACCUMULATOR RIGHT CIRCULAR | 6-124 |
| RRD — ROTATE ONE BCD DIGIT RIGHT BETWEEN THE ACCUMULATOR AND MEMORY LOCATION | 6-125 |
| RST n — RESTART | 6-126 |
| SBC A,data — SUBTRACT IMMEDIATE FROM ACCUMULATOR WITH BORROW | 6-127 |
| SBC A,reg — SUBTRACT REGISTER WITH BORROW FROM ACCUMULATOR | 6-128 |
| SBC A,(HL) — SUBTRACT MEMORY AND CARRY FROM ACCUMULATOR | 6-129 |
| SBC A,(IX+disp) | |
| SBC A,(IY+disp) | |
| SBC HL,rp — SUBTRACT REGISTER WITH CARRY FROM H AND L | 6-130 |
| SCF — SET CARRY FLAG | 6-131 |
| SET b,reg — SET INDICATED REGISTER BIT | 6-132 |
| SET b,(HL) — SET BIT B OF INDICATED MEMORY POSITION | 6-133 |
| SET b,(IX+disp) | |
| SET b,(IY+disp) | |
| SLA reg — SHIFT CONTENTS OF REGISTER LEFT ARITHMETIC | 6-134 |
| SLA (HL) — SHIFT CONTENTS OF MEMORY LOCATION LEFT ARITHMETIC | 6-135 |
| SLA (IX+disp) | |
| SLA (IY+disp) | |
| SRA reg — ARITHMETIC SHIFT RIGHT CONTENTS OF REGISTER | 6-136 |
| SRA (HL) — ARITHMETIC SHIFT RIGHT CONTENTS OF MEMORY POSITION | 6-137 |
| SRA (IX+disp) | |
| SRA (IY+disp) | |
| SRL reg — SHIFT CONTENTS OF REGISTER RIGHT LOGICAL | 6-138 |
| SRL (HL) — SHIFT CONTENTS OF MEMORY LOCATION RIGHT LOGICAL | 6-139 |
| SRL (IX+disp) | |
| SRL (IY+disp) | |
| SUB data — SUBTRACT IMMEDIATE FROM ACCUMULATOR | 6-140 |
| SUB reg — SUBTRACT REGISTER FROM ACCUMULATOR | 6-141 |
| SUB (HL) — SUBTRACT MEMORY FROM ACCUMULATOR | 6-142 |
| SUB (IX+disp) | |
| SUB (IY+disp) | |

# TABLE OF CONTENTS (Continued)

| CHAPTER | | PAGE |
|---|---|---|
| | XOR data — EXCLUSIVE-OR IMMEDIATE WITH ACCUMULATOR | 6-143 |
| | XOR reg — EXCLUSIVE-OR REGISTER WITH ACCUMULATOR | 6-144 |
| | XOR (HL) — EXCLUSIVE-OR MEMORY WITH ACCUMULATOR | 6-145 |
| | XOR (IX+disp) | |
| | XOR (IY+disp) | |
| 7 | SOME COMMONLY USED SUBROUTINES | 7-1 |
| | MEMORY ADDRESSING | 7-1 |
| |   INDIRECT ADDRESSING | 7-2 |
| |   INDIRECT, POST-INDEXED ADDRESSING | 7-2 |
| | DATA MOVEMENT | 7-3 |
| |   MOVING SIMPLE DATA BLOCKS | 7-3 |
| |   MULTIPLE TABLE LOOKUPS | 7-4 |
| |   SORTING DATA | 7-5 |
| | ARITHMETIC | 7-6 |
| |   BINARY ADDITION | 7-7 |
| |   BINARY SUBTRACTION | 7-9 |
| |   DECIMAL ADDITION | 7-9 |
| |   DECIMAL SUBTRACTION | 7-9 |
| | MULTIPLICATION AND DIVISION | 7-9 |
| |   8-BIT BINARY MULTIPLICATION | 7-10 |
| |   AN 8-BIT BINARY MULTIPLICATION PROGRAM | 7-12 |
| |   16-BIT BINARY MULTIPLICATION | 7-13 |
| |   BINARY DIVISION | 7-14 |
| | PROGRAM EXECUTION SEQUENCE LOGIC | 7-15 |
| |   THE JUMP TABLE | 7-15 |

## LIST OF FIGURES

| FIGURE | | PAGE |
|---|---|---|
| 2-1 | Configuration for Memory-Mapped I/O Addressing | 2-8 |
| 2-2 | Configuration for I/O Space I/O Addressing | 2-10 |
| 3-1 | Printwheel Control Logic | foldout |
| 3-2 | Printwheel Control Logic Timing Diagram | 3-4 |
| 3-3 | The Complete Simulation Program | 3-68 |
| 4-1 | Timing for Figure 3-1, from the Programmer's Viewpoint | 4-2 |
| 4-2 | Z80 Microcomputer Configuration | 4-4 |
| 4-3 | First Attempt at Program Flowchart | 4-8 |
| 4-4 | Program Flowchart to Compute Printhammer Firing Pulse Length | 4-19 |
| 4-5 | A Simple Print Cycle Instruction Sequence Without Initialization or Reset | 4-21 |
| 4-6 | A Simple Print Cycle Program | 4-32 |
| 5-1 | Z80 Microcomputer Configuration Using a PIO to Generate an Interrupt | 5-28 |

## LIST OF TABLES

| TABLE | | PAGE |
|---|---|---|
| 2-1 | Comparing Memory Utilization and Program Execution Speed for 7411 AND Gates' Simulation | 2-33 |
| 5-1 | The Shortest Economic Subroutine Length as a Function of the Number of Times the Subroutine Is Called | 5-15 |
| 6-1 | A Summary of the Z80 Instruction Set | 6-5 |
| 6-2 | A Summary of Instruction Object Codes and Execution Cycles | 6-20 |

# QUICK INDEX

| INDEX | | PAGE |
|---|---|---|
| A | AMPLIFIER | 2-15 |
| | ASSEMBLY LANGUAGE VERSUS DIGITAL LOGIC | 3-67 |
| | ASYNCHRONOUS LOGIC | 2-13 |
| B | BASE RELATIVE ADDRESSING | 2-30 |
| | BIT CONTROL | 3-12 |
| | BIT DATA | 2-6,2-7 |
| | BIT INVERSION USING XOR | 2-12 |
| | BIT MASKING | 2-11 |
| | BRANCH ON CONDITION | 4-29 |
| | BUFFER | 2-15 |
| C | CARRY STATUS | 3-29 |
| | CHIP SELECT IN LARGER SYSTEMS | 4-5 |
| | CHIP SELECT IN SIMPLE SYSTEMS | 4-5 |
| | CH READY | 3-5 |
| | CLOCK SIGNAL | 2-35 |
| | COMBINATORIAL LOGIC | 1-1 |
| | COMPARE IMMEDIATE | 4-29 |
| | COMPLEMENTING A BYTE OF MEMORY | 2-16 |
| | CONDITIONAL INSTRUCTION EXECUTION PATHS | 4-29 |
| | CONDITIONAL RETURN | 5-16 |
| | CPU REGISTERS | 2-5 |
| | CPU REGISTER UTILIZATION, CONFLICTS IN | 2-28 |
| D | DATA MEMORY ADDRESS COMPUTATION | 3-60 |
| | DATA SOURCE AND DESTINATION | 2-6 |
| | DIGITAL LOGIC DESIGN CYCLE | 2-1 |
| | DIRECT VERSUS IMPLIED ADDRESSING | 2-32 |
| | D-TYPE FLIP-FLOP | 2-35 |
| E | EVENT SEQUENCE | 3-58 |
| | EVENT TIMING IN MICROCOMPUTER SYSTEM | 3-31 |
| | EXTERNAL LOGIC AS THE SOURCE OR DESTINATION | 2-7 |
| F | FAN IN | 2-15,2-16 |
| | FAN IN IN MICROCOMPUTER PROGRAMS | 2-19 |
| | FAN OUT | 2-15,2-16 |
| | FAN OUT IN MICROCOMPUTER PROGRAMS | 2-21 |
| | FFA | 3-8 |
| | FLIP-FLOP CLEAR | 2-36 |
| | FLIP-FLOP PRESET | 2-36 |
| | FLIP-FLOP SIMULATION USING I/O PORTS | 3-27 |
| | FLOWCHART | 2-5 |
| G | GATE SETTLING TIME | 2-13 |
| H | HIGHER LEVEL LANGUAGES | 4-5 |
| | H IN OPERAND FIELD | 2-11 |

# QUICK INDEX (Continued)

**INDEX**                                                    **PAGE**

| | | |
|---|---|---|
| **I** | IMPLIED ADDRESSING | 2-28 |
| | IMPLIED MEMORY ADDRESSING | 2-25 |
| | INITIATING INTERRUPTS VIA THE PIO | 5-29 |
| | INPUT/OUTPUT | 2-7 |
| | INPUT SIGNAL PULSE WIDTH | 3-17 |
| | INPUT SIGNALS | 4-1 |
| | INTERRUPT ACKNOWLEDGE | 5-25 |
| | INTERRUPT ECONOMICS | 5-37 |
| | INTERRUPT ENABLE | 5-24 |
| | INTERRUPT PRIORITY ARBITRATION | 5-35 |
| | INTERRUPT PROGRAM ORIGIN | 5-31 |
| | INTERRUPTS, WHEN TO USE | 5-24 |
| | INTERRUPT TIMING CONSIDERATIONS | 5-37 |
| | INVERTER SIMULATION | 3-28 |
| | I/O IN MEMORY ADDRESS SPACE | 2-7 |
| | I/O PORT ADDRESS DETERMINATION | 3-12 |
| | I/O PORT ADDRESSING | 3-12 |
| | I/O PORT MODES | 3-9 |
| | I/O PORT MODE SELECT INSTRUCTION SEQUENCE | 3-14 |
| | I/O PORT MODE SELECTION | 3-12 |
| | I/O PORT PIN SELECT | 2-11 |
| | I/O VIA I/O PORTS | 2-9 |
| **J** | JK FLIP-FLOP | 2-35 |
| | JUMP ON NO CARRY | 3-49 |
| **L** | LEADING ZERO | 2-11 |
| | LEAKAGE CURRENT | 2-16 |
| | LIMIT CHECKING | 4-27 |
| | LOGIC EXCLUDED FROM MICROCOMPUTER | 3-56 |
| **M** | MACRO ASSEMBLER DIRECTIVES | 5-22 |
| | MACRO DEFINITION | 5-21 |
| | MACRO DEFINITION LOCATION IN A SOURCE PROGRAM | 5-23 |
| | MASTER-SLAVE FLIP-FLOP | 2-38, 2-43 |
| | MEMORY ADDRESSES | 4-7 |
| | MICROCOMPUTER LOGIC DESIGN CYCLE | 2-2 |
| | MONOSTABLE MULTIVIBRATOR | 2-38 |
| **N** | NEGATIVE EDGE TRIGGER | 2-35 |
| | NESTED SUBROUTINES | 5-17 |
| **O** | OBJECT CODE INTERPRETATION | 2-9, 2-10 |
| | OBJECT PROGRAM | 2-4 |
| | ONE-SHOT | 2-38 |
| | ONE-SHOT INITIATION | 2-47 |
| | ONE-SHOT TIME DELAY SIMULATION | 3-51 |
| | ONE-SHOT TIME OUT USING STATUS | 2-48 |

## QUICK INDEX (Continued)

| INDEX | | PAGE |
|---|---|---|
| | ONE-SHOT VARIABLE PULSE | 3-56 |
| | OR GATE SIMULATION | 3-28 |
| P | PARALLEL INPUT/OUTPUT INTERFACE | 3-9 |
| | PIN ASSIGNMENTS | 4-3 |
| | PIO INTERRUPT CONTROL WORD | 5-29 |
| | POSITIVE EDGE TRIGGER | 2-34 |
| | PRINTHAMMER FIRING DELAY | 4-17 |
| | PRINTWHEEL POSITION OF VISIBILITY | 3-7 |
| | PRINTWHEEL READY | 3-5 |
| | PRINTWHEEL REPOSITIONING PRINT CYCLE | 3-15,3-35 |
| | PROGRAM IMPLEMENTATION SEQUENCE | 4-9 |
| | PROGRAMMED SIGNAL PULSE | 4-24 |
| | PROGRAM TIMING | 2-6 |
| | PROGRAM VARIATIONS RANKED | 2-32 |
| | PROGRAMS MADE SHORTER | 3-44,3-47 |
| | PULSE WIDTH CALCULATION | 3-46 |
| | PW STROBE | 3-5 |
| R | RAM | 4-7 |
| | RESET | 3-26 |
| | RESET LOGIC | 4-6 |
| | RESET THE CPU | 3-18 |
| | ROM ADDRESSES | 4-6 |
| | ROM SELECT IN SIMPLE SYSTEMS | 4-7 |
| S | SAVING REGISTERS AND STATUS | 5-33 |
| | SETTLING DELAYS | 3-6 |
| | 7474 FLIP-FLOP | 3-25 |
| | SIGNAL BUFFERING | 2-16 |
| | SIGNAL ENABLE | 3-57 |
| | SIGNAL LEVEL CHANGES SENSED WITHOUT INTERRUPTS | 3-31 |
| | SIGNAL PULSE WIDTH | 3-18 |
| | SIMPLE I/O | 3-12 |
| | SIMULTANEOUS TIME DELAYS | 2-47 |
| | SORTING DATA | 7-5 |
| | SOURCE PROGRAM | 2-4 |
| | SOURCE PROGRAM LABEL ASSIGNMENTS | 2-24 |
| | STACK MANIPULATION | 5-19 |
| | START RIBBON PULSE | 3-8 |
| | STATUS CHANGES WITH INSTRUCTION EXECUTION | 6-4 |
| | STATUS DETERMINATION BY ANDING A REGISTER WITH ITSELF | 2-20 |
| | STATUS FLAGS USED TO REPRESENT LOGIC | 3-28 |
| | STATUS TESTING USING DEC INSTRUCTION | 2-45 |
| | SUBROUTINE CALL USING RST | 6-126 |
| | SUBROUTINE PARAMETER | 5-17 |

# QUICK INDEX (Continued)

| INDEX | | PAGE |
|---|---|---|
| | SWITCHING BITS OFF | 3-36 |
| | SWITCHING BITS ON | 3-31, 3-36 |
| | SYNCHRONOUS LOGIC | 2-13 |
| **T** | TABLES POSITIONED TO SIMPLIFY ACCESS INSTRUCTION SEQUENCE | 5-5 |
| | TIME DELAY | 3-63 |
| | TIME DELAY BASED ON INPUT SIGNAL | 3-19 |
| | TIME DELAY COMPUTATION | 3-63 |
| | TIME DELAY INITIATION | 2-45 |
| | TIME DELAY OF VARIABLE LENGTH | 3-49, 4-24 |
| | TIME DELAYS, EXECUTING PROGRAMS WITHIN | 2-45 |
| | TIMING AND LIMITS OF SIMULATION | 3-47 |
| | TIMING AND LOGIC SEQUENCE | 3-32, 3-39, 3-43 |
| | TIMING LONG TIME INTERVALS | 2-44 |
| | TIMING SHORT TIME INTERVALS | 2-44 |
| | TRANSFER FUNCTION | 4-1 |
| | TTL LOADS | 2-16 |
| **U** | USING Z80 CPU AUXILIARY REGISTERS | 5-34 |
| **Z** | Z80 CPU INTERRUPT MODE 0 | 5-25 |
| | Z80 CPU INTERRUPT MODE 1 | 5-25 |
| | Z80 CPU INTERRUPT MODE 2 | 5-25 |
| | Z80 PIO BIDIRECTIONAL DATA TRANSFERS WITH HANDSHAKING | 3-11 |
| | Z80 PIO INPUT WITH HANDSHAKING | 3-11 |
| | Z80 PIO INTERRUPT ACKNOWLEDGE RESPONSE | 5-27 |
| | Z80 PIO OUTPUT WITH HANDSHAKING | 3-10 |
| | Z80 PIO RESET LOGIC | 4-6 |
| | ZERO STATUS | 3-28 |

# Chapter 1
# INTRODUCTION

This book explains how an assembly language program within a microcomputer system can replace combinatorial logic — that is, the combined use of "off-the-shelf", non-programmable logic devices such as standard 7400 series digital logic.

COMBINATORIAL LOGIC

If you are a logic designer, this book will teach you how to do your old job in a new way — by creating assembly language programs within a microcomputer system.

If you are a programmer, this book will show you how programming has found a new purpose — in logic design.

This is a "how to do it" book; as such, it has to become very specific, and so a particular type of microcomputer, the Z80, is referenced directly.

Companies manufacturing these microcomputers are:

ZILOG, INCORPORATED
10460 Bubb Road
Cupertino, California 95014

MOSTEK, INCORPORATED
1215 West Crosby Road
Carrollton, Texas 75006

## WHAT THIS BOOK ASSUMES YOU KNOW

This book is a sequel to <u>An Introduction to Microcomputers</u>, which was a single volume in its first edition but is two volumes in its second edition.

<u>An Introduction to Microcomputers</u> describes microprocessors and microcomputers conceptually; it does not address itself to the practical matter of implementing a concept. This book addresses the practical matter of implementation.

In that this book is a sequel, it makes a single assumption — that you have read, or otherwise understand, the material covered in <u>An Introduction to Microcomputers</u>. However, before launching into a real design project, you will need vendor literature that specifically describes the devices you have elected to use.

Note in particular that hardware and timing are not described in this book, either for the Z80 CPU or any other microcomputer device; sufficient information may be found in <u>An Introduction to Microcomputers, Volume II — Some Real Products</u>.

The Z80 instruction set is described in Chapter 6, since programming is what this book is all about.

## UNDERSTANDING ASSEMBLY LANGUAGE

Assembly language instructions are the transfer functions of a microcomputer system; taken together, they constitute an "instruction set", which describes the individual operations which the microcomputer can perform.

You define the events which must occur serially within the microcomputer system — as a sequence of instructions which, taken together, constitute an assembly language program.

In reality, understanding what individual instructions do within a microcomputer system is very straightforward; it is one of the simplest aspects of working with microcomputers. Yet, it unduly terrifies users who are new to programming. If that includes you, a word of advice — forget about mnemonics and instruction sets; take instructions one at a time as you encounter them in this book. When you do not understand what an instruction is doing, look it up in Chapter 6.

The spectre of "programming" will haunt you only if you let it.

## HOW THIS BOOK HAS BEEN PRINTED

**Notice that text in this book has been printed in boldface type** and lightface type. **This has been done to help you skip those parts of the book that cover subject matter with which you are familiar. You can be sure that lightface type only expands on information presented in the previous boldface type.** Therefore, only read boldface type until you reach a subject about which you want to know more, at which point start reading the lightface type.

# Chapter 2
# ASSEMBLY LANGUAGE AND DIGITAL LOGIC

## THE DESIGN CYCLE

Any product that is to be built out of discrete digital logic components will go through a well-defined design cycle.

**DIGITAL LOGIC DESIGN CYCLE**

**Let us assume that the product has been defined — from marketing management's point of view.**

You are presented with a product specification which identifies necessary product performance and characteristics; your job is to deliver a viable design to manufacturing. **The design cycle will proceed as follows:**

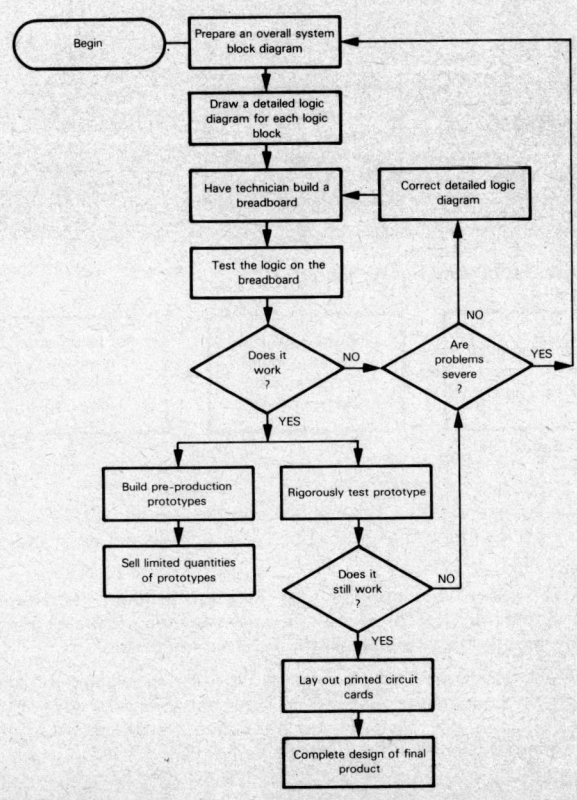

**There is an expensive and slow iterative loop in any digital logic design cycle;** as illustrated above, it consists of these steps:

- Redraw logic
- Build a new breadboard
- Test the breadboard for logic errors, technician errors, or faulty components

This iterative loop makes combinatorial logic design slow and expensive — not only during the initial design phase, but even more so when you subsequently decide to modify or enhance the product.

**What happens when you start using microcomputers? First of all, a portion of your logic vanishes into a "black box" — which is the microcomputer system:**

**MICROCOMPUTER LOGIC DESIGN CYCLE**

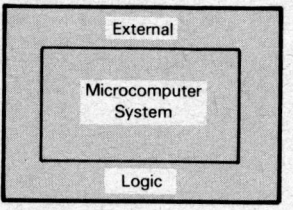

**Your first step:**

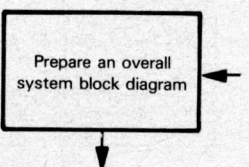

**must now be broken out as follows:**

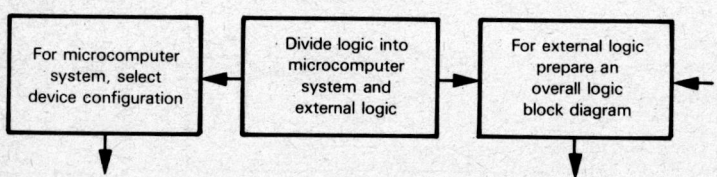

Partitioning your application into a microcomputer system and external digital logic may look like a difficult proposition — if you do not understand what the microcomputer system can do.

In fact, **once you have a microcomputer in your product, economics overwhelmingly favor making the "black box" assume as many tasks as possible; you must justify the existence of every single external logic gate.**

Remember, memory comes in finite increments. In order to expand the logic implemented within the microcomputer system, you may simply have to write additional instruction sequences that will reside in memory which would otherwise be wasted; adding program memory, for that matter, costs very little.

Also, compared to the cost of digital logic development, microcomputer logic development is quick and inexpensive. **A typical microcomputer system development cycle may be illustrated as follows:**

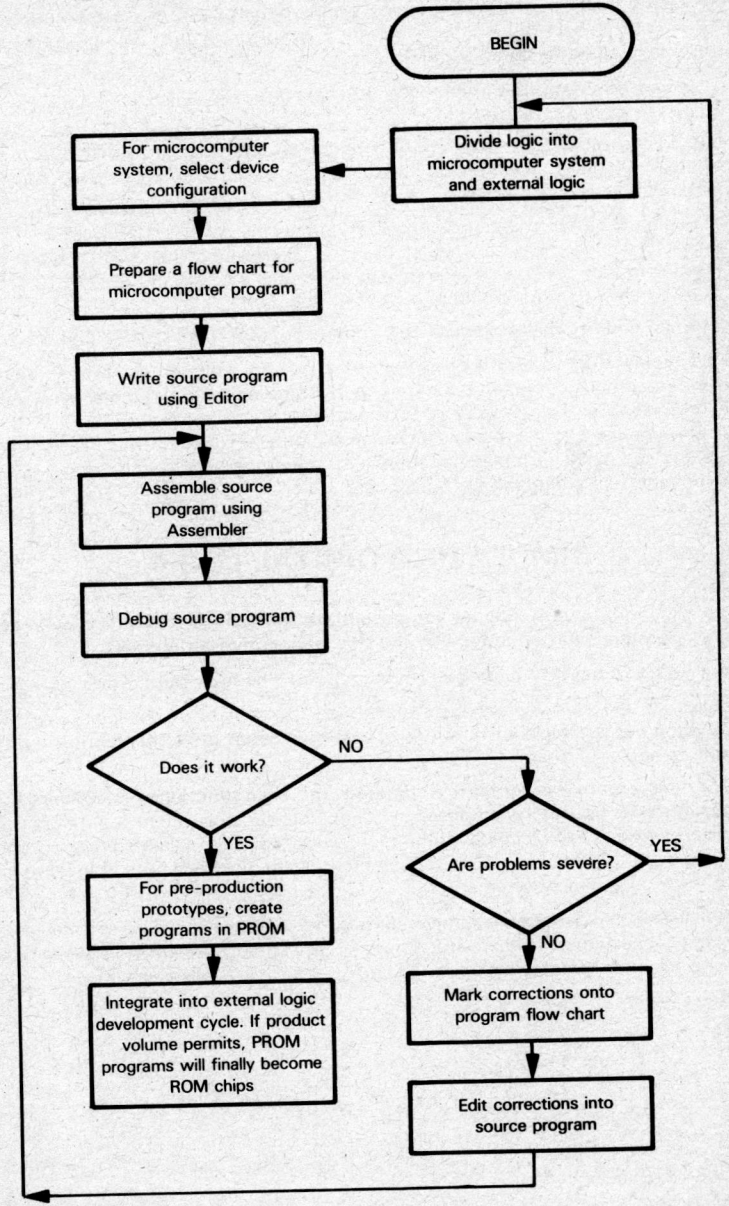

There are still iterative loops in the microcomputer development cycle illustrated above, but, compared to digital logic development, less time and expense are associated with microcomputer development cycle iterative loops.

**Every microcomputer is supported by a development system.** Characteristics and operation of these development systems vary markedly from one company to the next; however, they **all have these capabilities:**

1) You can **simulate the microcomputer system** you have configured without necessarily creating a breadboard.

2) You can execute a resident editor program to **create your source program.** Remember, a sequence of assembly language instructions is referred to as a "Source Program".

    > SOURCE PROGRAM

3) You can **assemble the source program** right at the development system to create an object program. Remember, the source program becomes a sequence of binary digits (referred to as an object program) before it can be executed.

    > OBJECT PROGRAM

4) You can **conditionally execute the object program** to make sure that it works.

**Using a typical microcomputer development system, you can go through several major development cycles in a single day, where each development cycle might have taken one or two weeks in a total digital logic implementation.** Within a single development cycle you can make many program corrections; in less than a minute you can make a simple correction, equivalent to adding or removing a gate (or MSI function) from a digital logic breadboard.

## SIMULATING DIGITAL LOGIC

**OK, so logic must eventually be separated into that which is within a microcomputer system and that which is beyond the microcomputer system.**

**We are going to have to address two aspects of this logic separation:**

1) Based on the ability of assembly language to simulate digital logic, **we must develop some simple criterion for estimating what a microcomputer system can do** and what it cannot do.

2) **We must create a program to implement the logic functions which have been assigned to the microcomputer system.** Unfortunately, there are innumerable ways of writing a microcomputer program. Once you have mastered the concept of using instructions to drive a microcomputer system, **the next step is to learn how to write efficient programs.**

**We will begin by describing simple digital logic simulation.** This is a necessary beginning because there are some fundamental conceptual differences between digital logic and microcomputer programming logic.

# MICROCOMPUTER SIMULATION OF A SIGNAL INVERTER

**Suppose you want to invert a single signal:**

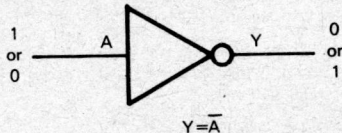

In the interests of developing good habits from the start, we will illustrate the signal inverter with the following logic flowchart:

**FLOW CHART**

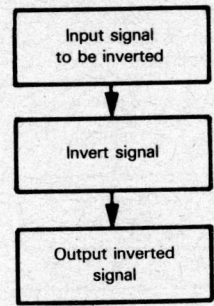

Although you would never use a microcomputer simply to replace a signal inverter, it is still worthwhile examining how it could be done.

## A MICROCOMPUTER EVENT SEQUENCE

**Recall that Z80 microcomputers have the following CPU registers:**

**CPU REGISTERS**

| | | | | | |
|---|---|---|---|---|---|
| | F | Program Status Words | | | F' |
| | A | Primary Accumulators | | | A' |
| B | C | Secondary Accumulators/Data Counters | B' | | C' |
| D | E | Secondary Accumulators/Data Counters | D' | | E' |
| H | L | Secondary Accumulators/Data Counters | H' | | L' |
| SP | | Stack Pointer | | | |
| PC | | Program Counter | | | |
| IX | | Index Register X | | | |
| IY | | Index Register Y | | | |
| | IV | Interrupt Vector | | | |
| | R | Memory Refresh Counter | | | |

**This single instruction:**

        CPL         :COMPLEMENT ACCUMULATOR

**when converted into object code and executed, inverts all eight bits of the primary Accumulator.** But that does not duplicate the inverter. First, one binary digit of the Accumulator must be selected to represent the signal being inverted. But which one?

| BIT DATA |

Having decided which binary digit, how does it reach the Accumulator in the first place? And, once inverted, how does the inverted bit become a signal again?

If the CPL instruction object code must be executed in order to perform the actual inversion, how and when does the object code reach the CPU? Clearly, execution of this instruction must be timed to occur after the binary digit to be inverted has reached the Accumulator.

| DATA SOURCE AND DESTINATION |
| PROGRAM TIMING |

**Steps needed to implement an inverter using a microcomputer may be illustrated by expanding our flowchart as follows:**

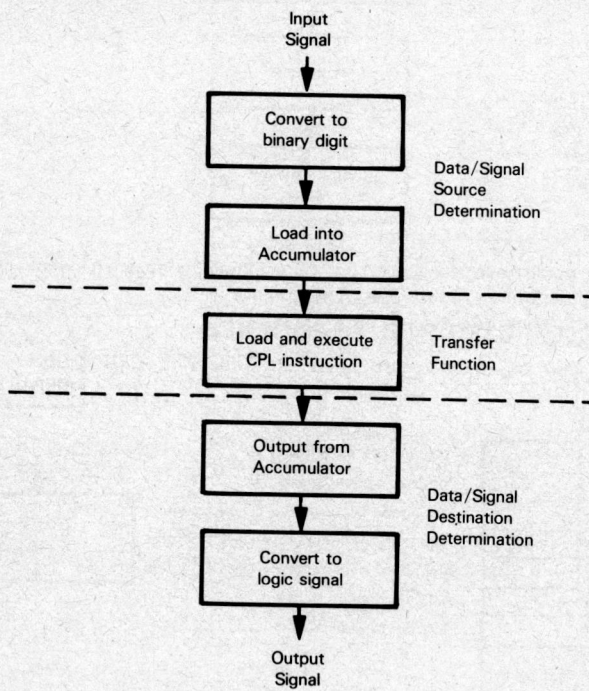

In the illustration above, pay most attention to the division of the problem into these three phases:

1) **Data/signal source determination.** We identify the data which is to be operated on. This data is transferred to a location out of which it can be accessed by the microcomputer Central Processing Unit (CPU).
2) **Transfer function execution.** The actual operation which must be performed on the source data will be referred to as a "transfer function".
3) **Data/signal destination determination.** The data or signals, having been subject to the transfer function, must now be transferred to some destination.

We will now generate an instruction sequence to implement the three phases of the inverter simulation illustrated above.

## IMPLEMENTING THE TRANSFER FUNCTION

The CPL instruction inverts every bit of the Accumulator.   | BIT DATA |

The CPL instruction, therefore, does not specify which bit of the Accumulator represents the signal to be inverted. This specification is implied by the way in which data is input to and output from the microcomputer system.

## DETERMINING DATA SOURCES AND DESTINATIONS

How will Accumulator data be input to and output from the microcomputer system? In answering this question, we touch on one of the fundamental strengths (and complexities) of microcomputers — their flexibility.

The input signal and the inverted output signal are just what their names imply — they are signals. But, to the microcomputer system they are "external logic". Information transfers between external logic and the microcomputer system are referred to generically as Input/Output (or I/O). During any programmed I/O operation, recall that the microcomputer is master and external logic is slave. This means that the microcomputer must indicate the direction of the I/O operation (input or output), and must identify the external logic being accessed.

| EXTERNAL LOGIC AS THE SOURCE OR DESTINATION |
| INPUT/OUTPUT |

External logic might decode a specific memory address as an enable strobe, so that I/O is handled as though it were a memory read or write. **Suppose the label INVD is being used in the assembly language source program to identify the signal being inverted.** This is the instruction sequence which will reproduce the signal inverter:

| I/O IN MEMORY ADDRESS SPACE |

```
LD      A,(INVD)      ;LOAD ACCUMULATOR FROM INVD
CPL                   ;COMPLEMENT THE ACCUMULATOR
LD      (INVD),A      ;STORE ACCUMULATOR CONTENTS TO INVD
```

In terms of microcomputer devices, Figure 2-1 shows the microcomputer configuration implied.

When the LD A,(INVD) instruction is executed, "Address Decode Logic" causes "Select Logic" to transmit the "Data In" signal to the Data Bus.

**There are eight Data Bus lines; the number of the line to which the "Data In" signal is connected becomes the significant bit number within the Accumulator.** When the LD A,(INVD) instruction has completed execution, the contents of the Data Bus will be in the Accumulator.

Next, the CPL instruction is executed. This instruction causes every bit of the Accumulator to be complemented.

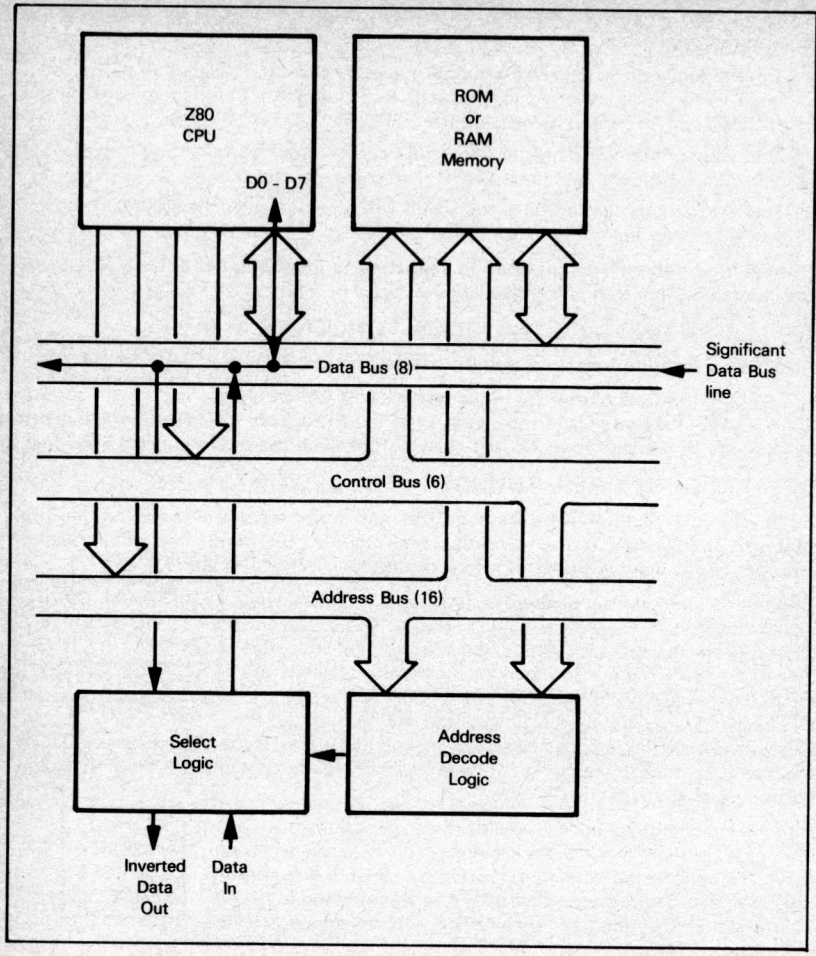

Figure 2-1. Configuration for Memory-Mapped I/O Addressing

**When the LD (INVD),A instruction is executed, the contents of the Accumulator are output to the Data Bus.** "Address Decode Logic" then causes "Select Logic" to output the contents of a single Data Bus line — which becomes the inverted "Data Out" signal.

**Because the "Select Logic" has "Data In" and "Data Out" signals connected to the same line of the Data Bus, "Data Out" is the complement of "Data In", and the signal inverter has been simulated.**

**ROM or RAM memory must be present in the microcomputer system, because the object codes for the three instructions must be stored in and fetched out of memory.**

**Consider object code in detail.** The three source program instructions become object code as follows:

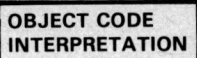

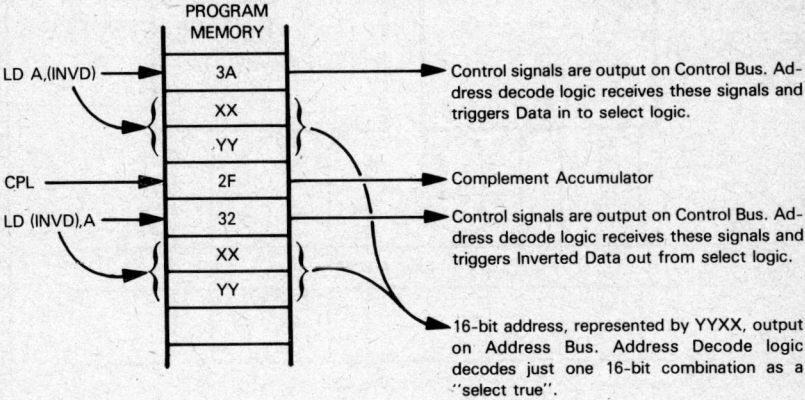

The program memory addresses of the bytes within which the object codes are stored are not important. However, no memory byte, ROM or RAM, can have the address represented by YYXX, since external logic is selected by this address.

Observe that the two bytes of the 16-bit address YYXX are reversed when stored in memory. There is nothing very significant about this inversion, it is just the way Z80 devices were designed.

**Now suppose that communication with external logic occurs via an I/O peripheral interface device.**

**I/O VIA I/O PORTS**

**In assembly language source program instructions, the label INVD will now identify an I/O port. This is the instruction sequence which reproduces the signal inverter:**

```
        IN      A,(INVD)    ;INPUT TO ACCUMULATOR FROM PORT INVD
        CPL                 ;COMPLEMENT THE ACCUMULATOR
        OUT     (INVD),A    ;OUTPUT ACCUMULATOR TO PORT INVD
```

**In terms of hardware, Figure 2-2 shows the microcomputer configuration implied.**

**All we have done by adding the Z80 Parallel I/O device is provide the "Address Decode" and "Select Logic" needed by the "Data In" and inverted "Data Out" signals.** Now the particular bit which is significant will be determined by the Z80 PIO pin to which the "Data In" and inverted "Data Out" signals are connected. In turn, these pins will be determined by the mode in which the Z80 PIO is used.

The fact that quite a few options are available to you when using the Z80 PIO is of no immediate consequence, in that it will confuse your early understanding of what assembly language programming is all about. We will therefore ignore Z80 PIO mode-control instructions, and simply assume that the appropriate mode control has been selected.

Z80
CPU
D0 - D7
ROM
or
RAM
Memory
Data Bus (8)
Control Bus (6)
Address Bus (16)
Z80
PIO
Data in
Inverted Data out

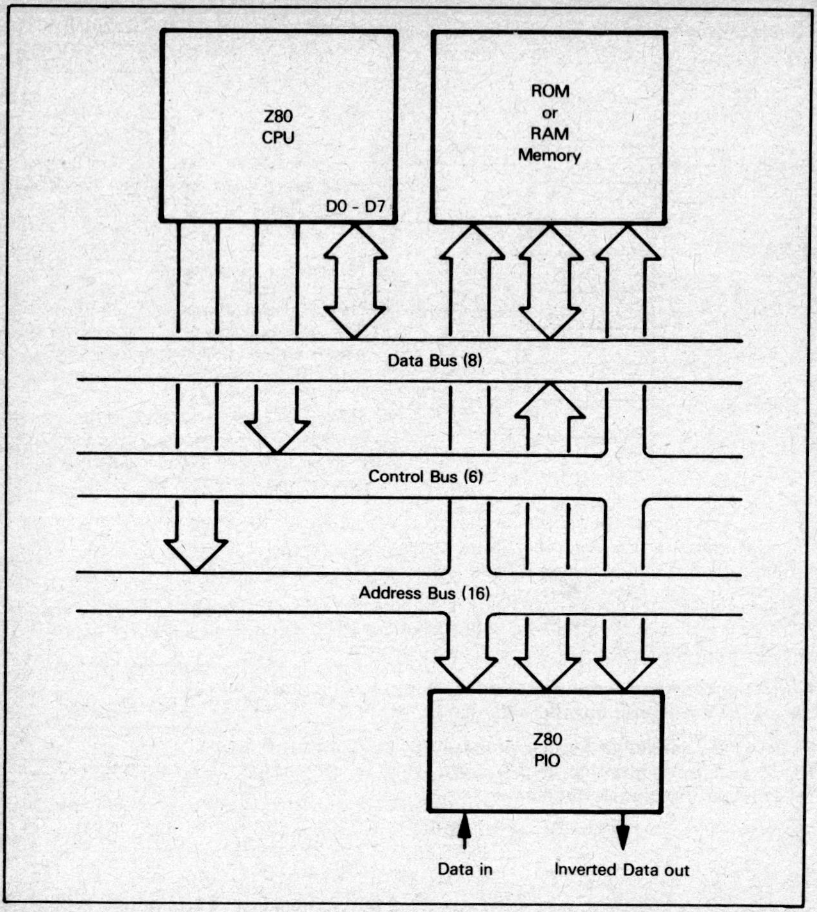

Figure 2-2. Configuration for I/O Space I/O Addressing

**OBJECT CODE INTERPRETATION**

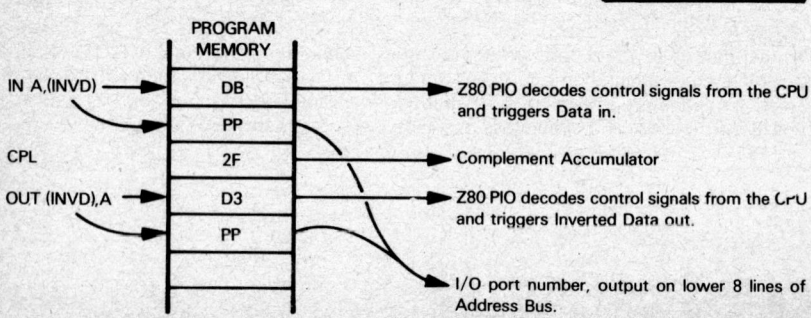

**In this case, the object code for the three instructions is interpreted as follows:**

PROGRAM
MEMORY
IN A,(INVD)
DB
PP
CPL
2F
OUT (INVD),A
D3
PP
Z80 PIO decodes control signals from the CPU and triggers Data in.
Complement Accumulator
Z80 PIO decodes control signals from the CPU and triggers Inverted Data out.
I/O port number, output on lower 8 lines of Address Bus.

Once again, addresses of the program memory bytes within which the above object codes are stored will not be important.

Observe that we are complementing every bit of I/O port INVD, even though only one bit corresponds to the signal being inverted.

**I/O PORT PIN SELECT**

**Suppose pin 4 alone must be inverted:**

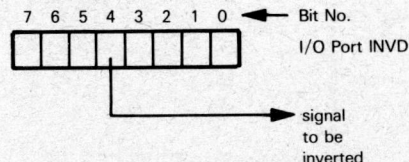

We can use a technique known as "masking" in order to invert a single I/O port pin, leaving all other pins alone. In this instance, masking may be illustrated as follows:

**BIT MASKING**

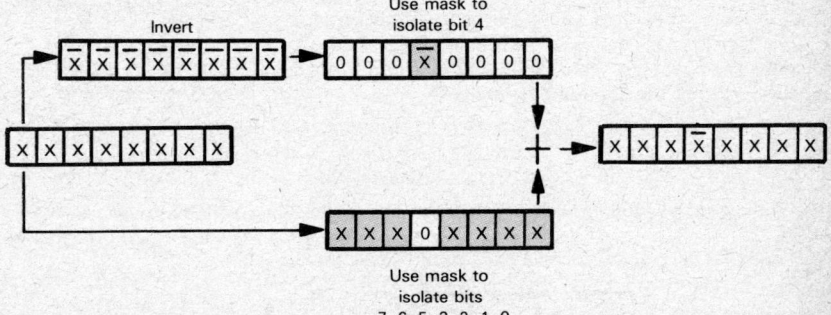

In the illustration above, X represents any binary digit; $\overline{X}$ represents its complement.

**The following instruction sequence will invert pin 4, leaving all other pins as they were:**

```
IN      A,(INVD)    ;INPUT TO ACCUMULATOR FROM I/O PORT INVD
CPL                 ;COMPLEMENT ACCUMULATOR
AND     10H         ;ISOLATE BIT 4
LD      B,A         ;SAVE IN REGISTER B
IN      A,(INVD)    ;INPUT TO ACCUMULATOR FROM I/O PORT INVD
AND     0EFH        ;CLEAR BIT 4
OR      B           ;OR A WITH B
OUT     (INVD),A    ;OUTPUT ACCUMULATOR TO I/O PORT INVD
```

**H, as the last character in the operand field, specifies a hexadecimal, immediate data value.** Thus, 0EFH represents the binary value:

**H IN OPERAND FIELD**

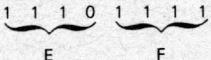

Hexadecimal numbers beginning with the characters A through F are preceded by a 0 to prevent the assembler from mistaking the numbers for variable names.

**LEADING ZERO**

In terms of registers' contents, this is what happens when the above instruction sequence is executed (again X represents any binary digit):

|     |         | I/O Port  | Accumulator | Register B |
|-----|---------|-----------|-------------|------------|
|     |         | XXXXXXXX  |             | ?          |
| IN  | A,(INVD)| XXXXXXXX →| XXXXXXXX    | ?          |
| CPL |         | XXXXXXXX  | $\overline{X}\overline{X}\overline{X}\overline{X}\overline{X}\overline{X}\overline{X}\overline{X}$ | ? |
|     |         |           | ∧ 00010000  |            |
| AND | 10H     | XXXXXXXX  | 000$\overline{X}$0000 | ?    |
| LD  | B,A     | XXXXXXXX  | 000$\overline{X}$0000 → | 000$\overline{X}$0000 |
| IN  | A,(INVD)| XXXXXXXX →| XXXXXXXX    | 000$\overline{X}$0000 |
|     |         |           | ∧ 11101111  |            |
| AND | 0EFH    | XXXXXXXX  | XXX0XXXX    | 000$\overline{X}$0000 |
|     |         |           | ∨ 000$\overline{X}$0000 |     |
| OR  | B       | XXXXXXXX  | XXX$\overline{X}$XXXX | 000$\overline{X}$0000 |
| OUT | (INVD),A| XXX$\overline{X}$XXXX ←| XXX$\overline{X}$XXXX | 000$\overline{X}$0000 |

The procedure given above demonstrates a valuable technique — namely, bit masking. However, for the specified function it is much too complicated. **Here is a simpler instruction sequence which performs the same bit inversion:**

> **BIT INVERSION USING XOR**

```
IN    A,(INVD)    ;INPUT TO ACCUMULATOR FROM I/O PORT INVD
XOR   10H         ;COMPLEMENT BIT 4, SAVING ALL OTHER BITS
OUT   (INVD),A    ;OUTPUT ACCUMULATOR TO I/O PORT INVD
```

In this instruction sequence we use Exclusive-OR and the appropriate mask to invert the desired bit while preserving the others. The truth table for Exclusive-OR shows that XOR with 1 inverts the bit, while XOR with 0 saves the bit value.

| Y | X | X⩛Y |
|---|---|-----|
| 0 | 0 | 0   |
| 0 | 1 | 1   |
| 1 | 0 | 1   |
| 1 | 1 | 0   |

$X \veebar 0 = X$

$X \veebar 1 = \overline{X}$

In programming as in logic design with discrete components, there will often be more than one way to implement the same function.

# EVENT TIMING

Within any digital logic implementation, events may be time synchronously, based on a clock signal:

**SYNCHRONOUS LOGIC**

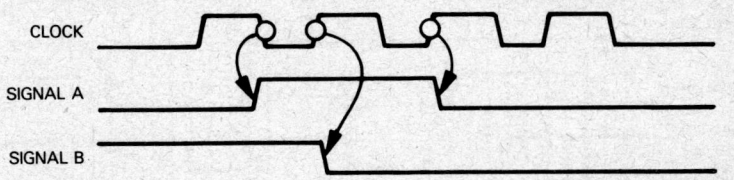

or asynchronously, based upon an output signal from one device changing state and thus triggering another device's state change:

**ASYNCHRONOUS LOGIC**

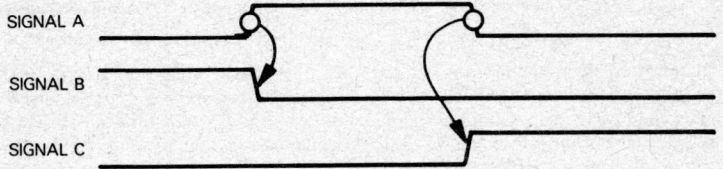

**Simple gates, however, are continuous devices.** Consider the following simple logic sequence:

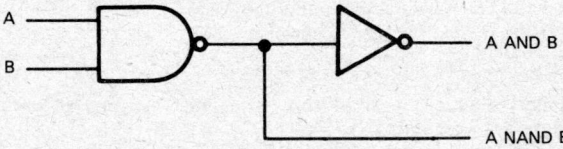

The signal inverter continuously inverts its input; a gate settling time of perhaps 10 nanoseconds is the only lag between input and output signal state changes.

**GATE SETTLING TIME**

Within a microcomputer system, however, three instructions must be executed before an output signal can reflect an input signal's state change.

In the unlikely event that the microcomputer system is emulating an inverter and doing nothing else, the inverter instruction sequence could be continuously re-executed as follows:

```
LOOP:   LD    A,(INVD)   ;LOAD ACCUMULATOR FROM INVD
        CPL              ;COMPLEMENT THE ACCUMULATOR
        LD    (INVD),A   ;STORE ACCUMULATOR CONTENTS AT INVD
        JP    LOOP       ;RE-EXECUTE THE SIGNAL INVERTER SEQUENCE
```

Depending on the microcomputer clock frequency, it will take approximately 20 microseconds to execute the signal inverter instruction loop once. Providing the period between input signal state changes is never less than 20 microseconds, the microcomputer implemented signal inverter will always work. But **there may be a delay of up to 30 microseconds between an input signal changing state and the output signal following suit.** This may be illustrated as follows:

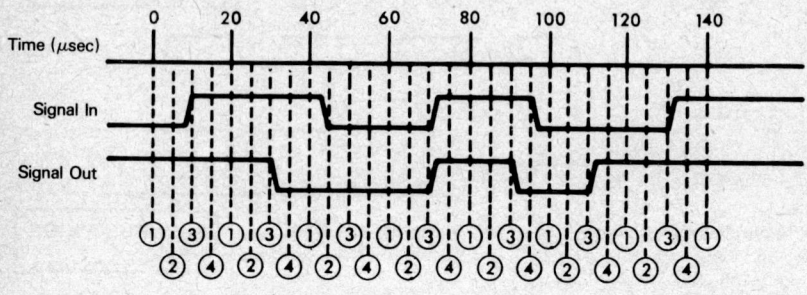

① = First LD instruction execution
② = CPL instruction execution
③ = Second LD instruction execution
④ = JP instruction execution

In the illustration above, the four instructions have been shown dividing twenty microseconds equally, so that each instruction is executed in five microseconds. In reality, this is not the case. Chapter 6 gives instruction execution times; you will see that the CPL instruction, for example, requires considerably less time to execute than any of the other three instructions. We will overlook this detail for the moment in order to concentrate on the concept at hand — which is that **we must pay careful attention to event sequences within the microcomputer system.**

Irrespective of when and how "Signal In" changes state, it is the state of "Signal In" at time ① (when the LD A,(INVD) instruction is executed) which is transported, as a binary digit, into the microcomputer system.

The actual binary digit inversion occurs at time ②

The inverted binary digit is converted into "Signal Out" at time ③ , when the LD (INVD),A instruction is executed.

Thus, "Signal Out" timing may differ considerably from "Signal In" timing.

**More serious problems arise when the signal inverter instruction sequence is just one small part of a larger microcomputer program.** Under these circumstances, many milliseconds may elapse between repeated executions of the inverter instruction sequence. If you leave it to chance, signal inversions may be completely missed. At very best, there may be considerable delays between the input signal changing state and the output signal following suit. This situation may be illustrated as follows:

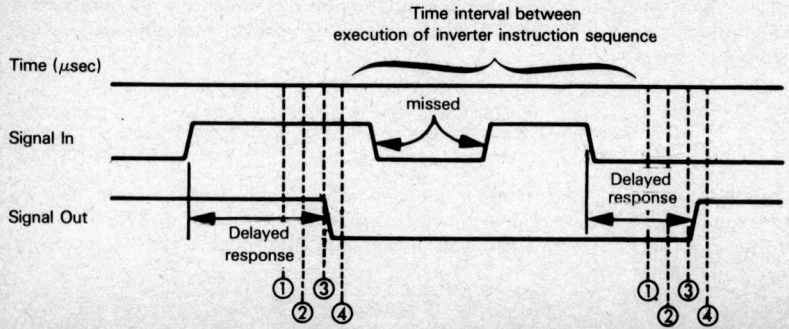

Again, ①, ②, ③ and ④ identify LD, CPL, LD and JP instructions' execution, respectively.

Having stressed the importance of timing in a microcomputer system plus the consequences of poor timing, we will drop the subject for the moment. This is because **timing problems largely evaporate when you simulate entire logic sequences as opposed to individual devices.** Therefore, solutions to timing problems should be looked at in the context of an entire logic simulation; we have not yet progressed that far.

## BUFFERS, AMPLIFIERS AND SIGNAL LOADS

Having looked at timing, we will now turn to some other fundamental digital logic concepts.

**A signal buffer increases the signal current level:**

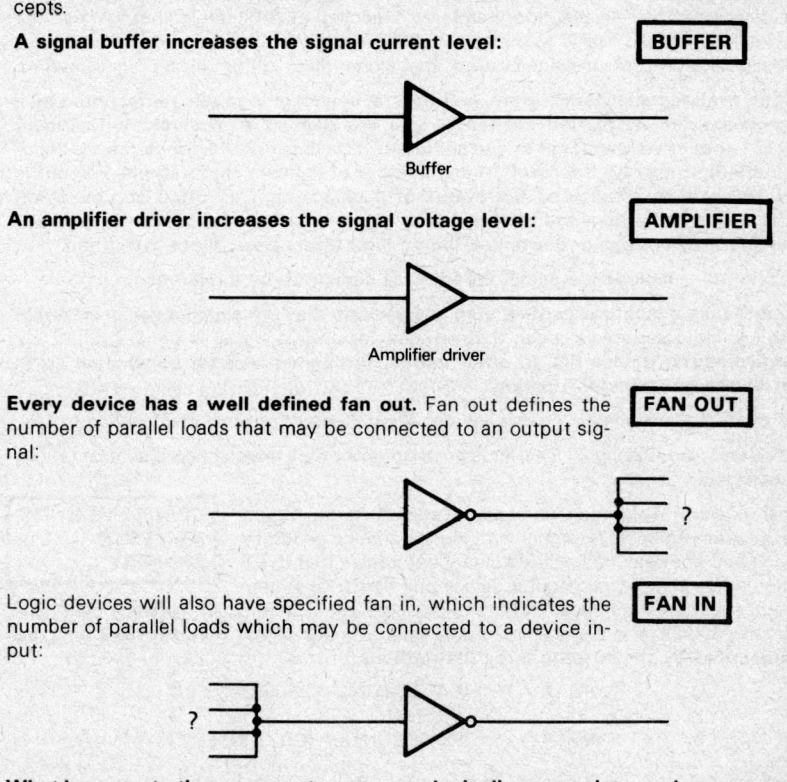

BUFFER

Buffer

**An amplifier driver increases the signal voltage level:**

AMPLIFIER

Amplifier driver

**Every device has a well defined fan out.** Fan out defines the number of parallel loads that may be connected to an output signal:

FAN OUT

Logic devices will also have specified fan in, which indicates the number of parallel loads which may be connected to a device input:

FAN IN

**What happens to these concepts once your logic disappears into a microcomputer program? The answer is simple: these concepts disappear — along with digital logic.**

Now, at the actual pins of a physical microcomputer device, fan in and fan out remain legitimate concepts; signals travelling between pins of individual microcomputer devices may need to be amplified and buffered. For example, a Z80 device's fan out may be as little as one or two Transistor-Transistor Logic (TTL) loads; that means that if more than one or two similar devices connect to an output signal, the output signal will have insufficient power to transmit usable signals to all connected devices. Therefore, for all but the simplest microcomputer configurations, bus lines will have to be buffered.

> FAN IN
> FAN OUT
> TTL LOADS
> SIGNAL BUFFERING

**When determining whether your bus lines need to be buffered, do not ignore leakage current.** For example, if you have sixteen ROM devices connected to the System Bus and only one device can be selected (and therefore connected) at any time, do not assume that the total signal load is due to the selected ROM. The fifteen unselected ROM devices will each tap off some leakage current; that alone may require System Bus buffering.

> LEAKAGE CURRENT

**Within a microcomputer program, however, when logic is totally represented by a microcomputer instruction sequence, you are dealing exclusively with binary digits — never with voltage or current levels. Fan in is infinite, since the status of a binary digit may be the result of any number of logical computations. Fan out is infinite since you can read the status of a binary digit as often as you want. Buffers and amplifiers are meaningless, since a binary digit has no qualities equivalent to voltage or current. A binary digit offers pure, finite resolution.**

**Take another look at the signal inverter, as simulated by a microcomputer.**

**We will take a giant conceptual step and assume that the signal inverter is buried within a logic sequence, such that no input or output signal is generated at any microcomputer device pin. In other words, the signal inverter becomes a small part of a larger transfer function.**

The input to the signal inverter is a binary digit created by some previous logic.

The output from the signal inverter is another binary digit which becomes input to subsequent logic.

Logic external to the microcomputer system does not supply the inverter input as a signal arriving at a microcomputer device pin, nor does the inverted signal get transmitted to external logic via a microcomputer device pin. Rather, the interface between external logic and the microcomputer system occurs at some point significantly before and beyond the signal inverter. **Our signal inverter may now be represented by these same three instructions:**

> COMPLEMENTING A BYTE OF MEMORY

```
        LD      A,(INVD)    ;LOAD ACCUMULATOR FROM INVD
        CPL                 ;COMPLEMENT
        LD      (INVD),A    ;STORE ACCUMULATOR CONTENTS AT INVD
```

**The source and destination become data memory bits; this may be illustrated as follows:**

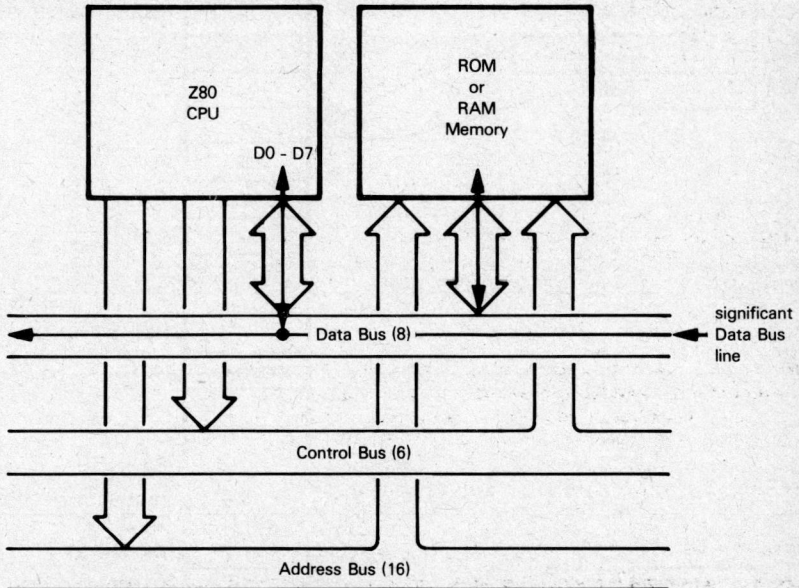

In terms of memory and CPU register contents, the signal inverter sequence proceeds as follows:

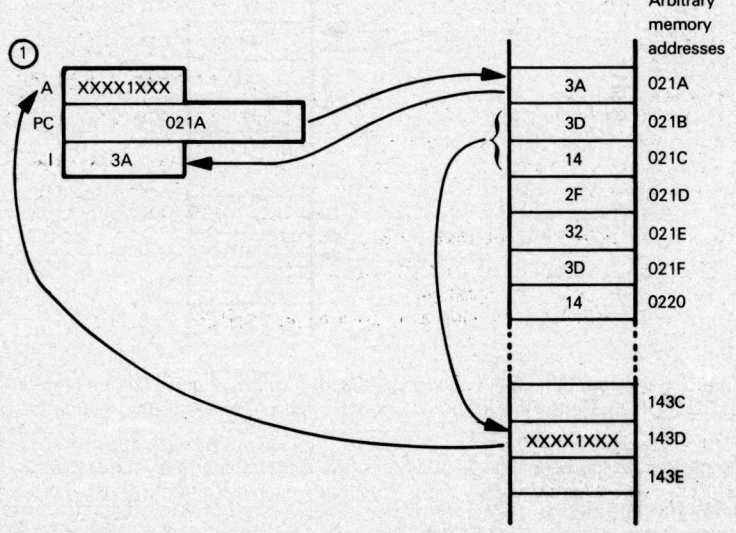

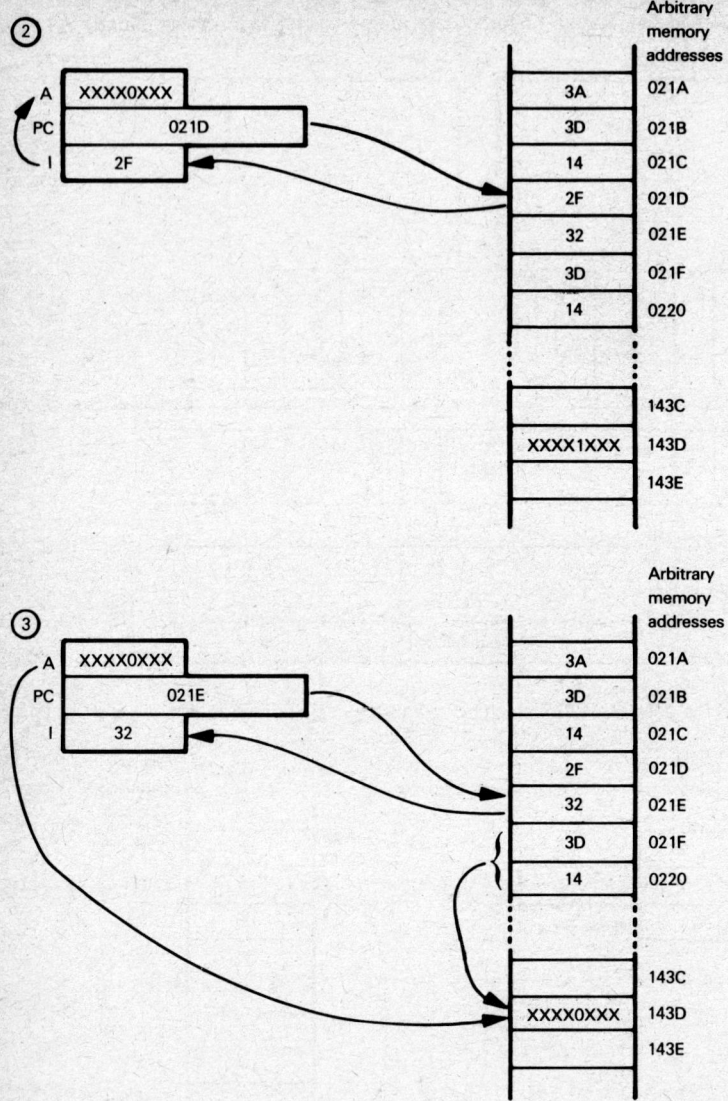

With regard to the illustration above, the letter A identifies the primary Accumulator of the Z80 CPU. PC represents the Program Counter, and I represents the Instruction register.

The contents of data memory byte $143D_{16}$ and the Accumulator are represented in binary format. X represents any binary digit. Note that we have arbitrarily selected bit 3 to be the significant bit.

In step ①, the LD A,(INVD) instruction is executed. This instruction causes the contents of data memory byte $143D_{16}$ to be loaded into the Accumulator.

During step ②, the CPL instruction is executed. This causes the contents of the Accumulator to be complemented.

During step ③, the contents of the Accumulator are loaded back into memory byte $143D_{16}$.

**Signal inversion has been simulated by inverting the contents of bit 3 (along with every other bit) of data memory byte $143D_{16}$.**

Where does the inverter's input come from? The answer is: from a data memory bit. **Let us suppose, to illustrate a point, that the inverter input is the OR of eight signals.** We could not wire-OR these eight signals to create an inverter input as follows:

**FAN IN IN MICROCOMPUTER PROGRAMS**

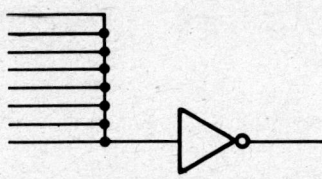

But, **presuming the eight signals are represented by the eight binary digit contents of the Accumulator, we would have no trouble generating the inverter input via the following logic sequence:**

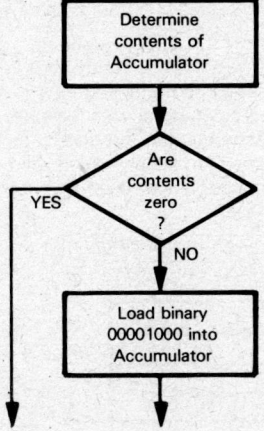

The fan in logic is implemented by this instruction sequence:

```
;ASSUME THE EIGHT SIGNALS ARE IN THE ACCUMULATOR
;EACH REPRESENTED BY ONE ACCUMULATOR BIT
         AND     A           ;AND ACCUMULATOR WITH ITSELF TO SET STATUS
                             ;FLAG
         JR      Z,NEXT      ;ACCUMULATOR HOLDS ZERO. SIGNAL IN
                             ;MUST BE 0
         LD      A,8         ;ACCUMULATOR HOLDS NONZERO SIGNAL IN
                             ;MUST BE 1
NEXT     LD      (INVD),A    ;SAVE INVERTER INPUT
```

**The above instruction sequence is a direct microcomputer program implementation of the eight-signal wire-OR. Let us examine how the instruction logic works.**

We are going to assume that the eight input signals are initially represented by the status of the eight Accumulator binary digits:

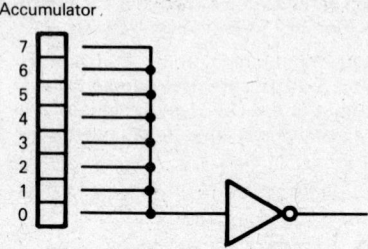

We are further going to assume that, in keeping with the prior illustration, bit 3 of the data byte will ultimately be the significant inverter signal bit.

Since the inverter input is the wire-OR of eight signals, program logic must set bit 3 of the Accumulator to 1 if any Accumulator bit is non-zero; bit 3 of the Accumulator must be set to 0 if all Accumulator bits are zero. The contents of the Accumulator are then stored in the data memory byte represented by label INVD. With regard to the previous illustration, INVD would be a label representing memory byte $143D_{16}$.

This is how the four-instruction sequence illustrated above works:

We do not know what the Accumulator initially contains, so we must determine its contents by setting CPU status flags appropriately. To do this, we AND the Accumulator contents with itself. ANDing the contents of the Accumulator with itself does not change the contents of the Accumulator, but status flags are set. We are only interested in the Zero status, which will be set to 1 if the AND of the Accumulator with itself generates a zero result; the Zero status flag will be set to 0 otherwise.

**STATUS DETERMINATION BY ANDING A REGISTER WITH ITSELF**

But the AND of the Accumulator with itself will only be zero if the Accumulator contains zero:

| X | Y | X ∧ Y |
|---|---|---|
| 0 | 0 | 0 |
| 0 | 1 | 0 |
| 1 | 0 | 0 |
| 1 | 1 | 1 |

→ $0 \wedge 0 = 0$

} not applicable

→ $1 \wedge 1 = 1$

Thus, after execution of the AND instruction, if the Zero status is 1 then bit 3 of the Accumulator must already be 0, which is what we want it to be. No operation is required, and we jump to the LD (INVD),A instruction.

If the Zero status bit was 0, then one or more bits of the Accumulator are non-zero. The LD A,8 instruction loads a 1 into bit 3 of the Accumulator:

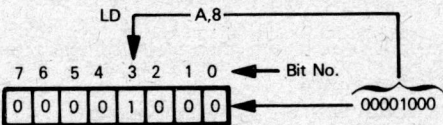

Finally, the LD (INVD),A instruction is executed to load the inverter input signal into the appropriate data memory byte.

**Now suppose the inverter output is distributed to numerous subsequent devices.**

**The following logic represents fan out that is not feasible:**

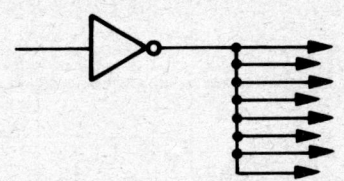

Within a microcomputer program, the whole concept of fan out disappears. The inverter output may be accessed an indefinite number of times by the simple re-execution of an LD instruction:

| **FAN OUT IN MICROCOMPUTER PROGRAMS** |

```
        LD      A,(INVD)    ;LOAD INVERTER OUTPUT INTO ACCUMULATOR
        -
        -
        -
        LD      A,(INVD)    ;LOAD INVERTER OUTPUT INTO ACCUMULATOR
        -
        -
        -
        LD      A,(INVD)    ;LOAD INVERTER OUTPUT INTO ACCUMULATOR
        -
        -
        -
        LD      A,(INVD)    ;LOAD INVERTER OUTPUT INTO ACCUMULATOR
        -
        -
        -
        LD      A,(INVD)    ;LOAD INVERTER OUTPUT INTO ACCUMULATOR
```

**What about amplifiers and buffers? Clearly, within the context of binary data stored in memory they have no meaning.** If amplifiers and buffers are present because of the electrical characteristics of the memory and processor chips, that has nothing to do with the logic function being implemented by a microcomputer program.

# MICROCOMPUTER SIMULATION OF 7404/05/06 HEX INVERTERS

**These three hex inverters differ only in their electrical characteristics:**

- **The 7404 is a simple hex inverter.**
- **The 7405 is a hex inverter with open collector outputs.**
- **The 7406 is a hex inverter buffer/driver with open collector, high voltage outputs.**

**Since these three devices differ only in their electrical characteristics, within a microcomputer assembly language simulation they are identical. Let us look at the 7404.** It consists of six independent signal inverters, which may be illustrated as follows:

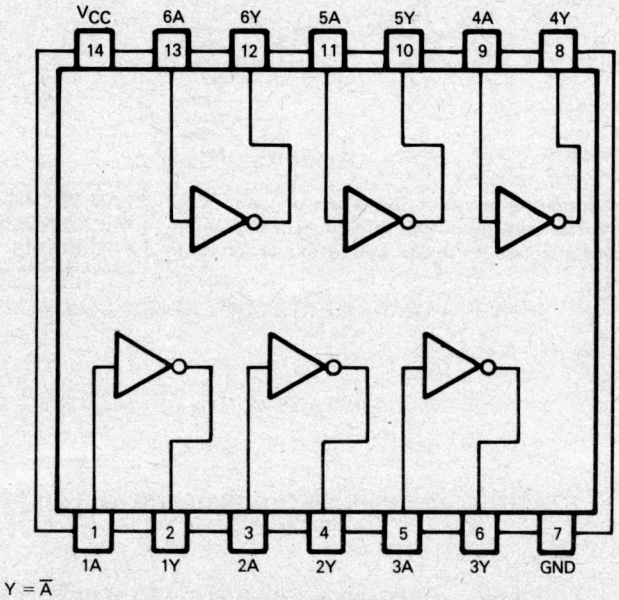

$Y = \overline{A}$

**The instruction sequence to represent a hex inverter is identical to the three-instruction, single-signal inverter instruction sequence, because Z80 microcomputers are eight-bit parallel devices.** Whether you like it or not, this inverter instruction sequence inverts eight independent binary digits. Hex inverters may therefore be represented within a microcomputer instruction sequence as follows:

```
LD      A,(INVD)      ;LOAD ACCUMULATOR FROM INVD
CPL                   ;COMPLEMENT
LD      (INVD),A      ;STORE ACCUMULATOR CONTENTS TO INVD
```

We will arbitrarily identify significant bits, as implied by the hex inverter, as follows:

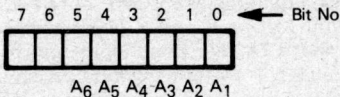

Note that the above selection of significant bits is completely arbitrary. There is absolutely no practical or philosophical argument favoring any one bit assignment as compared to any other.

## MICROCOMPUTER SIMULATION OF 7408/09 QUADRUPLE, TWO-INPUT POSITIVE AND GATES

**These two devices provide four independent two-input one-output AND gates, which may be illustrated as follows:**

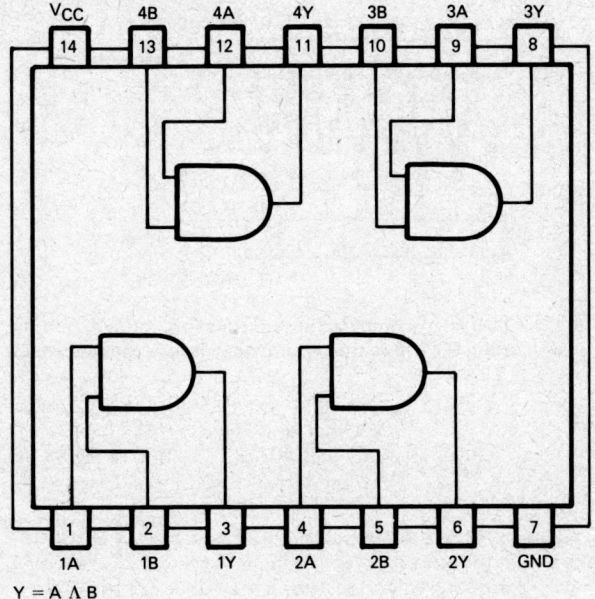

$Y = A \wedge B$

The 7409 has open collector outputs, which differentiates it from the 7408. This difference has no meaning in a microcomputer program simulation; therefore, the two devices can be looked on as being identical.

# TWO-INPUT FUNCTIONS

**From the microcomputer programmer's point of view, the most significant difference between a 7408 AND gate and a 7404 inverter is not the logic function; rather, it is the fact that a 7408 is a two-input device.** Conceptually, we might imagine a 7404 being simulated in one of the two following ways:

1) The eight input signals are loaded into the CPU Accumulator register. Each even-numbered bit is ANDed with the bit to its right. The result is deposited in the even-numbered bit for each bit pair:

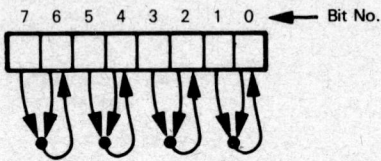

2) The two sets of four inputs are loaded into the CPU Accumulator and one other register. The result is returned in the Accumulator:

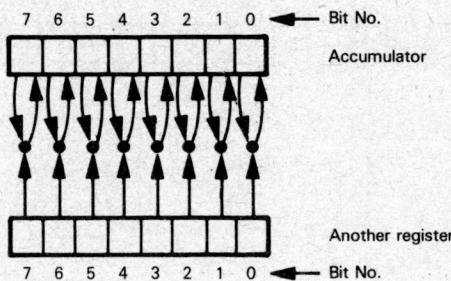

Upon examining the Z80 microcomputer instruction set, you will find that the second method of simulating a 7408 is the natural one. This is the required instruction sequence:

```
LD      A,(SRCA)    ;LOAD FIRST SET OF INPUTS FROM SRCA
LD      B,A         ;SAVE IN THE B REGISTER
LD      A,(SRCB)    ;LOAD SECOND SET OF INPUTS FROM SRCB
AND     B           ;AND B WITH A
LD      (DST),A     ;SAVE RESULT IN DST
```

**If the use of labels SRCA, SRCB and DST still confuses you, let us take a minute to clarify them.** Eventually, you will have some amount of memory, which may vary from as little as 256 bytes to as much as 65,536 bytes. Each of the labels SRCA, SRCB and DST identifies one memory byte. At the time

**SOURCE PROGRAM LABEL ASSIGNMENTS**

you are writing the source program, the exact memory byte identified by each label is unimportant. When you eventually assemble your source program, the assembler listing will print a memory map. The memory map will identify the exact memory byte associated with each label you have used. By examining the memory map, you will be able to determine whether or not all label assignments are valid. If any label assignments are invalid, you will have to take appropriate action. Appropriate action may involve adding more memory to your microcomputer configuration, or you may have to rewrite your source program so that it makes more effective use of the memory you have.

The problem of labels and memory allocations is irrelevant at the present level of discussion. Simply imagine every label as addressing one specific memory byte. Do not worry about which memory byte will eventually be addressed, and your problem will disappear.

**The 7408 simulation instruction sequence illustrated above by no means represents the only way in which a 7408 may be simulated.**

**First consider some minor variations.** CPU Registers C, D, E, H or L could be used instead of Register B to hold the second data input. Here is one example:

| | | |
|---|---|---|
| LD | A,(SRCA) | ;LOAD FIRST SET OF INPUTS FROM SRCA |
| LD | C,A | ;SAVE IN THE C REGISTER |
| LD | A,(SRCB) | ;LOAD SECOND SET OF INPUTS FROM SRCB |
| AND | C | ;AND C WITH A |
| LD | (DST),A | ;SAVE RESULT IN DST |

**Using Registers H or L to hold the second input is not encouraged. The primary use for these two registers is to hold a data memory address.** For example, the instructions LD A,(SRCA); LD A,(SRCB); and LD (DST),A could be replaced as follows:

**IMPLIED MEMORY ADDRESSING**

| | | |
|---|---|---|
| LD | HL,SRCA | ;LOAD ADDRESS FOR FIRST SET OF INPUTS INTO H,L |
| LD | A,(HL) | ;LOAD FIRST SET OF INPUTS INTO A |
| LD | HL,SRCB | ;LOAD ADDRESS OF SECOND SET OF INPUTS INTO ;H,L |
| AND | (H,L) | ;AND SECOND SET OF INPUTS WITH A |
| LD | HL,DST | ;LOAD ADDRESS OF DESTINATION INTO H,L |
| LD | (HL),A | ;STORE RESULT IN DST |

# THE MICROCOMPUTER SIMULATION OF A 7411 TRIPLE, THREE-INPUT POSITIVE AND GATE

The principal difference between the 7411 AND gate and the 7408 AND gate is the number of input signals. The 7411 generates three output signals, each of which is the AND for three inputs:

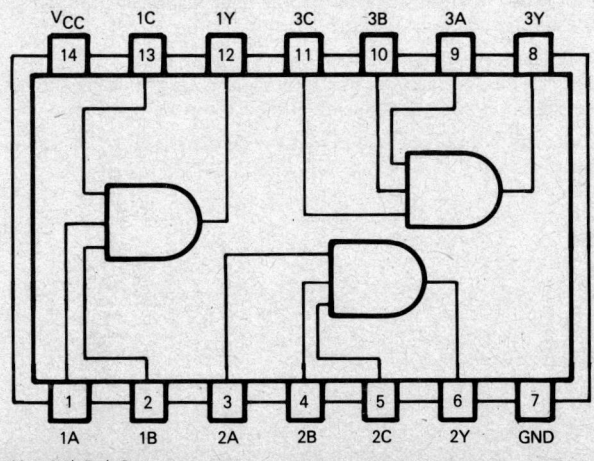

$Y = A \wedge B \wedge C$

## THREE INPUT FUNCTIONS

**Again we are faced with choices. We may load the three sets of inputs into three CPU registers (the Accumulator and two other registers), then perform two ANDs before restoring the result:**

```
ONE     LD    A,(SRCA)    ;LOAD FIRST SET OF INPUTS FROM SRCA
TWO     LD    B,A         ;SAVE IN B REGISTER
THREE   LD    A,(SRCB)    ;LOAD SECOND SET OF INPUTS FROM SRCB
FOUR    LD    C,A         ;SAVE IN C REGISTER
FIVE    LD    A,(SRCC)    ;LOAD THIRD SET OF INPUTS FROM SRCC
SIX     AND   B           ;AND B WITH A
SEVEN   AND   C           ;AND C WITH A
EIGHT   LD    (DST),A     ;SAVE THE RESULT IN DST
```

The instructions in the above sequence have been given labels so as to make the description which follows easier to understand. The instructions do not need labels in order to satisfy the needs of an assembly language source program.

When instruction ONE executes, an 8-bit value is loaded into the Accumulator from the memory byte addressed by label SRCA. We will assume that AND gate inputs are represented as follows:

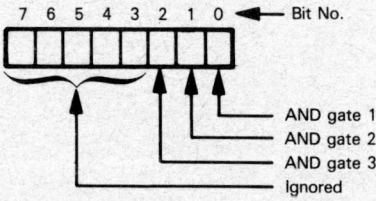

Understand that the assignment of data bits illustrated above is completely arbitrary. It is only necessary that all subsequent inputs be consistent.

After instruction ONE has executed, the first set of inputs is in the Accumulator. The Accumulator is the only CPU register into which data may be loaded if you use direct addressing. The first set of inputs must therefore be saved in another register, so that the Accumulator is free for a second set of inputs to be loaded. Instruction TWO moves the contents of the Accumulator to the B register.

Instructions THREE and FOUR load the second set of inputs into the Accumulator, then move it to the C register. We assume that bit assignments of this second set of inputs are identical to the bit assignments illustrated above for the first input.

The third and last set of inputs is loaded into the Accumulator by instruction FIVE.

The AND instruction ANDs the contents of the CPU register with the contents of the Accumulator, leaving the result in the Accumulator. Instruction SIX performs the first AND as follows:

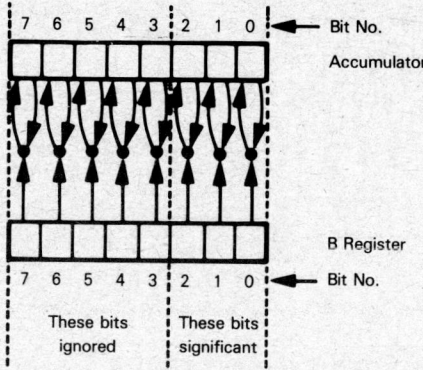

Instruction SEVEN performs the second AND operation. This time the AND occurs between the Accumulator and Register C. The Accumulator initially holds the result of the AND with B, illustrated above. After instruction SEVEN has executed, the AND of three inputs is in the Accumulator.

Instruction EIGHT returns the final result to a memory byte addressed by the label DST. The 7411 AND gate simulation is complete.

**Now consider an alternative simulation of the 7411 AND gates.** We may load the first input into the Accumulator and the second input into another register. After ANDing these two inputs, we may load the third input into the same "other" register, AND it with the result of the first AND, then return the result:

| ONE | LD | A,(SRCA) | ;LOAD FIRST SET OF INPUTS FROM SRCA |
| TWO | LD | B,A | ;SAVE IN B REGISTER |
| THREE | LD | A,(SRCB) | ;LOAD SECOND SET OF INPUTS FROM SRCB |
| FOUR | AND | B | ;AND B WITH A, THE RESULT IS IN A |
| FIVE | LD | B,A | ;SAVE THE RESULT IN B |
| SIX | LD | A,(SRCC) | ;LOAD THIRD SET OF INPUTS FROM SRCC |
| SEVEN | AND | B | ;AND B WITH A |
| EIGHT | LD | (DST),A | ;SAVE THE RESULT IN DST |

Let us compare this second simulation of the 7411 AND gate with the first simulation. Instructions ONE, TWO and THREE are identical to the first simulation. After these three instructions have executed, one set of inputs is in Register B and a second set of inputs is in the Accumulator. This is the situation:

> Inputs A are in Register B
>
> Inputs B are in the Accumulator

Now, instead of bringing the third set of inputs immediately into a CPU register, we execute instruction FOUR, which generates the AND of the first two inputs. Since this AND is generated in the Accumulator, we save the result in Register B by executing instruction FIVE. This is the net effect:

> A Λ B in Register B

Instruction SIX now loads the third set of inputs into the Accumulator. Instruction SEVEN ANDs the third set of inputs with the result of the first AND, as follows:

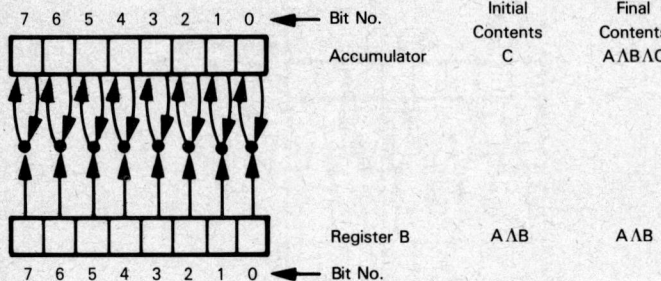

|  | Initial Contents | Final Contents |
|---|---|---|
| Accumulator | C | A∧B∧C |
| Register B | A∧B | A∧B |

Instruction EIGHT saves the result from the Accumulator in the memory byte addressed by label DST.

## MINIMIZING CPU REGISTER ACCESSES

**Which is the "better" 7411 AND gates' simulation? Clearly, the second option is better.** There is a non-obvious problem associated with the indiscriminate use of CPU registers. We have arbitrarily decided that Register B will hold a second input. So long as we are simulating 7411 AND gates without regard to what precedes or follows, the selection of Register B is arbitrary; its selection carries no rewards or consequences.

Invariably, an instruction sequence such as the 7411 AND gates' simulation is just a small part of a larger whole. Now we must worry about whether using Register B to house the second input will interfere with prior or subsequent use of Register B. A very common programming error involves CPU register utilization conflicts.

**CONFLICTS IN CPU REGISTER UTILIZATION**

For example, what if some prior logic step uses Register B to hold an intermediate data value? Now the 7411 simulation will wipe out the data which was being temporarily stored in this register.

**In order to reduce CPU register conflicts, it is always preferable to choose an instruction sequence that uses as few CPU registers as possible, providing there is no significant penalty. In this case, there is no significant penalty. It takes no more instructions to simulate 7411 AND gates using CPU Register B only than it does using CPU Registers B and C. Using CPU Register B only is therefore the better method.**

**Now let us consider a 7411 AND gate's simulation using implied addressing.** Assume that the three inputs to the AND gates are stored in sequential bytes of data memory and that the destination follows the last source byte, as follows:

**IMPLIED ADDRESSING**

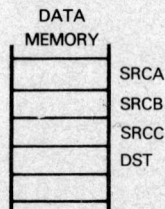

Now, using implied addressing, we have the following instruction sequence:

```
ONE     LD    HL,SRCA   ;LOAD THE FIRST SOURCE ADDRESS INTO HL
TWO     LD    A,(HL)    ;LOAD THE FIRST SOURCE INTO THE ACCUMULATOR
THREE   INC   HL        ;INCREMENT THE IMPLIED ADDRESS
FOUR    AND   (HL)      ;AND ACCUMULATOR WITH SECOND SOURCE
FIVE    INC   HL        ;INCREMENT THE IMPLIED ADDRESS
SIX     AND   (HL)      ;AND ACCUMULATOR WITH THIRD SOURCE
SEVEN   INC   HL        ;INCREMENT THE IMPLIED ADDRESS
EIGHT   LD    (HL),A    ;SAVE THE RESULT
```

**This is how the instruction sequence will be executed:**

Instruction ONE loads the address of the first source byte into the H and L registers.

Instruction TWO moves the contents of the memory byte addressed by H and L into the Accumulator.

Instruction THREE increments the 16-bit address in the H and L registers, which now addresses SRCB.

Instruction FOUR ANDs the contents of the Accumulator with the second source, as addressed by the H and L registers. The result is saved in the Accumulator. This may be illustrated as follows:

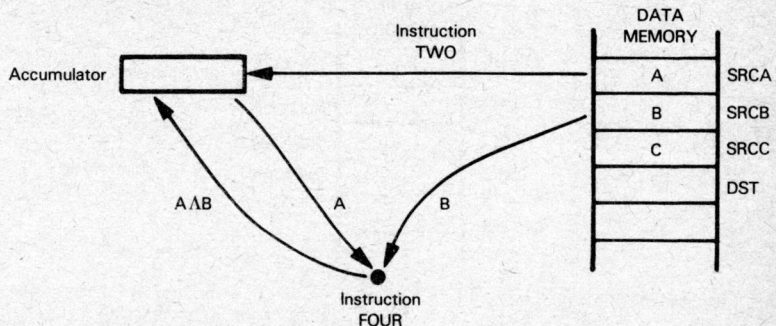

Instructions FIVE and SIX increment the implied address and repeat the AND operation, this time ANDing the third input with the AND of the first two inputs. This may be illustrated as follows:

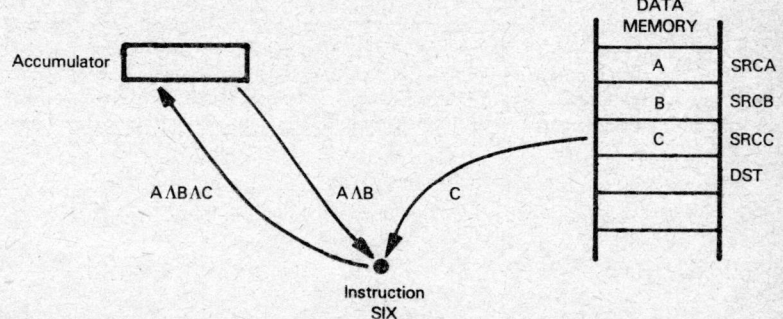

Instruction SEVEN increments the address in H and L again so that it now points to DST. Instruction EIGHT saves the result in the destination, as follows:

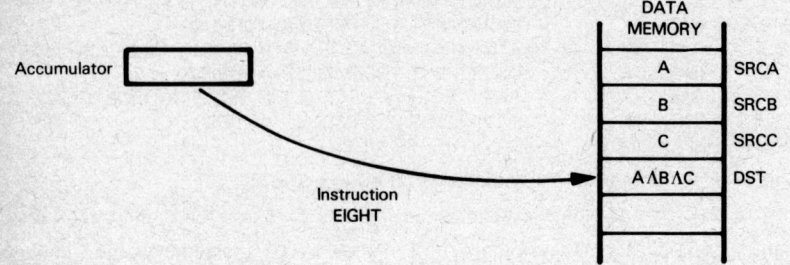

**We can use base relative addressing in the simulation of a 7411 AND gate.** As in the previous example, we assume that the three inputs are stored in sequential bytes of data memory and the destination follows the last source byte. We can think of each of these locations in terms of its relative distance from SRCA:

**BASE RELATIVE ADDRESSING**

Here is the instruction sequence:

```
ONE     LD    IX,SRCA     ;LOAD THE FIRST SOURCE ADDRESS INTO IX
TWO     LD    A,(IX+0)    ;LOAD THE FIRST SOURCE INTO THE ACCUMULATOR
THREE   AND   A,(IX+1)    ;AND ACCUMULATOR WITH SECOND SOURCE
FOUR    AND   A,(IX+2)    ;AND ACCUMULATOR WITH THIRD SOURCE
FIVE    LD    (IX+3),A    ;SAVE THE RESULT
```

As far as the Accumulator and data memory are concerned, this sequence operates exactly as the previous one. The Address register, however, is used in a different way: instead of incrementing the register before the next memory access, an index is added to the base address, leaving the register contents unchanged. This is base relative addressing, as described in <u>An Introduction to Microcomputers: Volume I — Basic Concepts</u>.

Here is the execution of the sequence, step-by-step:

Instruction ONE loads the address of the first source byte into Index Register X.

When instruction TWO is executed, the index 0 is added to the contents of Index Register X to obtain the address of the first source byte. That byte is then moved into the Accumulator. This may be illustrated as follows:

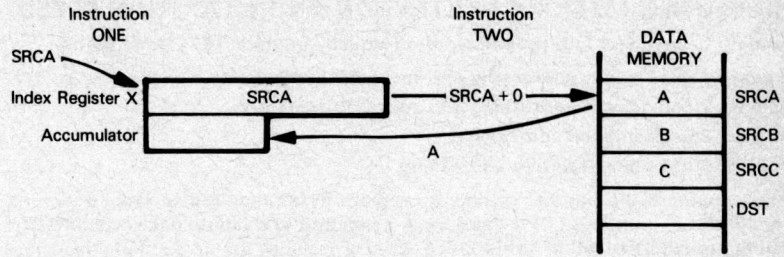

Instruction THREE ANDs the Accumulator contents with the second source byte, which is addressed by adding the index 1 to the contents of Index Register X. This may be illustrated as follows:

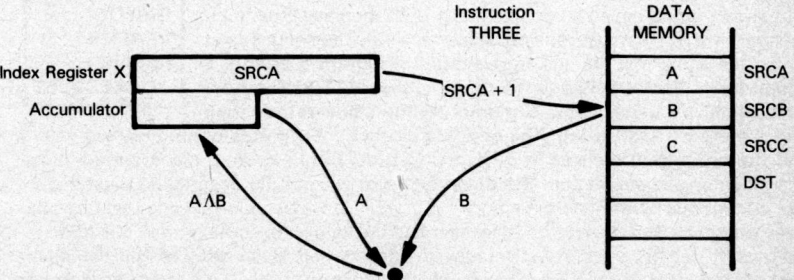

Instruction FOUR ANDs the Accumulator contents with the third source byte, which is addressed by adding the index 2 to the contents of Index Register X.

Instruction FIVE loads the Accumulator contents in the location addressed by adding 3 to the contents of Index Register X. Thus, the AND of the three source bytes is saved in the destination byte. This may be illustrated as follows:

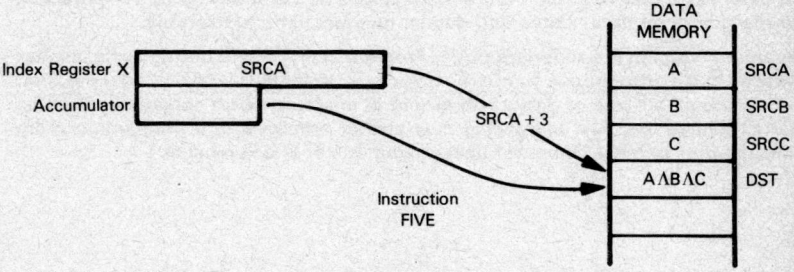

Although this instruction sequence has fewer lines of code than the preceding ones, it is actually less efficient, as we will demonstrate below. This is not an appropriate use for base relative addressing.

## COMPARING MEMORY UTILIZATION AND EXECUTION SPEED

We now have these four programs, all of which simulate 7411 AND gates:

Program 1 uses direct addressing and three CPU registers.

Program 2 uses direct addressing and two CPU registers.

Program 3 uses implied addressing.

Program 4 uses base relative addressing.

Let us compare the number of object program bytes required to store each program with the number of CPU clock cycles required to execute each program. The results are summarized in Table 2-1. Table 2-1 includes the instruction mnemonics for each program, to help you follow how total object program bytes and execution cycles have been computed. See Chapter 6 for the data you will need in order to verify Table 2-1.

Programs 1 and 2 have identical memory utilization and execution speeds — which is not surprising, since they vary the sequence in which the same instructions are executed. **Program 3 adopts a completely different philosophy towards the 7411 AND gates' simulation, by using implied memory addressing rather than direct memory addressing. The result is dramatic. Six bytes of memory are saved, and the program executes in 82% of the time.** But Program 3 places an additional restriction on the simulation; the three data sources and the destination must occupy four contiguous bytes of data memory. Program 4 has fewer lines of code than the other three programs, but it saves no bytes and has the longest execution time. In addition, it restricts the location of the data sources and destination. **Base relative addressing** is a sophisticated feature which can save time and program space, but it **is not appropriate for this particular program.**

`DIRECT VERSUS IMPLIED ADDRESSING`

How are we going to rank the three simulation options?

**The sophisticated addressing scheme of Program 4 is not suited to this application. We have already concluded that Program 2 beats Program 1, because Program 1 makes gratuitous use of an extra CPU register. Program 3 is clearly better than Program 2, providing the restriction on data source and destination locations is tolerable.**

`PROGRAM VARIATIONS RANKED`

**Regarding Program 3's superiority over Program 2, it is worth noting again, as was stressed in <u>An Introduction to Microcomputers: Volume I — Basic Concepts</u>, that the indiscriminate use of direct addressing in microcomputer applications can be costly. Implied memory addressing may appear primitive to a programmer with minicomputer or large computer background, but it is economical.**

| PROGRAM 1 | | | PROGRAM 2 | | | PROGRAM 3 | | | PROGRAM 4 | | |
|---|---|---|---|---|---|---|---|---|---|---|---|
| MNEMONIC | BYTES | CYCLES | MNEMONIC | BYTES | CYCLES | MNEMONIC | BYTES | CYCLES | MNEMONIC | BYTES | CYCLES |
| LD[1] | 3 | 13 | LD[1] | 3 | 13 | LD[1] | 3 | 10 | LD[1] | 4 | 14 |
| LD[2] | 1 | 4 | LD[2] | 1 | 4 | LD[1] | 1 | 7 | LD[1] | 3 | 19 |
| LD[1] | 3 | 13 | LD[1] | 3 | 13 | INC | 1 | 6 | AND[1] | 3 | 19 |
| LD[2] | 1 | 4 | AND[2] | 1 | 4 | AND[1] | 1 | 7 | AND[1] | 3 | 19 |
| LD[1] | 3 | 13 | LD[2] | 1 | 4 | INC | 1 | 6 | LD[1] | 3 | 19 |
| AND[2] | 1 | 4 | LD[1] | 3 | 13 | AND[1] | 1 | 7 | | | |
| AND[2] | 1 | 4 | AND[2] | 1 | 4 | INC | 1 | 6 | | | |
| LD[1] | 3 | 13 | LD[1] | 3 | 13 | LD[1] | 1 | 7 | | | |
| TOTAL | 16 | 68 | TOTAL | 16 | 68 | TOTAL | 10 | 56 | TOTAL | 16 | 90 |

[1] Register-memory version of instruction
[2] Register-register version of instruction

Table 2-1. Comparing Memory Utilization and Program Execution Speed for 7411 AND Gates' Simulation

# THE MICROCOMPUTER SIMULATION OF A 7474 DUAL, D-TYPE POSITIVE EDGE TRIGGERED FLIP-FLOP WITH PRESET AND CLEAR

Before looking at the 7474 flip-flop in particular, let us consider flip-flops in general. First, a few definitions.

## A DIGITAL LOGIC DESCRIPTION OF FLIP-FLOPS

**A flip-flop is a bi-stable logic device,** that is, a device which may exist in one of two stable conditions. 7474-type flip-flops have two outputs, Q and $\overline{Q}$; thus, the two bi-stable conditions may be represented as follows:

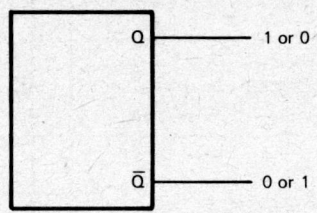

A clock signal causes the flip-flop to change from one bi-stable condition to the other. A positive edge triggered flip-flop changes upon sensing a zero-to-one transition of the clock signal:

**POSITIVE EDGE TRIGGER**

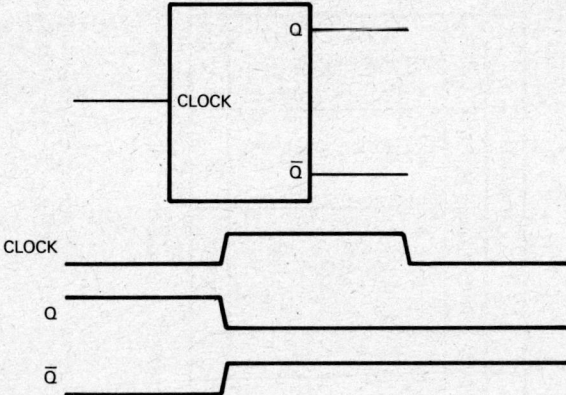

A negative edge triggered flip-flop changes state upon sensing a one-to-zero clock signal transition:

**NEGATIVE EDGE TRIGGER**

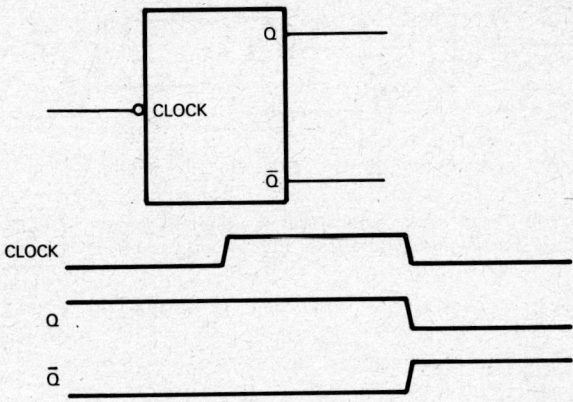

A JK flip-flop preconditions the Q and $\bar{Q}$ outputs which will be generated by the next clock edge trigger, as follows:

**JK FLIP-FLOP**

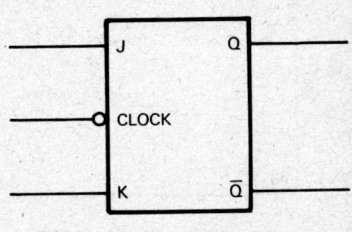

| Status of J and K at clock signal | | Outputs generated at clock signal | |
|---|---|---|---|
| J | K | Q | $\bar{Q}$ |
| 1 | 0 | 1 | 0 |
| 0 | 1 | 0 | 1 |
| 0 | 0 | Remain in previous state | |
| 1 | 1 | Change state regardless of previous state | |

In the table above, "clock signal" will be a zero-to-one transition for a positive edge triggered device; it will be a one-to-zero transition for a negative edge triggered device. This definition of "clock signal" also applies to the D-type flip-flop described next.

**CLOCK SIGNAL**

By inverting a J input in order to generate the K input, a D-type flip-flop is created. These are the D-type flip-flop characteristics that result:

**D-TYPE FLIP-FLOP**

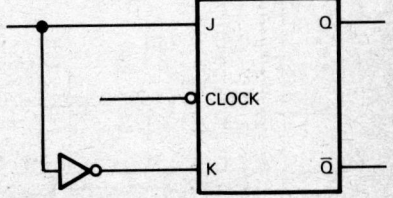

| Status of J and K at clock signal | | Outputs generated at clock signal | |
|---|---|---|---|
| J=D | K=$\bar{J}$ | Q | $\bar{Q}$ |
| 1 | 0 | 1 | 0 |
| 0 | 1 | 0 | 1 |

Here is a positive edge triggered, D-type flip-flop timing diagram:

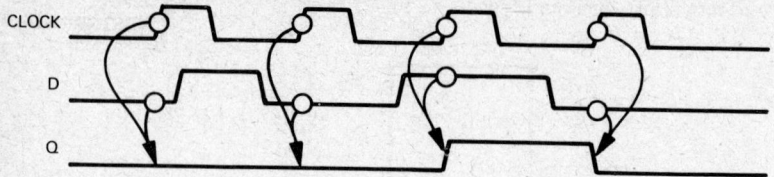

A D-type flip-flop, therefore, will always output the input conditions that existed at the previous clock pulse.

The presence of a Preset input means that the flip-flop may be forced to output $Q = 1$ and $\overline{Q} = 0$. Preset true forces this condition. **FLIP-FLOP PRESET**

A Clear input is the opposite of a Preset input. When true, the Clear input forces $Q = 0$ and $\overline{Q} = 1$. **FLIP-FLOP CLEAR**

Combining the definitions given above, this is what we get for a 7474 type flip-flop:

### FUNCTION TABLE

| INPUTS | | | | OUTPUTS | |
|---|---|---|---|---|---|
| 1PR or 2PR | 1CLR or 2CLR | 1CK or 2CK | 1D or 2D | 1Q or 2Q | $1\overline{Q}$ or $2\overline{Q}$ |
| L | H | X | X | H | L |
| H | L | X | X | L | H |
| L | L | X | X | H* | H* |
| H | H | ↑ | H | H | L |
| H | H | ↑ | L | L | H |
| H | H | L | X | $Q_0$ | $\overline{Q}_0$ |

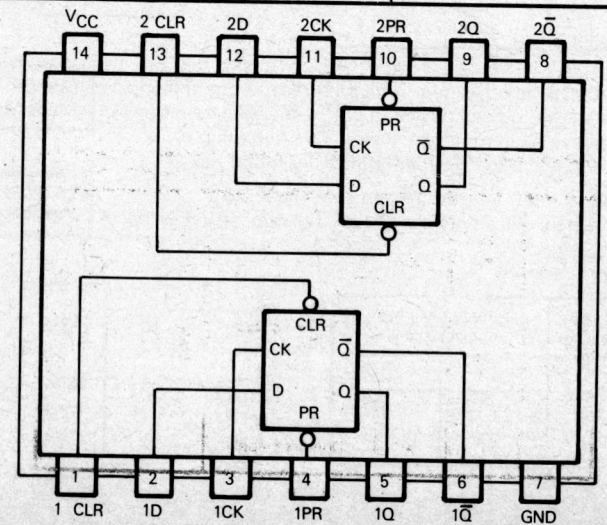

In the function table above, ↑ represents a clock zero-to-one transition. H* signifies an unstable state. $Q_0$ is the previous state for Q. X signifies "don't care"

# AN ASSEMBLY LANGUAGE SIMULATION OF FLIP-FLOPS

**Now, our first problem, when trying to simulate a 7474 flip-flop, is the fact that there is no clock signal within a microcomputer instruction set. Instead, we must assume that events are triggered by execution of an appropriate instruction rather than a clock signal transition.**

**How will we represent outputs Q and $\overline{Q}$? Two bits of memory could be used to represent these two outputs:**

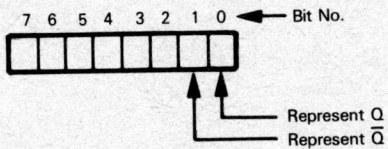

**Since we are dealing with data and not signals, $\overline{Q}$ is redundant. The single flip-flop therefore reduces to one memory bit.** A 7474 device, since it contains two flip-flops, reduces to two memory bits, one for each flip-flop implemented on the chip.

There is nothing surprising about this conclusion. Each bit of a microcomputer's read/write memory is a simple, bi-stable element; it could, indeed, be a flip-flop.

**The logic of a 7474 flip-flop may be represented by instructions that clear a memory bit, set the memory bit to 1, or store an unknown binary digit in the memory bit.**

Suppose memory bits are assigned as follows:

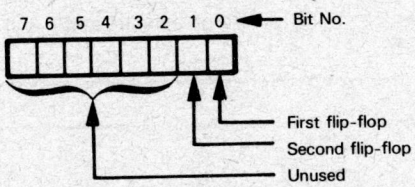

The 7474 function table now becomes these instructions:

| Preset | Clear | D | First flip-flop | | Second flip-flop | |
|--------|-------|---|---|---|---|---|
| L<br>H | H<br>H | X<br>H | LD<br>SET<br>LD | A,(FLP)<br>0,A<br>(FLP),A | LD<br>SET<br>LD | A,(FLP)<br>1,A<br>(FLP),A |
| H<br>H | L<br>H | X<br>L | LD<br>RES<br>LD | A,(FLP)<br>0,A<br>(FLP),A | LD<br>RES<br>LD | A,(FLP)<br>1,A<br>(FLP),A |
| L | L | X | Does not apply | | | |

With regard to the table above, bits 0 and 1 of the memory word identified by FLP are presumed equivalent to the two flip-flops of the 7474 device. The LD instructions move the word between memory and the Accumulator. In the Accumulator, the SET instruction sets the appropriate bit to 1; the RES instruction sets the specified bit to 0.

## MICROCOMPUTER SIMULATION OF FLIP-FLOPS IN GENERAL

**In conclusion, a flip-flop becomes a single bit of read/write memory within a microcomputer system.**

Within a microcomputer system, all flip-flops are the same. Flip-flop logic reduces to these four questions:

1) When do I execute an instruction to set a memory bit to 1?
2) When do I execute an instruction to set a memory bit to 0?
3) When do I execute an instruction to store a binary digit in a memory bit?
4) When do I execute an instruction to read the contents of a memory bit?

## THE MICROCOMPUTER SIMULATION OF REAL TIME DEVICES

**There are two types of real time devices that we will look at: the one-shot (including monostable multivibrators) and the master-slave flip-flop. Specifically, these devices will be described:**

- The Signetics 555 monostable multivibrator
- The 74121 monostable multivibrator
- The 74107 dual J-K master-slave flip-flop with Clear

**A one-shot is a device which generates a signal pulse with a specific time period:** | **ONE-SHOT**

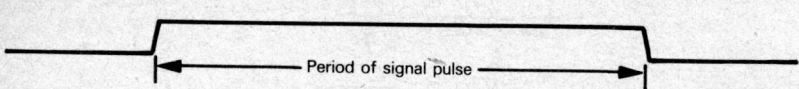

A monostable multivibrator is a device with one stable, or passive, state. It produces one-shot output signals, as illustrated above, where the pulse is in the unstable, or active, state: | **MONOSTABLE MULTIVIBRATOR**

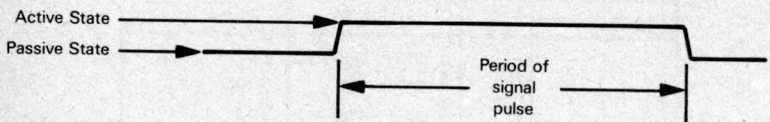

The device is a "multivibrator" because it can output a continuous stream of signals — much like a clock signal. In other words, a multivibrator output consists of a continuous stream of one-shot signals.

**The time period of the signal pulse is a real time value — it is a finite number of microseconds, or milliseconds, or even seconds.**

**A master-slave flip-flop is a flip-flop which generates output signals based on the condition of input signals at some earlier time.** Again we encounter a real time value — the delay between inputs and outputs. | **MASTER-SLAVE FLIP-FLOP**

# THE 555 MONOSTABLE MULTIVIBRATOR
**The Signetics 555 monostable multivibrator may be illustrated as follows:**

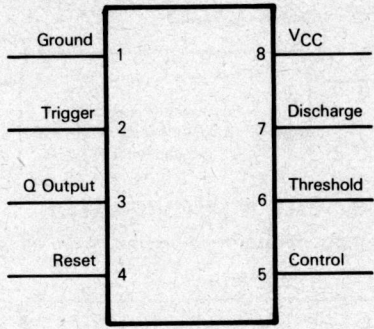

The negative edge of a clock signal at the Trigger input (pin 2) causes a negative-to-positive transition at the Output Q. The duration of the high-level output at Q is controlled by a resistor/capacitor circuit connected to the Discharge and Threshold pins (7 and 6, respectively).

Reset is a standard reset input; a low input will hold the Q output low.

The Control pin is used to control voltage within the multivibrator; it is not significant to an overall understanding of how the 555 device works.

The ground and power pins (1 and 8, respectively) are self-explanatory.

**Here is one way in which the 555 monostable multivibrator may be configured:**

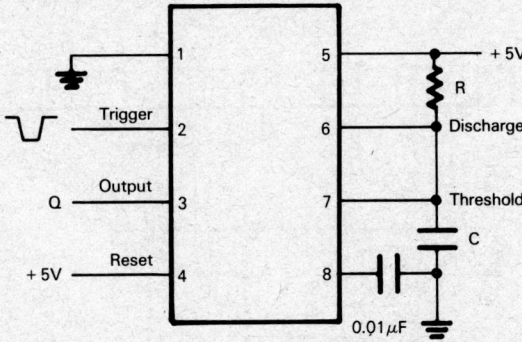

As soon as a high-to-low signal level is sensed at the Trigger input, the capacitor between pin 6 and ground charges. Signal levels at the Threshold and Discharge pins, as controlled by the resistor R and the capacitor C, control the period for which Q will output high. This time period is given by the following equation:

$$T = 1.1RC$$

Where:

    T is time in seconds
    R is resistance in megohms
    C is capacitance in microfarads

An output signal pulse is generated as follows:

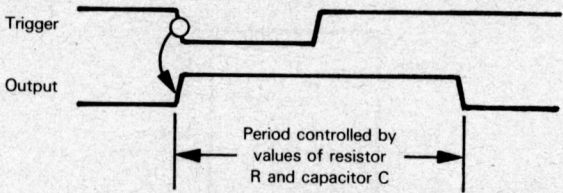

## THE 74121 MONOSTABLE MULTIVIBRATOR

The 74121 monostable multivibrator may be illustrated as follows:

**FUNCTION TABLE**

| INPUTS | | | OUTPUTS | | |
|---|---|---|---|---|---|
| A1 | A2 | B | Q | $\bar{Q}$ | |
| L | X | H | L | H | Monostable outputs |
| X | L | H | L | H | |
| X | X | L | L | H | |
| H | H | X | L | H | |
| H | ↓ | H | ⎍ | ⎎ | One-shot outputs |
| ↓ | H | H | ⎍ | ⎎ | |
| ↓ | ↓ | H | ⎍ | ⎎ | |
| L | X | ↑ | ⎍ | ⎎ | |
| X | L | ↑ | ⎍ | ⎎ | |

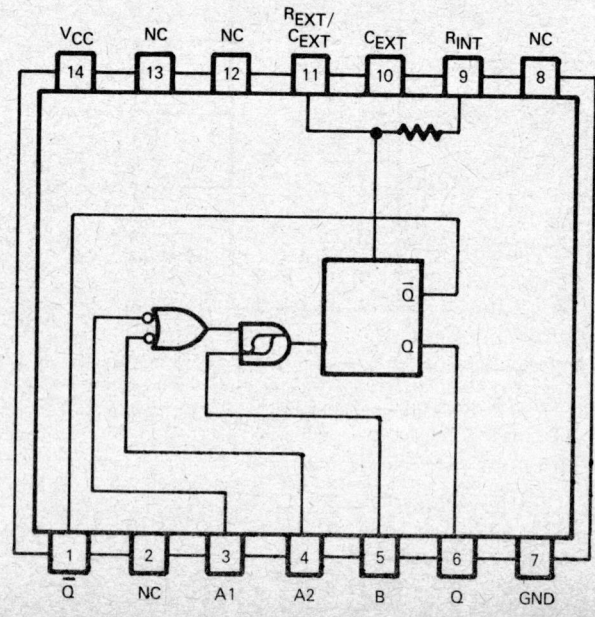

A constant low input at A1, A2 or B will hold the 74121 monostable multivibrator in its stable condition — with a low Q output and a high $\bar{Q}$ output. High inputs at A1 and A2 have the same effect.

There are five input signal combinations that will generate one-shot outputs. These input signal combinations are identified in the function table above.

With regard to the function table, symbols are used as follows:

X          represents a "don't care"
↓          represents a one-to-zero logic transition
↑          represents a zero-to-one transition
⊓          represents a one-shot with a zero monostable logic level and a one pulse level
⊔          is the NOT of ⊓

The duration of the one-shot output is determined by a resistor-capacitor network, just as described for the Signetics 555 monostable multivibrator; but, there are some differences. The 74121 provides an internal resistor which may be accessed by connecting $R_{INT}$ (pin 9) to $V_{CC}$ (pin 14). A variable external resistor may be connected between $R_{INT}$ (pin 9) or $R_{EXT}$ (pin 11) and $V_{CC}$ (pin 14).

An external timing capacitor, if present, will be connected between $C_{EXT}$ (pin 10) and $R_{EXT}$ (pin 11).

Here is one way in which a 74121 monostable multivibrator may be connected:

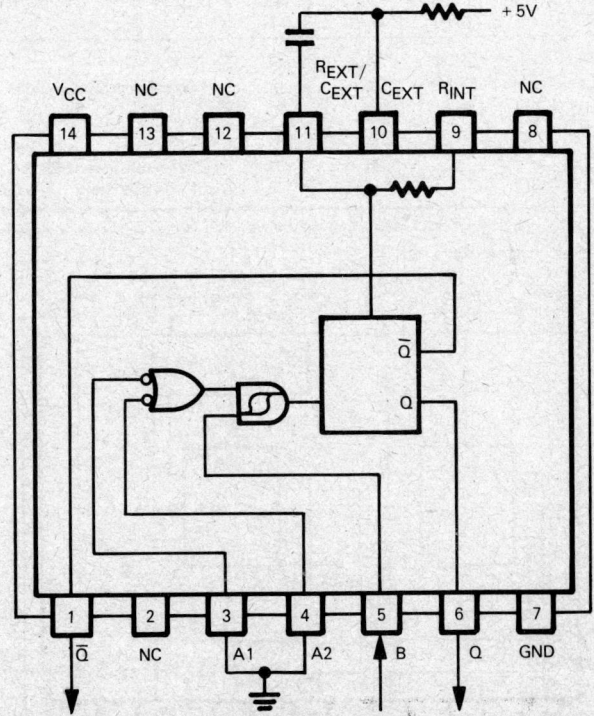

This use of the 74121 monostable multivibrator corresponds to the bottom two lines of the function table.

An external resistor/capacitor network controls one-shot pulse duration. Each one-shot pulse will be triggered by a low-to-high transition at pin 5 (B).

**From the programming point of view, there are only two significant features of the 74121 monostable multivibrator:**

1) **The monostable outputs are equivalent to binary digits of fixed value.** Any Immediate instruction which loads a zero or a one into any register bit simulates the monostable output. Here is an example:

   LD  B,4  SET BIT 3 OF REGISTER B TO 1, RESET ALL OTHER BITS

   Bit 3 of Register B is equivalent to a flip-flop; so is every other bit of Register B and all other registers.

2) **A one-shot output becomes a time delay of fixed value.** We will show how this time delay may be computed within a microcomputer system, but first let us examine the 74107 master-slave flip-flop.

## THE 74107 DUAL J-K MASTER-SLAVE FLIP-FLOP WITH CLEAR

Consider the master-slave flip-flop. This flip-flop is illustrated as follows:

| INPUTS | | | | OUTPUTS | |
|---|---|---|---|---|---|
| 1CLR or 2CLR | 1CK or 2CK | 1J or 2J | 1K or 2K | 1Q or 2Q | 1$\overline{Q}$ or 2$\overline{Q}$ |
| L | X | X | X | L | H |
| H | ⊓ | L | L | Remain in previous state | |
| H | ⊓ | H | L | H | L |
| H | ⊓ | L | H | L | H |
| H | ⊓ | H | H | Change state regardless of previous state | |

⊓  identifies a clock pulse; the way in which it is used is described below.
X  means "don't care".

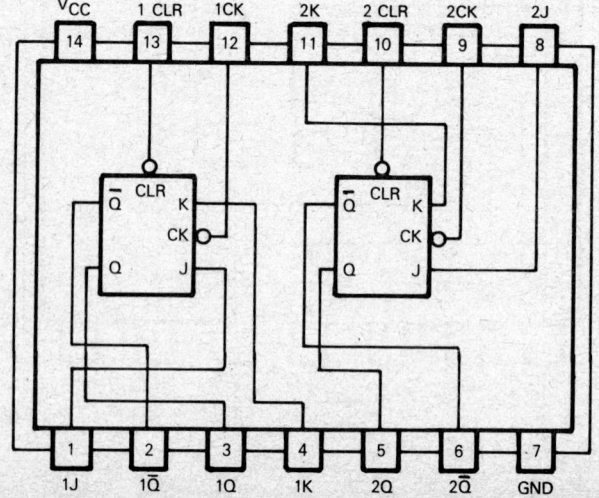

Let us examine the function table illustrated above. Unless you are familiar with this type of logic device, its features are not self-evident.

The connotation "master-slave" identifies a circuit which is, in fact, two flip-flops. Therefore, there are four flip-flops in the 74107 device illustrated above.

**MASTER-SLAVE FLIP-FLOPS**

The flip-flops in each master-slave pair respond to a clock signal, as follows:

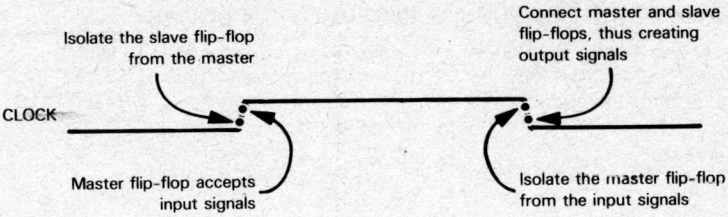

The significance of this clock signal's response is that the flip-flop inputs must be present at the positive edge of the clock signal; these inputs must remain steady while the clock signal is high. The flip-flop outputs, however, do not change state until the negative edge of the clock signal.

The clock signal may be used to create time delays. The 74107 flip-flop output is determined by input signal levels as they existed some time period earlier. This may be illustrated as follows:

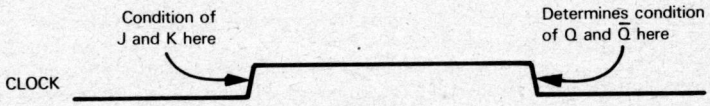

Here is a specific example:

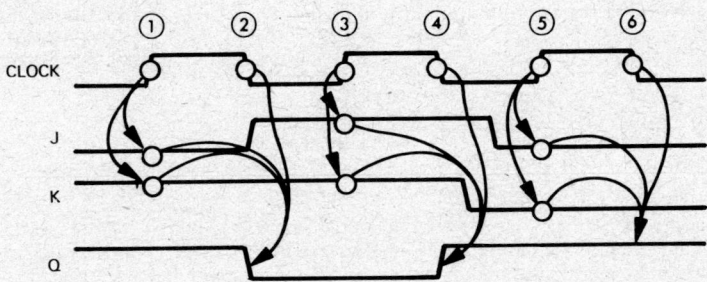

The following description of the timing diagram illustrated above is keyed to the circled numbers above the clock signal.

At ②, the Q output goes low, because at ① J was low and K was high.
At ④, Q changes state, because at ③ J and K were both high.
At ⑥, Q remains unaltered, because at ⑤ J and K were both low.

# MICROCOMPUTER SIMULATION OF REAL TIME

**What is the significance of the 555 monostable multivibrator and the master-slave flip-flops? When it comes to microcomputer simulation of these devices,**

there is only one feature that is important to our present discussion — and that is the concept of real time.

The 555 monostable multivibrator creates high logic level pulses at its output, where the duration of the high logic level is a controllable real time function.

The 74107 master-slave flip-flop allows an output signal to be generated based on input conditions as they existed some real time earlier.

## MICROCOMPUTER TIMING INSTRUCTION LOOPS

It is simple enough to create a time delay using a microcomputer system — providing the microcomputer system is not being called upon to perform any simultaneous operations. Consider the following instruction sequence:

**TIMING SHORT TIME INTERVALS**

```
Cycles
                LD    A,TIME    ;LOAD TIME CONSTANT INTO
                                ;ACCUMULATOR
  4     LOOP:   DEC   A         ;DECREMENT ACCUMULATOR
 10             JP    NZ,LOOP
                                ;REDECREMENT IF NOT ZERO
```

The above instruction sequence loads a data value, represented by the label TIME, into the Accumulator. The Accumulator is decremented until it reaches zero, at which time program execution continues. Let us assume that a 500 nanosecond clock is being used by the microcomputer system. The DEC and JP instructions, taken together, execute in 14 cycles — which is equivalent to seven microseconds. This means that the program sequence illustrated above can cause a delay with a minimum value of seven microseconds (when TIME equals 1), increasing in seven microsecond steps to a maximum of 1792 microseconds, which is equivalent to 7 x 256. This maximum time delay will result when TIME has an initial value of zero, since TIME is decremented before being tested to see if it is zero; therefore, the time out occurs when 1 decrements to 0, not when 0 decrements to $FF_{16}$.

Longer time delays may be generated by having a 16-bit counter. Here is the appropriate instruction sequence:

**TIMING LONG TIME INTERVALS**

```
Cycles
                LD    DE,T16    ;LOAD TIME CONSTANT INTO D
                                ;AND E
  6     LOOP:   DEC   DE        ;DECREMENT DE
  4             LD    A,D       ;TEST FOR ZERO BY ORING D
  4             OR    E         ;AND E CONTENTS VIA ACCUMULATOR
 12             JP    NZ,LOOP
```

The first LD instruction loads a 16-bit value, represented by the label T16, into the DE register pair. The LD instruction, being an immediate instruction, creates three bytes of object code. When the LD instruction executes, this is what happens:

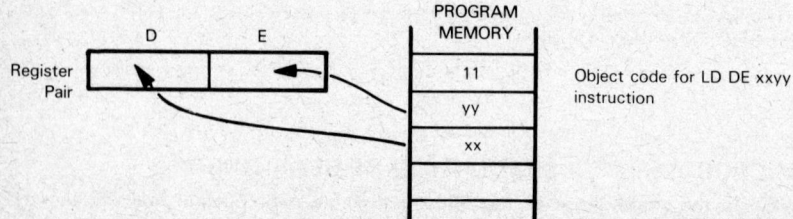

The DEC instruction decrements the 16-bit value in the DE registers as a single data entity. However, a quirk of the Z80 instruction set neglects to set status bits based on the result of the 16-bit decrement. This means that we have no immediate way of knowing whether the DE registers now contain a zero or non-zero value. To make this test, we load the contents of the D register into the Accumulator, then OR with the contents of the E register. If the result in the Accumulator is 0, then both D and E registers must contain 0. If the result is not zero, we return and redecrement the 16-bit value.

**STATUS TESTING USING DEC INSTRUCTION**

Observe that 26 cycles are required to travel once through the long time interval instruction loop. Again, assuming that the microcomputer is being driven by a 500-nanosecond clock, it will take 13 microseconds to execute the instruction loop once. The minimum value that T16 may have is 1. The maximum value is again 0 because a decrement occurs before the test for 0; should 0 initially be loaded into D and E, it will be decremented to $FFFF_{16}$ before the first test for zero is made. Thus, the long time interval instruction loop will generate delays that vary in 13-microsecond increments, from a minimum of 13 microseconds to a maximum of 0.851968 seconds.

$$FFFF_{16} = 65{,}535_{10}$$
$$13 \times 65{,}536 = 851{,}968 \text{ microseconds}$$

Now, the actual simulation of a one-shot is complicated by the fact that we may compute time delays, but when does the time delay begin? For digital logic devices, the answer is simple—the time delay begins when an input signal changes state:

**TIME DELAY INITIATION**

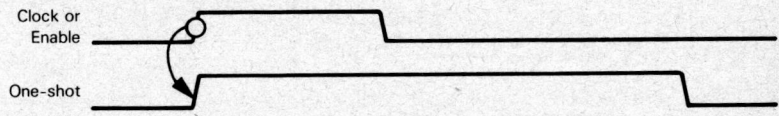

To parallel this concept within a microcomputer program, we must initiate a time delay upon completing some other program sequence's execution. This concept may be illustrated as follows:

```
        ≡
        JP      DELAY       ;LAST INSTRUCTION OF SOME PRIOR SEQUENCE
        ≡
DELAY:  LD      A,TIME      ;SHORT TIME INTERVAL INSTRUCTION
LOOP:   DEC     A           ;SEQUENCE
        JR      Z,LOOP
```

**There is another problem associated with creating time delays within a microcomputer system by executing instruction loops, as we have described: the microcomputer is, in essence, doing no useful work during the time delay.** There may be a simple remedy to this problem, providing we can define a program for the microcomputer to execute during the period of the time delay. This may be illustrated as follows:

**EXECUTING PROGRAMS WITHIN TIME DELAYS**

| Start of desired time delay | An instruction sequence whose execution time is known exactly executes during this time period. This is a coarse time interval | The remaining time is timed out using a time delay instruction loop This is a fine tuning time interval |

We must assume that we can calculate the exact time it will take for our program to execute within the one-shot time delay; also, the computed time must be less than or equal to the time delay. Not many programs are going to fit this description. If, for example, more than one instruction sequence may get executed depending on current conditions, then there may be many different times required for a program to execute. Still, so long as there is a fixed number of identifiable branches, the problem is tractable and may be illustrated as follows:

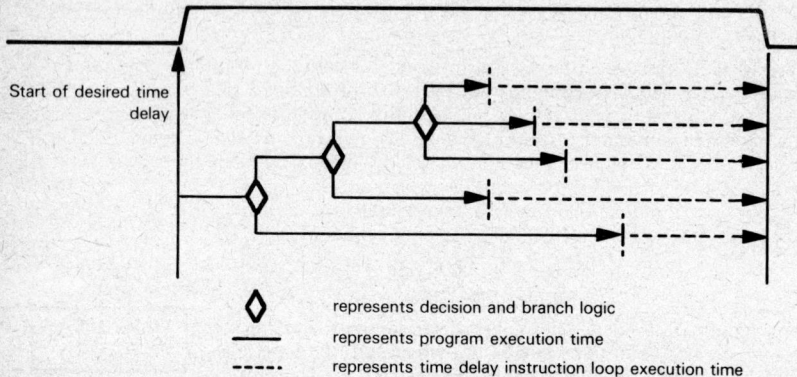

◇  represents decision and branch logic
—  represents program execution time
----  represents time delay instruction loop execution time

Now each "limb" of the program branches will end as follows:

```
            LD      A,DLY1      ;LOAD FIRST TIME DELAY
            JP      LOOP        ;START TIME DELAY LOOP
            ≡
            LD      A,DLY2      ;LOAD SECOND TIME DELAY
            JP      LOOP        ;START TIME DELAY LOOP
            ≡
            LD      A,DLY3      ;LOAD THIRD TIME DELAY
            JP      LOOP        ;START TIME DELAY LOOP
            ≡
            LD      A,DLY4      ;LOAD FOURTH TIME DELAY
            JP      LOOP        ;START TIME DELAY LOOP
            ≡
            LD      A,DLY5      ;LOAD FIFTH TIME DELAY
            JP      LOOP        ;START TIME DELAY LOOP
            ≡
LOOP:       DEC     A           ;SHORT TIME INTERVAL INSTRUCTION
            JR      NZ,LOOP     ;SEQUENCE
```

It is more common than not for a microcomputer program to contain numerous conditional branches; there may be hundreds of different possible execution times, depending on various combinations of current conditions. Executing a program within the time interval of the required delay now becomes impractical, because the logic needed to compute remaining time for the innumerable program branches is just too complicated.

## THE LIMITS OF DIGITAL LOGIC SIMULATION

**A Z80 microcomputer can compute time delays so long as no other program needs to be executed during the time delay, or providing a very simple instruction sequence with very limited branching is executed during the time delay.**

You cannot simulate simultaneous time delays, nor can you simulate a time delay which must occur in parallel to undefinable parallel program executions. External logic must handle all such time delays.

**SIMULTANEOUS TIME DELAYS**

## INTERFACING WITH EXTERNAL ONE-SHOTS

Note that, even though external logic may have to create time delays, it is very easy for the microcomputer system to trigger the start of the time delay and for the external logic to report the completion of the time delay.

**We can identify the start of a time delay by simply outputting an appropriate binary digit.** Look again at the way "Signal Out" was output to external logic by the signal inverter simulation. Outputting a signal to external logic is indeed very easy. Consider the following four instructions:

**ONE-SHOT INITIATION**

```
LD    A,0             ;LOAD 0 INTO THE ACCUMULATOR
OUT   (PORT B),A      ;OUTPUT VIA I/O PORT B
LD    A,2             ;LOAD 1 INTO THE ACCUMULATOR BIT 1
OUT   (PORT B),A      ;OUTPUT VIA I/O PORT B
```

A 1 is output at pin 1 of I/O Port B. Assuming that the pin associated with this I/O port is connected to the trigger of a multivibrator and that this connection was previously high, then the simple execution of the above instructions will trigger a one-shot.

**This may be illustrated as follows:**

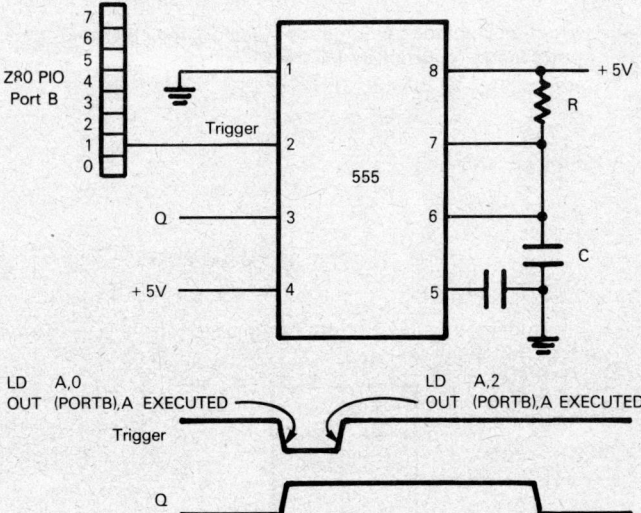

It is equally easy for external logic to signal the end of a time delay.

If we are dealing with "greater than or equal to" logic, all that is necessary is for the one-shot output to be connected to another pin of a microcomputer I/O port:

**ONE-SHOT TIME OUT USING STATUS**

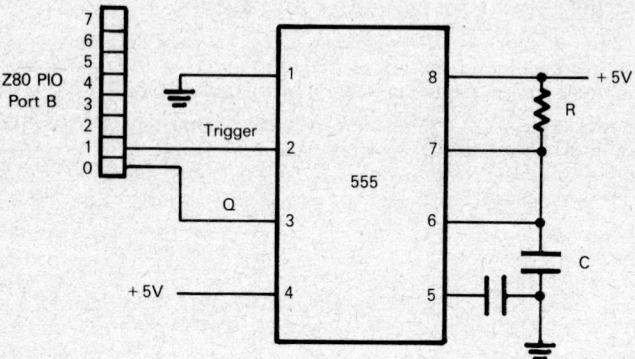

Signals arriving at pins of I/O ports are buffered. The program being executed by the microcomputer may, at any time, input the contents of the I/O port and test the condition of bit 0, which has been wired to the Q output. When this bit is found to equal 0, microcomputer program logic knows that the time interval has been surpassed.

**The following instruction sequence will test the I/O port and clear the "time interval complete" status being reported by I/O Port B, pin 0:**

```
            IN    A,(PORT B)    ;INPUT CONTENTS OF I/O PORT B TO ACCUMULATOR
            BIT   0,A           ;TEST BIT 0
            JP    NZ,NEXT       ;CONTINUE IF BIT IS 1
;TIME OUT PROGRAM BEGINS HERE
            -
            -
            -
NEXT:                           ;TIME NOT OUT PROGRAM BEGINS HERE
```

The IN instruction moves the current contents of I/O Port B to the Accumulator.

The following BIT instruction tests bit 0 of the Accumulator and sets the Zero flag to reflect the bit contents in the following way:

| Z | BIT |
|---|-----|
| 1 | 0   |
| 0 | 1   |

If the binary digit input from pin 0 of the I/O Port B is 1, then the Q output is still high. The JP NZ,NEXT instruction simply continues program execution.

If bit 0 of I/O Port B is 0, then the time delay is over; we branch to a program sequence which only gets executed immediately following a time out.

## TIME OUT AND INTERRUPTS
**The exact end of a time out can be signaled to the microcomputer using an interrupt.**

Now, as soon as the one-shot times out, it will force the microcomputer system to cease executing whatever program was currently being executed. A branch will be forced to some other program which has been specifically designed to respond to the time out.

The programming considerations associated with interrupts are more complicated than the level we have been dealing with in Chapter 2. We will therefore defer a detailed description of interrupt processing until later in this book. For the moment, it is sufficient to understand that the exact instant of a time out may be signaled to the microcomputer system using interrupt logic.

## INTERFACING WITH PROGRAMMABLE TIMERS
**Another type of external logic that can be used to create time delays is a programmable timer circuit such as the Z80 CTC (Counter/Timer Circuit). The CTC is a programmable device which contains four separate counter/timer circuits with associated control logic.** Each counter/timer can be accessed by the CPU as an I/O port or a memory location.

**Each of the four counter/timers can be programmed to operate as a timer, where it is decremented by the system clock, or as a counter, where it will be decremented upon reception of a clock/trigger signal.** There are several other operating options that can be established under program control — that is, by simply writing a control word to the appropriate counter/timer. We will not attempt to describe all of these options here; the Z80 CTC is described in detail in <u>An Introduction to Microcomputers: Volume II — Some Real Products</u>. Let us just briefly look at a typical sequence of events and at the flexibility and simplicity obtained by using a programmable timer.

Let us assume that the CTC is being accessed as though it were an I/O port — actually four I/O ports, since each timer/counter within the CTC operates independently and is selected individually. In order to initiate a time delay, we would perform the following steps:

1) Output a control word to the desired counter/timer, to specify that it is to operate in the timer mode. The same control word also specifies other mode information, such as the rate at which the timer is to be decremented, when the timer is to be started, and so on.
2) Output a constant representing the desired time delay to the timer/counter.

As soon as the time delay constant has been output, the timer will begin to count down. When the count reaches zero, a time out signal is generated. This signal can be used to inform the CPU that the time interval is complete. The information could be transmitted using an interrupt input to the CPU or via some intermediate logic.

**The use of a programmable timer offers obvious advantages over the external one-shot. The CTC can be programmed and reprogrammed to provide any desired time delay, whereas the external one-shot can only provide a single, fixed time delay. The CTC also provides four timer/counters so that simultaneous or overlapping time delays can be generated.**

In the design example we develop in this book only a few time delays are required, and there are no requirements for simultaneous delays. Therefore, we will use simple CPU instruction loops to generate the required delays. **If your application requires more than the most rudimentary timing sequences, however, you should investigate the use of programmable timers.**

# Chapter 3
# A DIRECT DIGITAL LOGIC SIMULATION

The discrete logic devices which we simulated in Chapter 2 were not selected at random; correctly sequenced, they will simulate the logic illustrated in Figure 3-1. This logic is a portion of the printer interface for the Qume Q-Series and Sprint Series printers. Figure 3-2 is the timing diagram that goes with Figure 3-1. We are going to describe both figures at a very elementary level.

The purpose of this chapter is to provide a one-for-one correlation between microcomputer assembly language programming and digital logic design. What you must understand is that, while such a one-for-one correlation can be forced, it is not natural — and that is where the problem in understanding lies. Microcomputer programs should be written to stress the nature of microcomputers, not the characteristics of digital logic.

The correct way to program a microcomputer is described beginning with Chapter 4.

Nevertheless, the juxtaposition of digital logic design and microcomputer programming is underscored in this chapter. This is the chapter that bridges two concepts; for that reason it is the most important chapter in this book. If you are a logic designer, this chapter is important because it will eliminate digital logic concepts which are inapplicable to microcomputers. If you are a programmer, this chapter is important because it will acquaint you with a new programming goal — efficient logic implementation.

To achieve the goal of this chapter, we will describe the logic illustrated in Figures 3-1 and 3-2; the description will be careful and detailed so that you can follow this chapter even if you are not a logic designer. As the logic description proceeds, we will blend in assembly language — in easy stages.

If you understand digital logic, it is particularly important that you confine your reading to the boldface type in this chapter. The logic of Figure 3-1 has been described in sufficient detail to meet the needs of a programmer or a reader with no logic background.

# HOW THE QUME PRINTER WORKS

The active Qume printing element is a 96-petal printwheel, with one character on each petal:

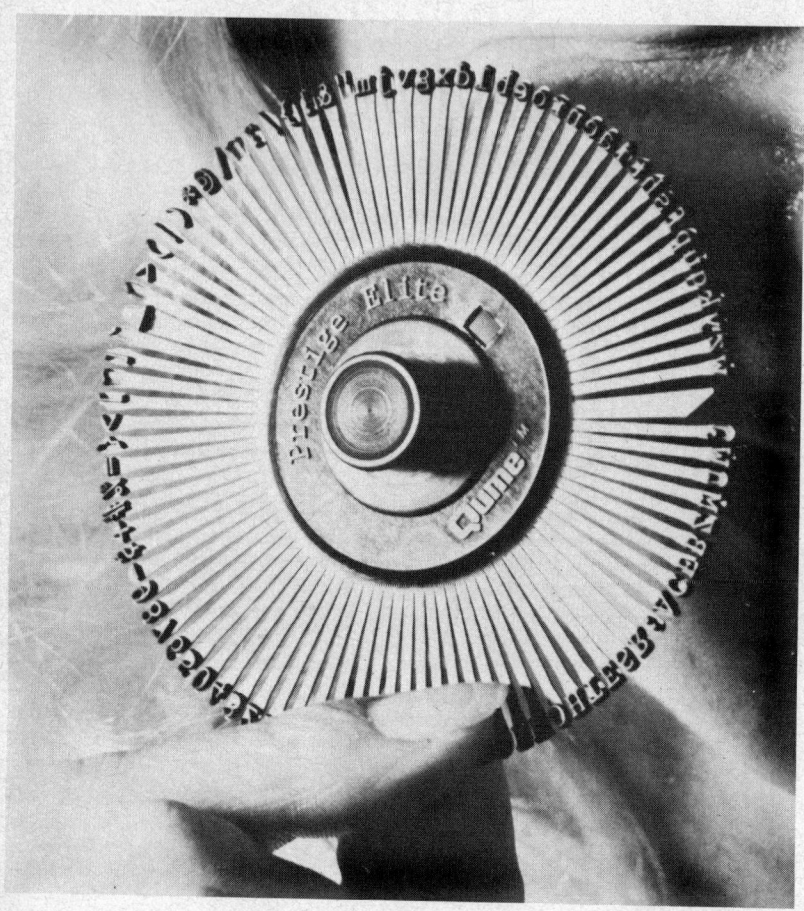

COURTESY OF QUME CORPORATION

A character is printed by moving the printwheel until the appropriate petal is in front of a solenoid-driven printhammer. The printhammer is then fired; it strikes the printwheel petal, which marks the paper:

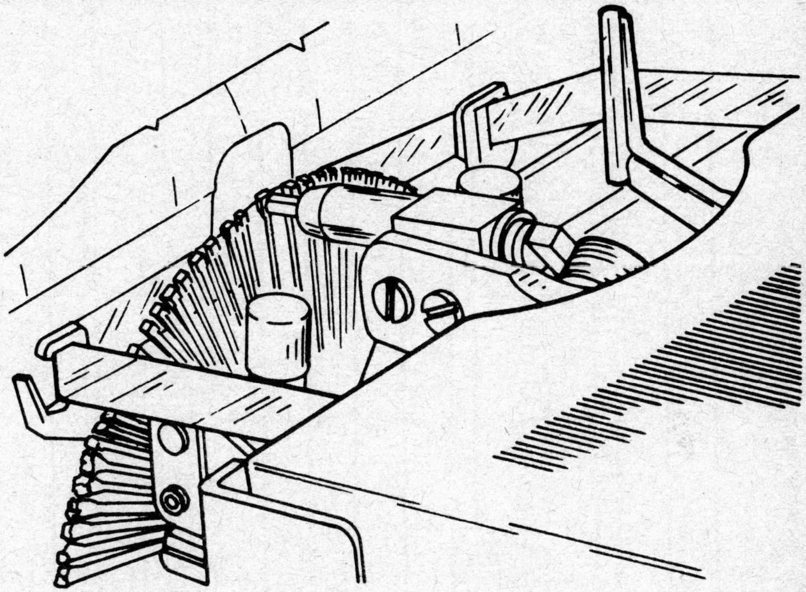

Whenever a character is not in the process of being printed, the printwheel is positioned with a short petal immediately vertical so that the character just printed is visible:

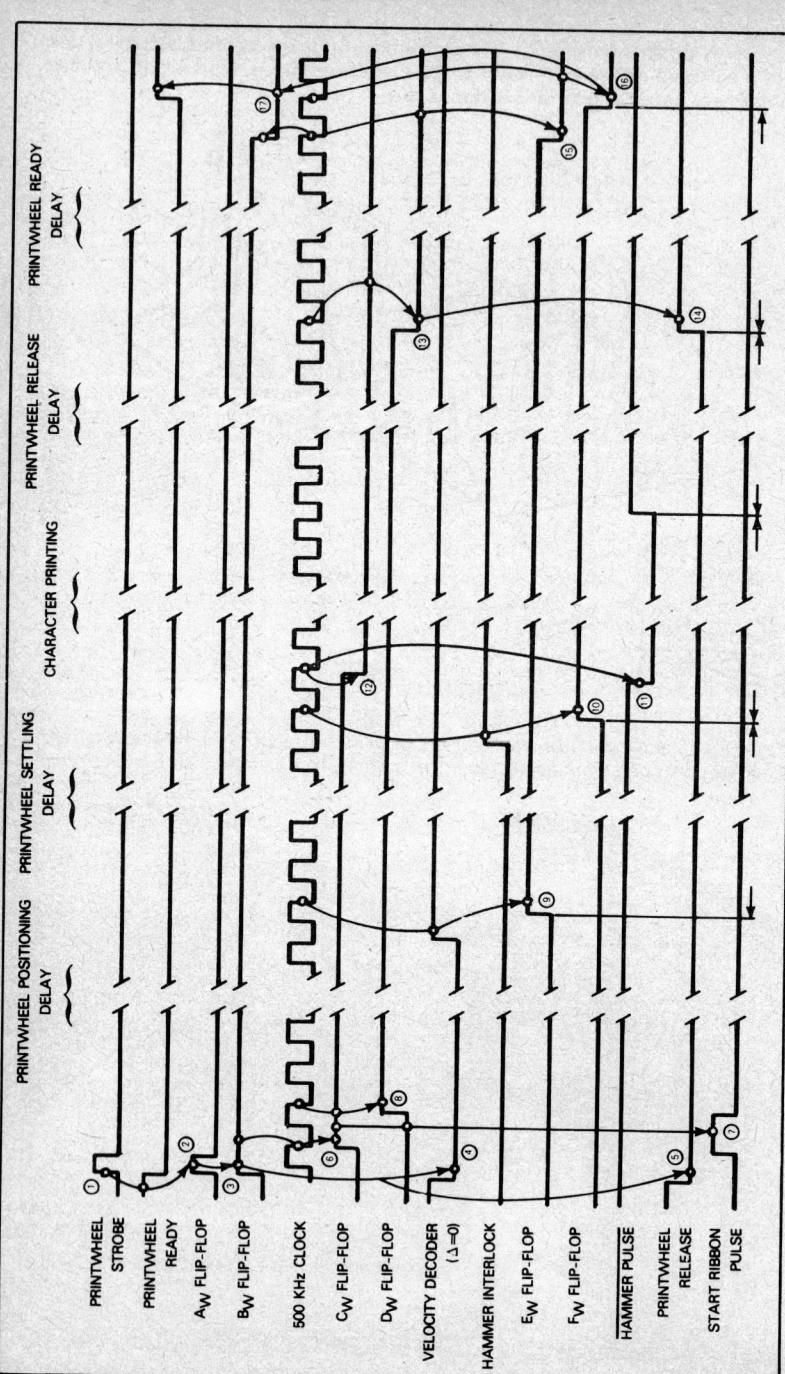

Figure 3-2. Printwheel Control Logic Timing Diagram

**As part of the print cycle, the printer ribbon and paper carriage must be moved.**

Every character is printed according to a definite sequence of events, collectively referred to as a "print cycle". The logic illustrated in Figure 3-1 controls the character print cycle. **These are the events which must occur within a print cycle:**

1) First, the print cycle must be initiated. A **signal (PW STROBE) is pulsed high to initiate the print cycle:**

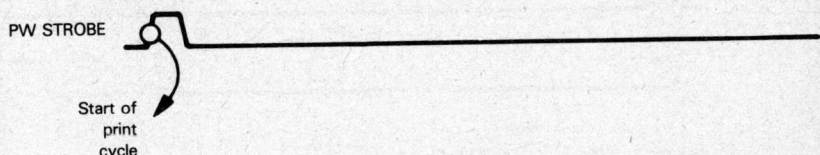

2) The print cycle will endure for a fixed time interval. Obviously, during this time interval another print cycle must not be initiated. Therefore, the **external logic** responsible for generating PW STROBE true **must be given a signal identifying the duration of the print cycle. This signal is PRINTWHEEL READY, also called CH RDY:**

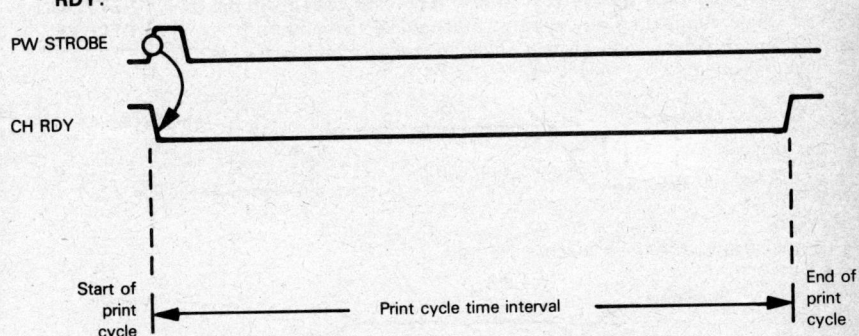

The sequence of events which actually cause a character to be printed can now proceed, with the assurance that external logic will not attempt to start printing the next character before the current print cycle has gone to completion.

3) **The printwheel is moved from its position of visibility until the appropriate character petal is in front of the printhammer:**

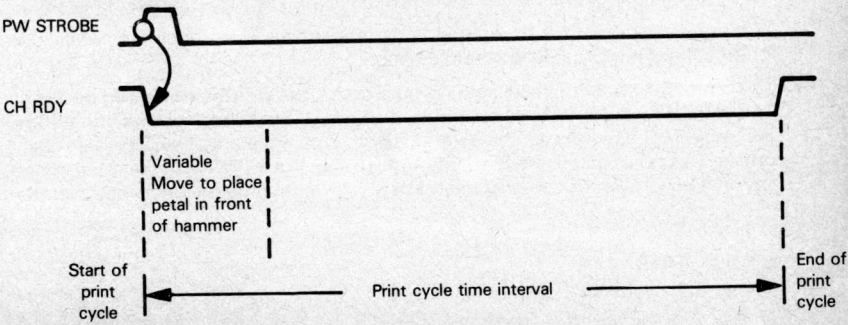

A variable time delay is needed by the printwheel positioning logic. Obviously it will take longer to position a petal that is far from the position of visibility than to position to an adjacent petal.

4) Before the printhammer is fired, **the printwheel must be given time to settle.** A fixed, two millisecond time delay is sufficient:

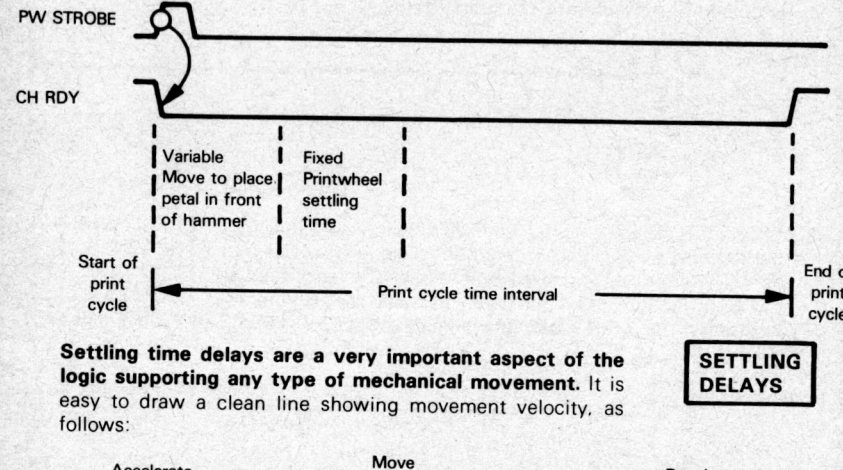

**Settling time delays are a very important aspect of the logic supporting any type of mechanical movement.** It is easy to draw a clean line showing movement velocity, as follows:

**SETTLING DELAYS**

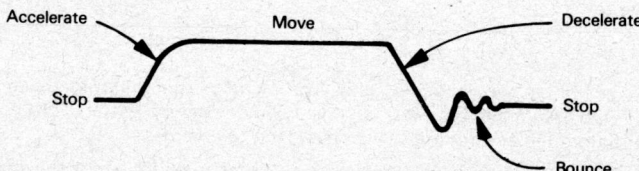

But in reality, movement occurs like this:

The bounce that follows deceleration must be passed over by a settling time delay.

A blurred character will be printed if the printwheel is still vibrating when the printhammer hits a petal against the paper.

5) At the end of the printwheel settling time delay, **the printhammer can be fired.** This is done by outputting an impulse to a solenoid. **Six firing impulse intensities are provided,** since some characters have a more substantial surface area than others. To strike a comparatively large surface area like a "W" with the same intensity that you strike a small character, like a ".", would produce unevenness in the

density of the printed text. The duration of the printhammer solenoid pulse is controlled by the next time delay:

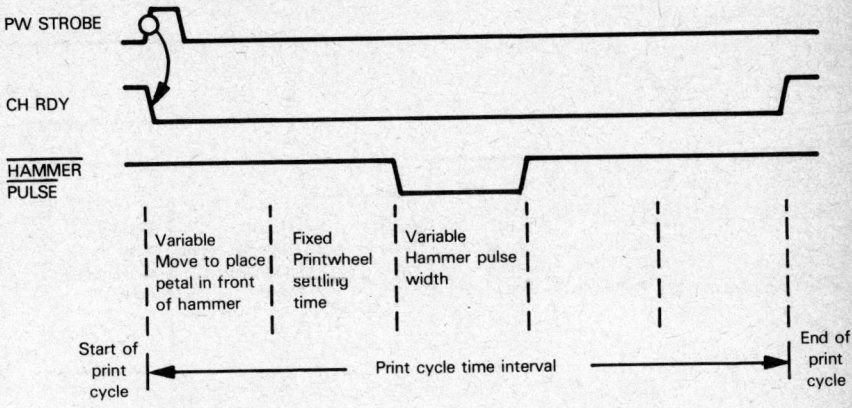

The bar over HAMMER PULSE identifies the signal as one which is low when active.

6) At the completion of the printhammer pulse time delay, the hammer has struck a petal and forced it onto the paper. Now **the hammer must be given time to return to its prefiring position.** A three millisecond delay is generated for this purpose:

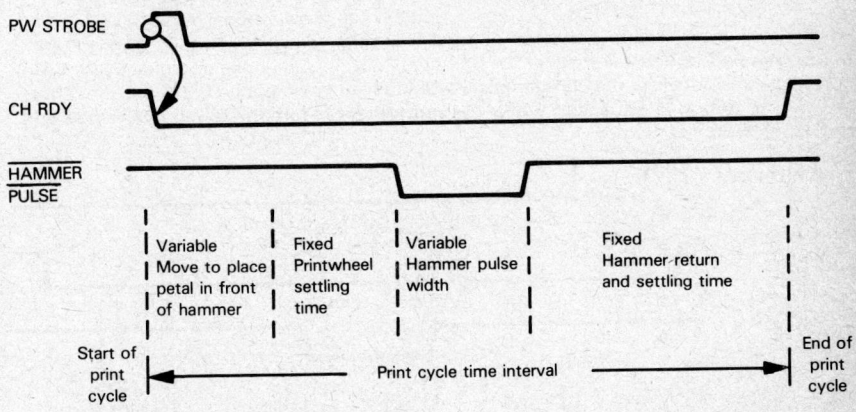

7) Now **the printwheel can be moved to its position of visibility** and the paper carriage can be advanced to the next character position. The printwheel's "position of visibility" is its normal inactive position. In this position, a short petal is in front of the printhammer, so the most recently printed character is visible above the short petal; hence the "position of visibility". Had we not given time for the printhammer to settle back before moving the printwheel to its position of visibility, a printwheel petal may have been broken striking the tip of the still protruding hammer. Also, the paper may have smudged moving against a bent petal. Since the printhammer has been given time to fully retract, none of these problems will arise.

> PRINTWHEEL
> POSITION OF
> VISIBILITY

**A final two millisecond time delay allows the printwheel and paper carriage to reposition themselves:**

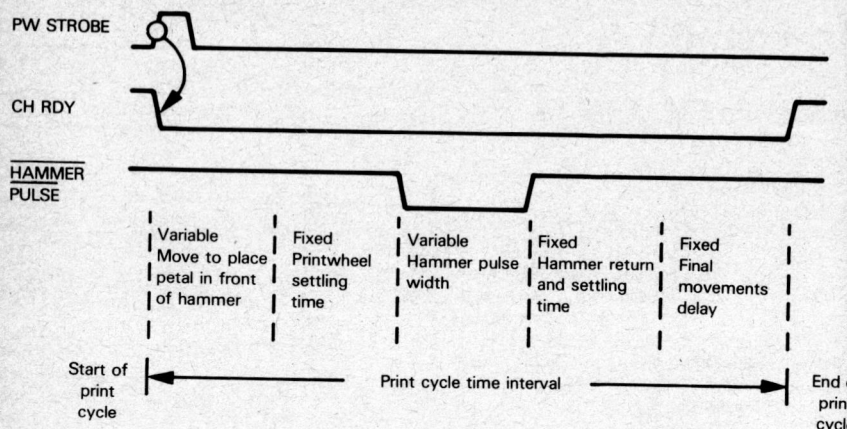

8) What about ribbon logic? **In order to get a clean impression on the paper, a fresh piece of ribbon must present itself between the character petal and the paper.** Shortly after the beginning of the print cycle, therefore, a signal (START RIBBON MOTION PULSE), which actually controls ribbon movement, is output to external logic. This external logic (it is not part of Figure 3-1) sends back a ribbon movement completed signal ($\overline{FFA}$), since we cannot allow the printhammer to be fired while the ribbon is still moving. Thus, **the ribbon is advanced while the printwheel is initially being positioned and settled:**

| START RIBBON PULSE |
| FFA |

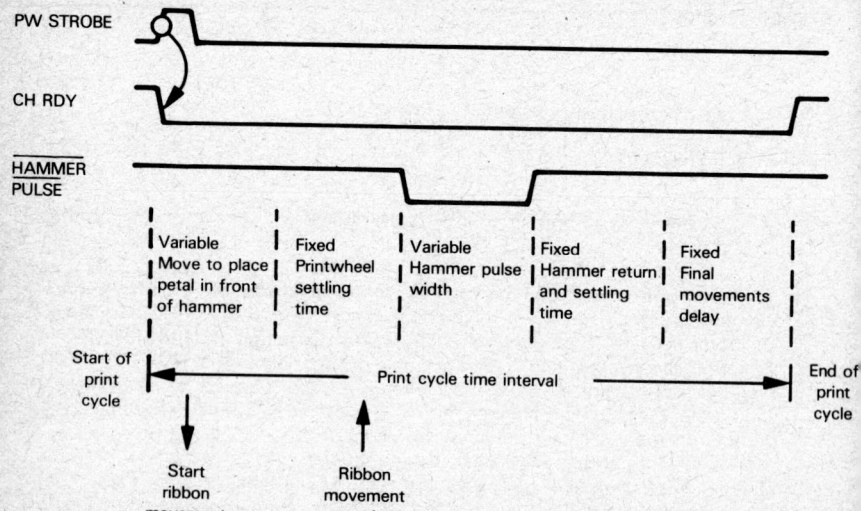

In summary, a print cycle consists of five time delays; each time delay starts out with a flurry of logical activity followed by a period of mechanical movement.

# INPUT AND OUTPUT SIGNALS

Now that you have a general understanding of the functions which are controlled by logic in Figure 3-1, **the next step is to take a closer look at input and output signals.**

In order to know what to do and when to do it, we must rely entirely upon input signals. Similarly, output signals represent the only way in which we can transmit control information to external logic.

Our limited goal, at this point, is to understand what function each input and output signal performs, and how — physically — we are going to handle the signals. We will discuss the "how" first.

## INPUT/OUTPUT DEVICES

**The principal device used to transmit signals and data between a Z80 microcomputer system and external logic is the Z80 Parallel Input/Output interface (PIO). We are going to use two Z80 PIO devices.**

| PARALLEL INPUT/OUTPUT INTERFACE |

Since this device has been described in <u>An Introduction to Microcomputers</u>, we are going to assume that you superficially understand its capabilities and organization; if you do not, see <u>An Introduction to Microcomputers: Volume II — Some Real Products</u> before continuing. Otherwise, you will not understand the discussion which follows.

## THE Z80 PARALLEL I/O INTERFACE (PIO)

**The Z80 Parallel I/O interface (PIO) provides 16 I/O pins which may be grouped into I/O ports as follows:**

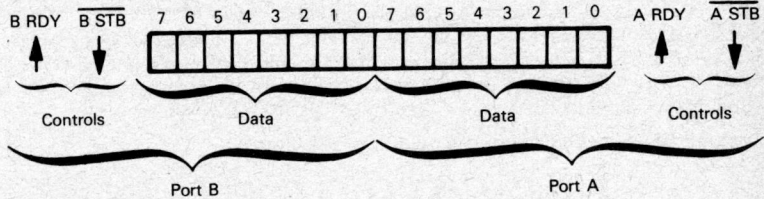

Each port has two associated control signals, RDY and $\overline{STB}$, for use in parallel data transfers with automatic handshaking.

RDY is output by the Z80 PIO to external logic; $\overline{STB}$ is input from external logic to the Z80 PIO.

**Each port may be programmed to operate in one of three modes; in addition, Port A may operate in a fourth mode which is not available on Port B. Port A and Port B do not have to operate in the same mode.**

| I/O PORT MODES |

**Let us now look at the Z80 PIO modes.**

**Output mode (Mode 0) allows Port A and/or Port B to be used as a conduit for transferring data to external logic. The handshaking works in the following way:**

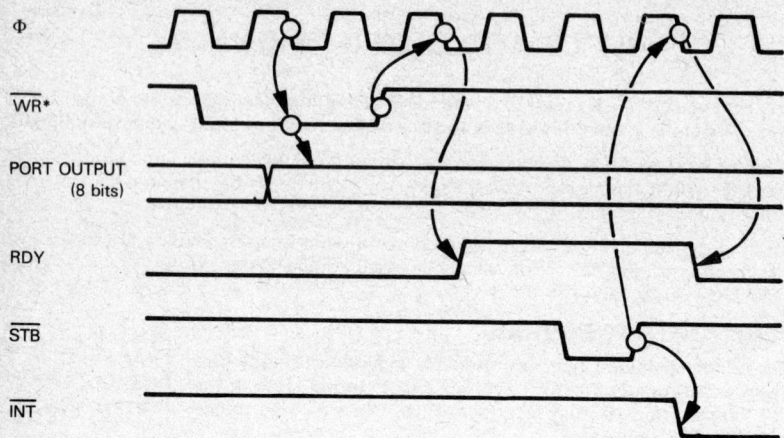

When the CPU executes an output instruction, it generates control signals which the Z80 PIO combines into an internal write pulse. This is shown as the signal $\overline{WR}^*$ in the diagram above. $\Phi$ is a system clock which Z80 PIO logic uses to synchronize its internal signal transitions.

An output cycle is initiated when the CPU executes any output instruction accessing the I/O port. The write pulse ($\overline{WR}^*$ above) is used to strobe data off the Data Bus and into the addressed I/O port's output register. After the write pulse, on the next high-to-low transition of the clock pulse $\Phi$, the RDY control signal is output high to external logic. RDY remains high until external logic returns a low pulse on the $\overline{STB}$ input. On the following high-to-low clock pulse $\Phi$ transition, RDY returns low. The low-to-high $\overline{STB}$ transition also generates an interrupt request — if interrupts have been enabled.

**Z80 PIO OUTPUT WITH HANDSHAKING**

**Timing for Input mode (Mode 1) is illustrated below:**

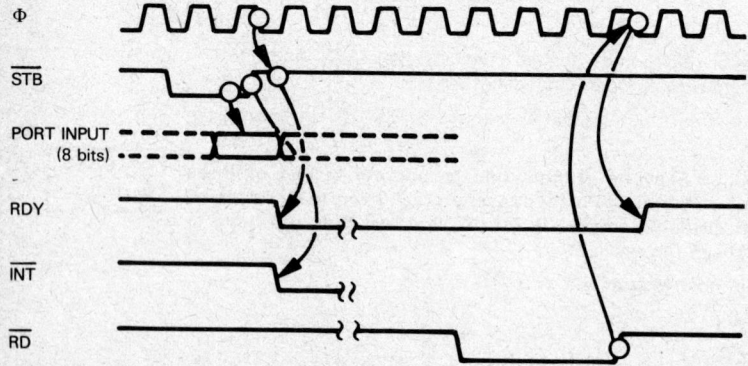

External logic initiates an input cycle by pulsing $\overline{STB}$ low. This low pulse causes the Z80 PIO to load data from the I/O port pins into the port input register. On the rising edge of the $\overline{STB}$ pulse an interrupt request will be triggered if interrupts have been enabled.

**Z80 PIO INPUT WITH HANDSHAKING**

On the falling edge of the Φ clock pulse which follows $\overline{STB}$ input high, RDY will be output low informing external logic that its data has been received but has not yet been read. RDY will remain low until the CPU has read the data, at which time RDY will be returned high.

**It is up to external logic to ensure that data is not input to the Z80 PIO while RDY is low.** If external logic does input data to the Z80 PIO while RDY is low, then the previous data will be overwritten and lost -- and no error status will be reported.

**In bidirectional mode (Mode 2), the control lines supporting I/O Ports A and B are both applied to bidirectional data being tranferred via Port A; Port B must be set to the bit control mode (Mode 3).**

**Z80 PIO BIDIRECTIONAL DATA TRANSFERS WITH HANDSHAKING**

Timing for bidirectional data transfers is simply a combination of input and output handshaking where A control lines apply to data output while the B control lines apply to data input. This may be illustrated as follows:

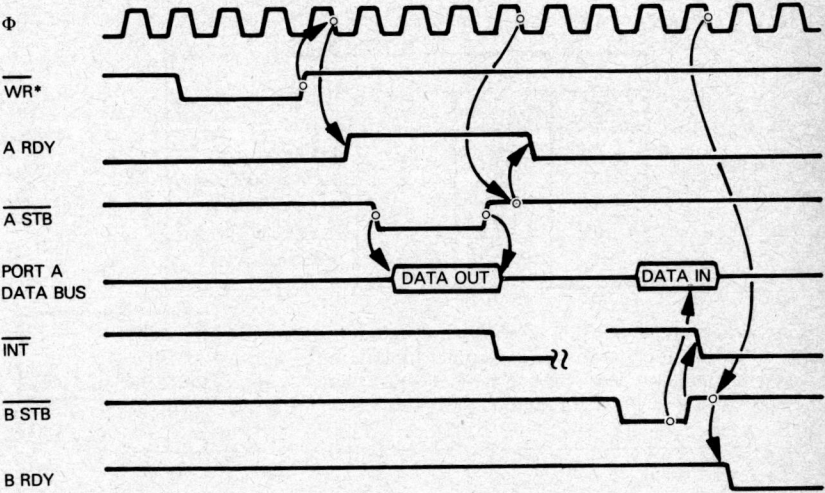

The only unique feature of the illustration above is that data being output via Port A is stable only for the duration of the $\overline{A\ STB}$ low pulse. This is necessary in bidirectional mode since the Port A pins must be ready to receive input data as soon as the output operation has been completed.

Once again, it is up to external logic to make sure that it conforms with the timing requirements of bidirectional mode operation. External logic must read output data while $\overline{A\ STB}$ is low. If external logic does not read data at this time the data will not be read, but the Z80 PIO will not report an error status to the CPU; there is no signal that external logic sends back to the Z80 PIO following a successful read.

Also, it is up to external logic to make sure that it transmits data to Port A only while B RDY is high and A RDY is low. If external logic tries to input data while the Z80 PIO is outputting data, input data will not be accepted. If external logic tries to input data before previously input data has been read, the previously input data will be lost and no error status will be reported.

**Control mode (Mode 3) does not use control signals. You must define every pin of an I/O port in Mode 3 as an input or an output pin.** Input and output are controlled by the CPU; there is no handshaking with external logic. If all the pins of a port are defined in the same direction, then the port can be used for simple parallel input or output.

BIT CONTROL

SIMPLE I/O

**You select port modes by writing an appropriate code into the port's control buffer.** A detailed discussion of control codes will not help you understand the subject matter of this chapter, so we leave that discussion to An Introduction to Microcomputers: Volume II — Some Real Products.

I/O PORT MODE SELECTION

**Every Z80 PIO has four I/O port addresses assigned to it.** Three Z80 PIO pins are used to select the device and a device port, as follows:

I/O PORT ADDRESSING

$\overline{CE}$: Input 0 to select the device. Input 1 to disconnect it.
B/$\overline{A}$ SEL: Input 0 to select Port A. Input 1 to select Port B.
C/$\overline{D}$ SEL: Input 0 to select data buffer. Input 1 to select control buffer.

Here is a summary of device select combinations:

| | SIGNAL | | SELECTED LOCATION |
|---|---|---|---|
| $\overline{CE}$ | B/$\overline{A}$ SEL | C/$\overline{D}$ SEL | |
| 0 | 0 | 0 | Port A data buffer |
| 0 | 0 | 1 | Port A control buffer |
| 0 | 1 | 0 | Port B data buffer |
| 0 | 1 | 1 | Port B control buffer |
| 1 | X | X | Device not selected |

**Now, when an IN or OUT instruction is executed by the Z80 CPU, the port number is output on the low-order eight Address Bus lines. We will use two Z80 PIO devices, and connect them to the Address Bus as follows:**

I/O PORT ADDRESS DETERMINATION

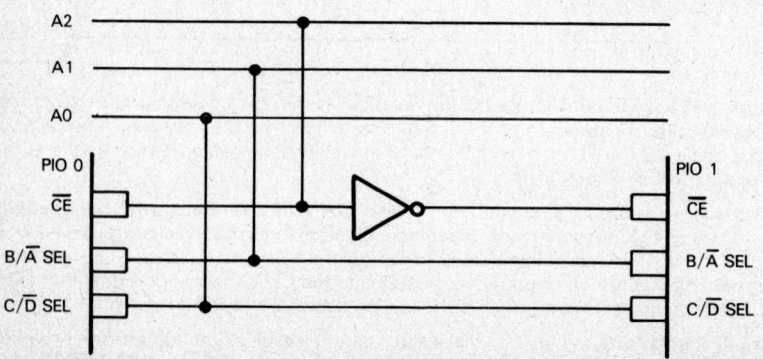

As a consequence of the connections shown above, the Z80 PIOs will respond to the following I/O port addresses:

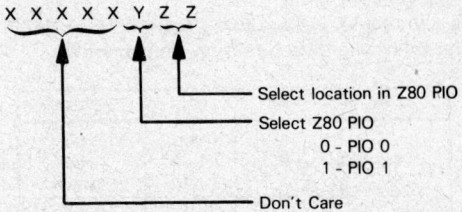

For the sake of consistency, we will always assign 0's to "don't care" bits. The Z80 PIO locations will thus be addressed as follows:

| ADDRESS | LOCATION | MNEMONIC |
|---------|----------|----------|
| 0 | PIO 0 Port A data | A0 |
| 1 | PIO 0 Port A control | AC0 |
| 2 | PIO 0 Port B data | B0 |
| 3 | PIO 0 Port B control | BC0 |
| 4 | PIO 1 Port A data | A1 |
| 5 | PIO 1 Port A control | AC1 |
| 6 | PIO 1 Port B data | B1 |
| 7 | PIO 1 Port B control | BC1 |

Since the "don't care" bits could have any value, **we have actually used all 256 I/O port addresses to access only eight separate locations.** Since we need only two Z80 PIOs for the program we are about to develop, **this addressing scheme is satisfactory for our limited purpose.**

**There are two ways to address more I/O ports:**

1) **Assign memory addresses to any further I/O ports,** as we demonstrated in Chapter 2.
2) **Reserve just the required eight I/O port addresses for the two Z80 PIOs.** This means that we must add more logic to decode a single enable signal from Address Bus lines A3 through A7.

Here is logic to reserve the addresses $F8_{16}$ through $FF_{16}$:

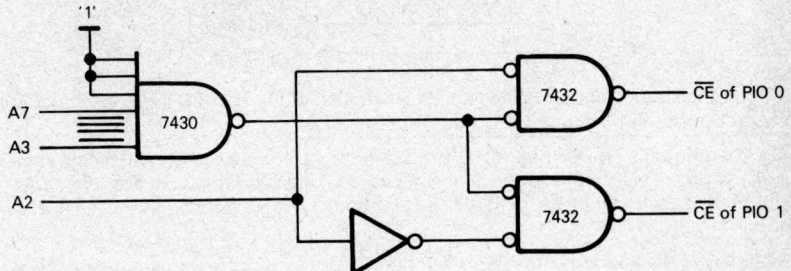

A 7430 is an 8-input positive NAND gate; a 7432 is a 2-input positive OR gate. When the upper five I/O port address lines are all 1's, the Z80 PIO selected by A2 will receive a 0 at its $\overline{CE}$ input.

**Initially, to keep things simple, we are going to program both Z80 PIOs to operate in Mode 3, with the following data direction assignments:**

| Z80 PIO | PORT | PINS | DIRECTION |
|---------|------|------|-----------|
| 0 | A | All | Input |
|   | B | 7 - 4 | Input |
|   |   | 3 - 0 | Output |
| 1 | A | All | Output |
|   | B | All | Input |

In order to understand the discussion at hand, you do not need to know how the Z80 PIO is programmed to meet our requirements; nevertheless, here is an example of the appropriate instruction sequence followed by an explanation of the control words:

**I/O PORT MODE SELECT INSTRUCTION SEQUENCE**

```
;INITIALIZE I/O PORTS
        LD      B,0CFH          ;PUT MODE 3 CONTROL WORD IN REGISTER B
;PIO 0, PORT B
        LD      C,3             ;PUT CONTROL ADDRESS IN REGISTER C
        OUT     (C),B           ;SET PORT IN MODE 3
        LD      A,0F0H          ;PUT PIN DIRECTION WORD IN ACCUMULATOR
        OUT     (C),A           ;SET DIRECTION: UPPER HALF INPUT, LOWER
                                ;OUTPUT
;PIO 0, PORT A
        -
        -
        -
```

**The following control word causes the addressed port to operate in Mode 3:**

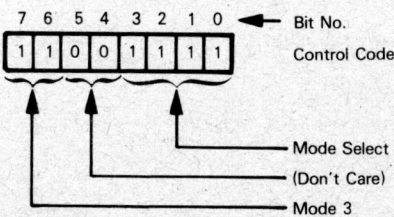

To verify the control word format, see the description of the Z80 PIO in <u>An Introduction to Microcomputers: Volume II — Some Real Products</u>.

We have arbitrarily chosen Register B to hold this control word, which will be the same for all I/O ports; thus, we load the control word into Register B only once, at the beginning of the sequence which initializes all the ports. The instruction LD B,0CFH does this.

We then load the address of the control port into Register C with the instruction LD C,3, and then output the control word via the instruction OUT (C),B.

**If a Mode Select control code is output specifying that an I/O port will operate in Mode 3, then the next byte output is assumed to be a pin direction mask.** We used the following mask in the example above:

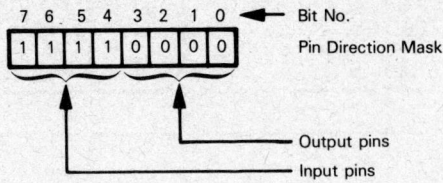

A 1 identifies an input pin, whereas 0 identifies an output pin. The instruction LD A,0F0H puts this pin direction mask in the Accumulator. The instruction OUT (C),A sends the pin direction mask to the port.

## INPUT SIGNALS

**Let us turn our attention to the input signals that appear on the left-hand side of Figure 3-1. We will describe each signal, assign it to an appropriate input pin, and include a rudimentary instruction sequence to access the signal at the most elementary level.**

## RETURN STROBE

**If the operator is to see the most recently printed character, two things must happen:**

**1) The printwheel must be moved to its position of visibility.**

**2) The ribbon must be dropped.**

External logic can take care of dropping and raising the ribbon, but logic in Figure 3-1 creates the signals that allow the printwheel to move.

In order to move the printwheel to its position of visibility, therefore, the ribbon control external logic inputs RETURN STROBE low while the ribbon is dropped.

Logic within Figure 3-1 uses RETURN STROBE as an alternative signal to start a print cycle; however, RETURN STROBE low is accompanied by HAMMER ENABLE FF low, which prevents the printhammer from firing. Therefore, **a print cycle initiated by**

**PRINTWHEEL REPOSITIONING PRINT CYCLE**

**RETURN STROBE low is a "dummy" print cycle which moves the printwheel back**

to its position of visibility but does not fire the printhammer; we refer to this as a **printwheel repositioning print cycle**:

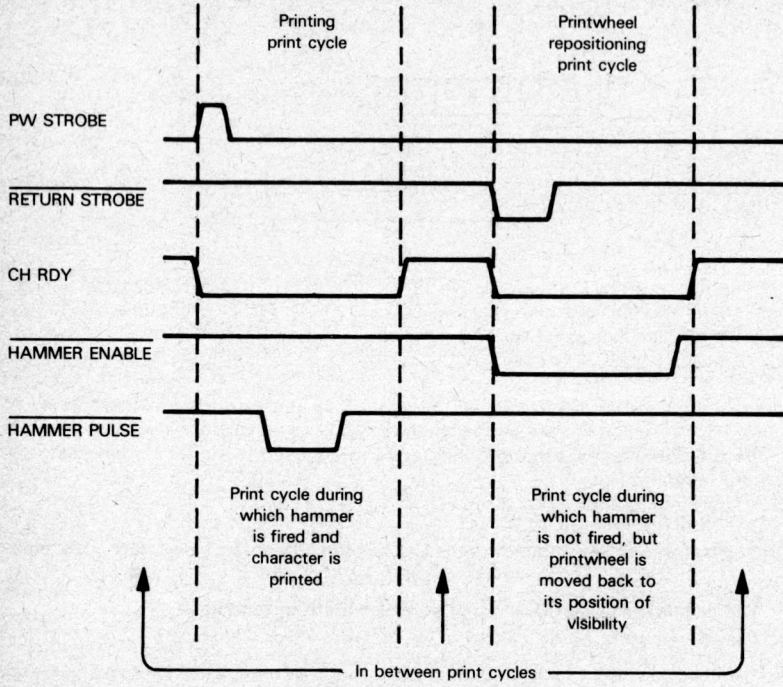

**We will assign I/O Port B0; pin 4 to RETURN STROBE.**

**In between print cycles, we can test this pin in order to trigger a new print cycle via the following instruction sequence:**

```
LOOP:      IN       A,(2)      ;INPUT I/O PORT B0 CONTENTS TO
                               ;ACCUMULATOR
           BIT      4,A        ;TEST VALUE OF BIT 4
           JR       NZ,LOOP    ;IF IT IS 1, RETURN AND RETEST
;NEW PRINT CYCLE INSTRUCTION SEQUENCE BEGINS HERE
```

## PFL REL

**The printhammer cannot be fired while the paper feed mechanism is moving, therefore, at such times external logic inputs PFL REL low.**

Logic within Figure 3-1 will delay firing the printhammer for as long as PFL REL is being input low.

**We will assign Pin 0 of I/O Port A0 to PFL REL.**

Before executing the instruction sequence which fires the printhammer, we will input the contents of Port A0 and test bit 0; so long as this bit contains zero, we will not execute the printhammer firing sequence.

The following instructions perform the required test:

```
LOOP:   IN    A,(0)     ;INPUT CONTENTS OF I/O PORT A0 TO
                        ;ACCUMULATOR
        BIT   0,A       ;TEST VALUE OF BIT 0
        JR    Z,LOOP    ;IF VALUE IS 0, DO NOT FIRE PRINTHAMMER
;PRINTHAMMER FIRING INSTRUCTION SEQUENCE BEGINS HERE
```

## RIB LIFT RDY

**This signal** is similar to PFL REL; it **is input low when ribbon lift logic is moving the ribbon.** Just as the printhammer cannot be fired while the paper feed mechanism is active, so it cannot be fired while the ribbon is being moved. By connecting RIB LFT RDY to Pin 1 of I/O Port A, we may adjust the printhammer firing initiation instruction sequence as follows:

```
LOOP:   IN    A,(0)     ;INPUT CONTENTS OF I/O PORT A0 TO
                        ;ACCUMULATOR
        OR    0FCH      ;MASK OUT ALL BITS EXCEPT 0 AND 1
        CPL             ;COMPLEMENT THE RESULT TO TEST FOR
                        ;ANY 0 BIT PRESENT
        JR    NZ,LOOP   ;ANY 0 BIT WILL NOW BE 1. IF ANY BIT
                        ;IS NOW 1, DO NOT FIRE PRINTHAMMER
;PRINTHAMMER FIRING INSTRUCTION SEQUENCE BEGINS HERE
```

## PW STROBE

We have already encountered **this signal; it is pulsed high by external logic to start a normal print cycle,** during which a character will be printed.

Remember, RETURN STROBE is input low to initiate a print cycle, during which the printwheel will be moved to its position of visibility but no character will be printed.

Assuming that **PW STROBE is connected to pin 5 of I/O Port B0,** this is the instruction sequence that will be executed between print cycles:

```
LOOP:   IN    A,(2)     ;INPUT I/O PORT B0 CONTENTS TO
                        ;ACCUMULATOR
        AND   30H       ;ISOLATE BITS 5 (PW STROBE) AND 4 (RETURN
                        ;STROBE
        CP    10H       ;TEST FOR PW STROBE = 0, RETURN STROBE = 1
        JR    Z,LOOP    ;IF TEST IS TRUE STAY IN LOOP
;PRINT CYCLE INSTRUCTION SEQUENCE STARTS HERE
```

Observe that either PW STROBE = 1 or RETURN STROBE = 0 can trigger the start of a print cycle; that is why only PW STROBE = 0 and RETURN STROBE = 1 keeps us in the testing instruction loop.

Now, the four instructions shown above execute in a combined total of 36 clock cycles. With a 500-nanosecond clock, the four instructions will execute in 18 microseconds — which becomes the minimum pulse width allowed for PW STROBE. **If PW STROBE is pulsed high for less than 18 microseconds, our instruction cycle may miss it.**

> **INPUT SIGNAL PULSE WIDTH**

## FFA

**This** is another printhammer warning **signal. It is set to 0 while external logic is advancing the ribbon. By connecting this signal to pin 2 of I/O Port A0, we can**

modify the instruction sequence which precedes printhammer firing as follows:

```
LOOP:   IN      A,(0)       ;INPUT CONTENTS OF I/O PORT A0 TO
                            ;ACCUMULATOR
        OR      0F8H        ;ISOLATE BITS 2,1, AND 0
        CPL                 ;COMPLEMENT THE RESULT TO TEST FOR
                            ;ANY 0 BIT
        JR      NZ,LOOP     ;ANY 0 BIT WILL NOW BE 1. IF ANY BIT IS
                            ;1, DO NOT FILE PRINTHAMMER.
;PRINTHAMMER FIRING SEQUENCE BEGINS HERE
```

All we have done is add one more test condition which must be met before the printhammer firing instruction sequence gets executed.

## RESET

**This is** a signal which is commonly seen in the most diverse types of logic. It is **an initializing signal.** Its purpose is to ensure that all logic is in a "beginning" state, which in our case is the condition which exists between printwheel cycles.

**The logic in Figure 3-1 connects the RESET signal to logic devices, such that RESET going high forces all logic to a "beginning" condition.**

There are many ways in which a microcomputer system can handle a RESET signal. **The simplest scheme is to input this signal to the $\overline{\text{RESET}}$ pin of the Z80 CPU.**

> RESET THE CPU

Another method of handling RESET is to test the signal in between print cycles and to prevent any print cycle from starting while RESET is high; this may be accomplished by connecting RESET to pin 6 of I/O Port B0 and then modifying our "in between print cycles" instruction sequence as follows:

```
LOOP:   IN      A,(2)       ;INPUT I/O PORT B0 TO ACCUMULATOR
        BIT     6,A         ;TEST BIT 6 (RESET)
        JR      NZ,LOOP     ;IF RESET IS HIGH, STAY IN LOOP
;RESET IS LOW. TEST PW STROBE AND RETURN STROBE
        AND     30H         ;ISOLATE BITS 5 (PW STROBE) AND
                            ;4 (RETURN STROBE)
        CP      10H         ;TEST FOR PW STROBE = 0, RETURN STROBE = 1
        JR      Z,LOOP      ;IF TEST IS TRUE STAY IN LOOP
;PRINT CYCLE INSTRUCTION SEQUENCE STARTS HERE
```

**This longer test loop will now require 51 cycles to execute. That means PW STROBE must pulse high for at least 25.5 microseconds, assuming a 500-nanosecond clock.**

> SIGNAL PULSE WIDTH

## PFR REL

**This** is yet another **signal** which must be tested before initiating printhammer firing. It **indicates when external logic is moving the paper feed.** Under such circumstances, we cannot fire the printhammer. **By connecting this signal to pin 3 of Input Port A0, we merely have to adjust the printhammer firing instruction initiation sequence as follows:**

```
LOOP:   IN      A,(0)       ;INPUT CONTENTS OF I/O PORT A0 TO
                            ;ACCUMULATOR
        OR      0F0H        ;ISOLATE BITS 3, 2, 1, and 0
        CPL                 ;COMPLEMENT THE RESULT TO TEST FOR
                            ;ANY 0 BIT
        JR      NZ,LOOP     ;ANY 0 BIT WILL NOW BE 1. IF ANY BIT IS
                            ;1, DO NOT FIRE PRINTHAMMER
;PRINTHAMMER FIRING SEQUENCE BEGINS HERE
```

## CA REL

**This signal** is almost identical to PFR REL. It **comes from external logic that controls carriage movement. We will connect this signal to pin 4 of Input Port A0** and modify the hammer firing instruction initiation sequence as follows:

```
LOOP:  IN    A,(0)      ;INPUT CONTENTS OF I/O PORT A0 TO
                        ;ACCUMULATOR
       OR    0E0H       ;ISOLATE BITS 4, 3, 2, 1 AND 0
       CPL              ;COMPLEMENT THE RESULT TO TEST FOR
                        ;ANY 0 BIT
       JR    NZ,LOOP    ;ANY 0 BIT WILL NOW BE 1. IF ANY BIT IS
                        ;1, DO NOT FIRE PRINTHAMMER
;PRINTHAMMER FIRING SEQUENCE BEGINS HERE
```

## FFI

**This is the signal which times the first delay in the print cycle -- the time during which the printwheel moves from its position of visibility until the required petal is in front of the printhammer.**

FFI is generated by external logic; it is low while the printwheel is moving and high while the printwheel is not moving.

**We will tie FFI to pin 7 of I/O Port A0.** The following instruction loop will create a delay which lasts until FFI goes high:

> TIME DELAY BASED ON INPUT SIGNAL

```
LOOP:  IN    A,(0)      ;INPUT PORT A0 TO ACCUMULATOR
       RLA              ;SHIFT BIT 7 INTO THE CARRY
       JR    NC,LOOP    ;IF CARRY = 0 STAY IN THE LOOP
```

Do you see how this loop works? After I/O Port A0 contents have been input to the Accumulator, we are only interested in bit 7, since this is the bit that corresponds to FFI.

This is what the RLA instruction does:

If the Carry status equals 1, the printwheel move delay is over. If Carry equals 0, program logic must continue the delay.

Why did we use an RLA instruction to test this bit instead of a BIT instruction? The BIT instruction uses two bytes of object code and eight clock cycles of execution time, whereas the RLA instruction is just one byte and executes in four clock cycles.

## EOR DET

**This signal indicates that the end of the ribbon has been reached.** Under these circumstances, character printing cannot continue.

When this signal is generated, there will still be fresh ribbon in front of the printhammer, so the signal is not used to inhibit printhammer firing; rather, it is used to prevent the end of the print cycle from ever being indicated. This effectively prevents a new print cycle from ever starting.

**We will connect the $\overline{\text{EOR DET}}$ signal to bit 7 of I/O Port B0.** Since $\overline{\text{EOR DET}}$ is a negative logic signal, we will test it prior to going into the "in between print cycle" loop, as follows:

```
;TEST FOR VALID END OF PRINT CYCLE
VALND:  IN    A,(2)        ;INPUT I/O PORT B0 TO ACCUMULATOR
        RLA                ;SHIFT BIT 7 INTO CARRY
        JR    NC,VALND     ;IF ZERO IN CARRY, STAY IN PRINT CYCLE
;START OF IN BETWEEN PRINT CYCLES LOOP
LOOP:   IN    A,(2)        ;INPUT I/O PORT B0 TO ACCUMULATOR
        BIT   6,A          ;TEST BIT 6 (RESET)
        JR    NZ,LOOP      ;IF RESET IS HIGH, STAY IN LOOP
;RESET IS LOW. TEST PW STROBE AND RETURN STROBE
        AND   30H          ;ISOLATE BITS 5 (PW STROBE) AND
                           ;4 (RETURN STROBE)
        CP    10H          ;TEST FOR PW STROBE = 0, RETURN STROBE = 1
        JR    Z,LOOP       ;IF TEST IS TRUE STAY IN LOOP
;PRINT CYCLE INSTRUCTION SEQUENCE STARTS HERE
```

Look at the instruction sequence above. There are some interesting aspects to it.

**The first three instructions above will be the last three instructions in the print cycle sequence.** The instruction labeled LOOP is the first instruction of a sequence which gets executed continuously until the start of the next print cycle. Thus, if $\overline{\text{EOR DET}}$ is low, program logic will hang up in the first three instructions listed above, constantly looping within these three instructions until $\overline{\text{EOR DET}}$ goes high. At that time, the print cycle ends and we go into the "in between print cycles" instruction loop. The program now hangs up indefinitely in this instruction loop until bit 6 (which corresponds to RESET) equals 0, while bit 5 (which corresponds to PW STROBE) equals 1, or bit 4 (which corresponds to RETURN STROBE) equals 0.

**There is another interesting feature of the instruction sequence above. We could, if we wished, eliminate the second IN instruction, as follows:**

```
;TEST FOR VALID END OF PRINT CYCLE
VALND:  IN    A,(2)        ;INPUT I/O PORT B0 TO ACCUMULATOR
        RLA                ;SHIFT BIT 7 INTO CARRY
        JR    NC,VALND     ;IF ZERO IN CARRY, STAY IN PRINT CYCLE
;START OF IN BETWEEN PRINT CYCLES LOOP
        BIT   7,A          ;TEST BIT 6 (RESET)
        JR    NZ,VALND     ;IF RESET IS HIGH, STAY IN LOOP
;RESET IS LOW. TEST PW STROBE AND RETURN STROBE
        AND   60H          ;ISOLATE BITS 5 (PW STROBE) AND 4
                           ;(RETURN STROBE)
        CP    20H          ;TEST FOR PW STROBE = 0, RETURN STROBE = 1
        JR    Z,VALND      ;IF TEST IS TRUE STAY IN LOOP
;PRINT CYCLE INSTRUCTION SEQUENCE STARTS HERE
```

By eliminating one instruction, we have saved two bytes of object code. The penalty is that we have added 11 clock cycles to the entire instruction loop, which means that the PW STROBE high pulse goes up from the 25.5 microseconds we calculated when discussing the RESET signal to 31 microseconds.

**Why does the condensed instruction sequence illustrated above work?** The reason is because external logic is not supposed to be moving the ribbon in between print cycles; therefore, $\overline{\text{EOR DET}}$ will always be high during the "in between print cycle" instruction execution loop. If this is so, the RLA instruction will always shift a 1 into the Carry, which will always cause execution to continue with the BIT instruction. Thus, the first three instructions become harmless. Notice that the BIT, AND, and CP instructions' operands have changed, since all the bits have been shifted one position to the left by the RLA instruction.

## HAMMER ENABLE FF

**This is the signal which prevents the printhammer from being fired after the printwheel is moved to its position of visibility,** as described in connection with the RETURN STROBE signal.

**We will connect HAMMER ENABLE FF to pin 6 of I/O Port A0,** then modify the instruction sequence which precedes printhammer firing as follows:

```
LOOP:   IN      A,(0)           ;INPUT CONTENTS OF I/O PORT A0 TO
                                ;ACCUMULATOR
        OR      0A0H            ;ISOLATE BITS 6, 4, 3, 2, 1, AND 0
        CPI                     ;COMPLEMENT THE RESULT TO TEST FOR
                                ;ANY 0 BIT
        JR      NZ,LOOP         ;ANY 0 BIT WILL NOW BE 1. IF ANY BIT IS
                                ;1, DO NOT FIRE PRINTHAMMER
;PRINTHAMMER FIRING SEQUENCE BEGINS HERE
```

## CLK

**This is the clock signal that synchronizes all logic in Figure 3-1. Try as we may, we cannot include this signal in our simulation of Figure 3-1,** since events within the microcomputer program are going to be synchronized by the sequence in which instructions are executed -- not by a clock. **Similarly, the next two signals, +5V and RV1, are power supplies. They are meaningless within a microcomputer program.**

## H1 - H6

**These are the six signals which select one of six time durations for the printhammer firing pulse. We will assign these signals to I/O Port B1.** Once the printhammer firing instruction sequence gets executed, it simply loads these signals into the Accumulator as follows:

```
        IN      A,(6)           ;INPUT FIRING PULSE TIME CODE TO
                                ;ACCUMULATOR
```

# INPUT SIGNAL SUMMARY

In summary, this is how input signals have been assigned:

Z80 PIO 0, Port A (Port A0) assigned to input:

| Bit | Signal |
|---|---|
| 7 | FFI |
| 6 | HAMMER ENABLE |
| 5 | |
| 4 | CA REL |
| 3 | PFR REL |
| 2 | FFA |
| 1 | RIB LIFT RDY |
| 0 | PFL REL |

Z80 PIO 0, Port B (Port B0) assigned to input:

| Bit | Signal |
|---|---|
| 7 | EOR DET |
| 6 | RESET |
| 5 | PW STROBE |
| 4 | RETURN STROBE |

Z80 PIO 1, Port B (Port B1) assigned to input:

| Bit | Signal |
|---|---|
| 7 | |
| 6 | |
| 5 | H6 |
| 4 | H5 |
| 3 | H4 |
| 2 | H3 |
| 1 | H2 |
| 0 | H1 |

# OUTPUT SIGNALS

**We will now turn our attention to the output signals listed on the right-hand side of Figure 3-1.** These signals are much easier to describe than the input signals. They consist of six flip-flop outputs — which are simply timing indicators used by external logic — plus four control signals. **We are going to output these signals to the B port of one Z80 PIO and the A port of the second Z80 PIO, as follows:**

Z80 PIO 1, Port A (Port A1) assigned to output:

| Bit | Signal |
|---|---|
| 7 | |
| 6 | FFF |
| 5 | FFE |
| 4 | FFE |
| 3 | FFD |
| 2 | FFC |
| 1 | FFB |
| 0 | FFA |

Z80 PIO 0, Port B (Port B0) assigned to output:

| Bit | Signal |
|---|---|
| 3 | START RIB MOTION |
| 2 | HAMMER PULSE |
| 1 | CH RDY |
| 0 | PW RELEASE |

We assign a pin for FFC even though it is not output, because I/O Port A1 is going to serve a double purpose — as a data storage location and as an output signals buffer. Simple routines to generate output signals cannot be concocted; that is the whole purpose of the logic in Figure 3-1. We will therefore simply define the four output control signals:

1) **PW REL. This signal marks the end of the fixed printhammer return and settling time delay, and the beginning of the fixed Final Movement's delay** during which external logic can move the paper feed and carriage.

2) **CH RDY.** This is also referred to as the **PRINTWHEEL READY** signal. This is the signal which defines the entire print-cycle time interval; it goes low at the start of the print cycle and stays low until the end of the print cycle.

3) **HAMMER PULSE. This signal must be output low for the time interval during which external logic is supposed to transmit a firing pulse to the printhammer solenoid.**

4) **START RIBBON MOTION PULSE. This signal is pulsed high early in the print cycle, telling external logic that it is safe to begin advancing the ribbon so that fresh ribbon will be in front of the printhammer when it is fired.**

## A DIGITAL-LOGIC ORIENTED SIMULATION

We are now ready to start simulating the logic illustrated in Figure 3-1 — but first, a brief overview of the logic.

### A LOGIC OVERVIEW

At the center of the logic sequence are four 74107 flip-flops, labeled $FFC_W$, $FFD_W$, $FFE_W$ and $FFF_W$. You will find these flip-flops in the center and to the left of Figure 3-1. **These four flip-flops form what is known as a "Johnson Counter".** Each flip-flop is controlled by the output of the previous flip-flop, coupled with a test for external conditions:

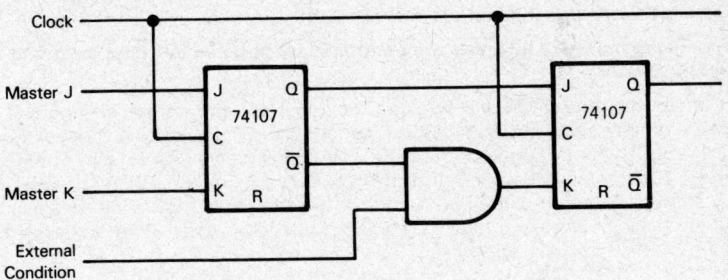

Thus, the four flip-flops may be visualized as initiating print cycle events in the following way:

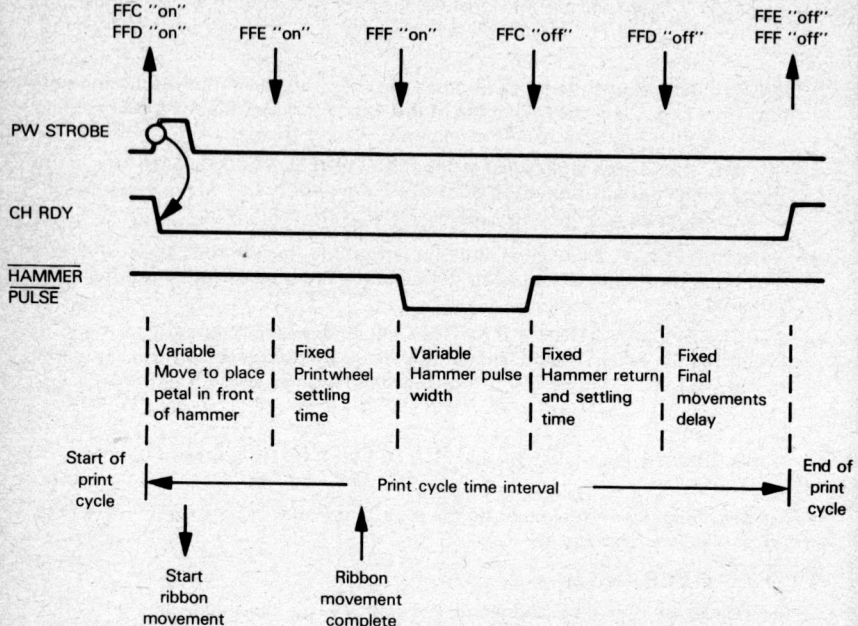

As illustrated above, the print-cycle time interval may be divided into five periods.

**During the first time interval, the printwheel is moved from its position of visibility until the required petal is in front of the printhammer. This time interval is controlled by external logic, via the FFI input.**

**The remaining four time intervals are controlled by three 74121 one-shots and the 555 multivibrator.**

**What about the two 7474 flip-flops at the top left-hand corner of Figure 3-1? These are simply cycle initiation logic.** Flip-flop FFA is triggered by a combination of signals necessary for a print cycle to begin. Flip-flop FFB acts as a switch for the four 74107 flip-flops, forcing them to turn "off" in between print cycles. Flip-flop FFB does this by tying its Q output to the reset inputs of the 74107 flip-flops. This results in the 74107 flip-flops always being turned off if FFB is turned off; later on we will explain in more detail how this happens.

**We are now going to follow a print cycle through Figure 3-1. As we progress, we will create a microcomputer assembly language program that simulates the logic, device-by-device.**

## FLIP-FLOP FFA_W

Our print cycle begins at the 7474 flip-flop designated FFA$_W$. You will find this flip-flop at the top left-hand corner of Figure 3-1. Let us isolate FFA$_W$, and illustrate it as follows:

| 7474 |
| FLIP-FLOP |

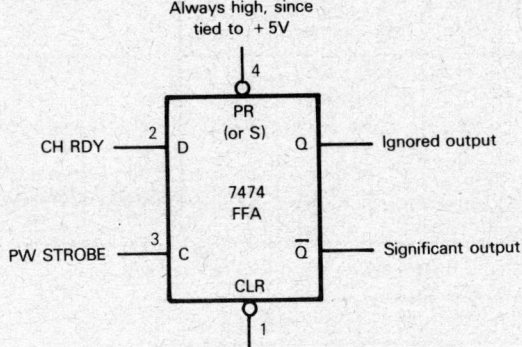

((CH RDY) OR (PW STROBE)) AND (NOT RESET)

Refer back to the general function table for a 7474 flip-flop given in Chapter 2.

Since PRESET (PR) is always high, being tied to +5V, a low CLEAR (CLR) input will force the flip-flop "off", at which time Q is output low and $\overline{Q}$ is output high.

Look at Figure 3-1 and you will see that CLR is generated as follows:

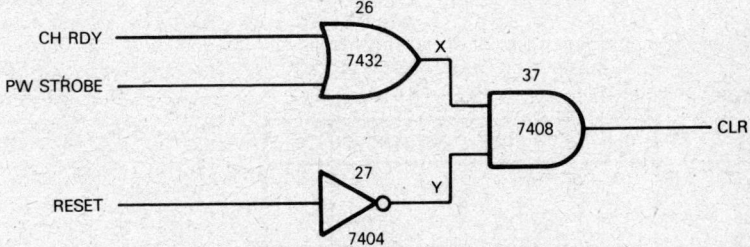

This is the truth table for CLR:

| CH RDY | PW STROBE | X | RESET | Y | CLR |
|--------|-----------|---|-------|---|-----|
| 0 | 0 | 0 | 0 | 1 | 0 |
|   |   |   | 1 | 0 | 0 |
| 0 | 1 | 1 | 0 | 1 | 1 |
|   |   |   | 1 | 0 | 0 |
| 1 | 0 | 1 | 0 | 1 | 1 |
|   |   |   | 1 | 0 | 0 |
| 1 | 1 | 1 | 0 | 1 | 1 |
|   |   |   | 1 | 0 | 0 |

For flip-flop FFA$_W$ to turn "on", CLR must be high; for CLR to be high, RESET must be low, and either CH RDY or PW STROBE must be high.

Now CH RDY provides FFA_W with its data (D) input, and PW STROBE provides the clock (C) input. Therefore the function table for flip-flop FFA_W may be illustrated as follows:

| | INPUTS | | | OUTPUTS | | |
|---|---|---|---|---|---|---|
| PRESET | CLR | CLOCK (PW STROBE) | D (CH RDY) | Q | $\bar{Q}$ | |
| 0 | 1 | 0 or 1 | 0 or 1 | 1 | 0 | PRESET=1 |
| 1 | 0 | 0 or 1 | 0 or 1 | 0 | 1 | |
| 0 | 0 | 0 or 1 | 0 or 1 | Unstable | | PRESET=1 |
| 1 | 1 | 0→1 | 1 | 1 | 0 | |
| 1 | 1 | 0→1 | 0 | 0 | 1 | |
| 1 | 1 | 0 | 0 or 1 | Previous Q | Previous $\bar{Q}$ | No change |

And this reduces to the following small function table:

| CLR | CH RDY | PW STROBE | $\bar{Q}$ | |
|---|---|---|---|---|
| 0 | | | 1 | "off" condition |
| 1 | 0 | 0→1 | 1 | } possible "on" conditions |
| 1 | 1 | 0→1 | 0 | |

It takes a zero-to-one transition of PW STROBE for flip-flop FFA_W to turn on. When FFA_W turns on, however, if CH RDY is 0 then the $\bar{Q}$ output is still 1, representing the "off" condition. Thus, to turn FFA_W "on", PW STROBE must go from 0 to 1 while CH RDY is 1.

Recall that CH RDY is a signal which is output high in between print cycles and is output low for the duration of a print cycle. This means that flip-flop FFA_W will only turn on if PW STROBE pulses high in between print cycles, as characterized by CH RDY being output high:

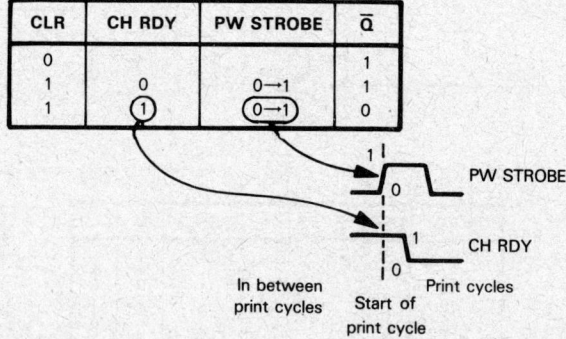

For the moment do not worry about how CH RDY goes to 0 shortly after flip-flop FFA_W turns on; we will explain how this happens later. The only important thing to note is that a PW STROBE high pulse will be ignored if it occurs while CH RDY is low.

**What about the RESET signal? What this signal does is override all other logic associated with flip-flop FFA_W; whenever RESET is input high, CLR is forced low which turns flip-flop FFA_W off irrespective of whatever else is going on.**

## SIMULATING FLIP-FLOP FFA$_W$

We concluded in Chapter 2 that a flip-flop is represented in a microcomputer system by a single bit of read/write memory. A single bit of a read/write buffer will do just as well.

**I/O Port A1 has been assigned to output signals.** This port has an 8-bit buffer to which port pins are connected; thus, **each bit of the port buffer will simulate the flip-flop whose output is transmitted via the port pin:**

| FLIP-FLOP SIMULATION USING I/O PORTS |
|---|

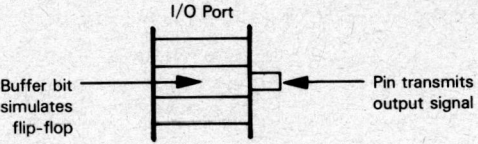

Recall that FFA has been assigned pin 0 of I/O Port A1.

O.K., we are ready to simulate flip-flop FFA$_W$.

At the same time, how about simulating the three gates below and to the left of FFA$_W$? These three gates are numbered 26, 27 and 37, and together they create the CLR input.

Simulating these three gates individually, the following instruction sequence applies:

```
;SIMULATE GATE 27
        IN      A,(2)           ;INPUT I/O PORT B0 CONTENTS TO REG A
        CPL                     ;COMPLEMENT ALL EIGHT BITS
        LD      B,A             ;SAVE COMPLEMENT IN REGISTER B
;SIMULATE GATE 26
        CPL                     ;RE-COMPLEMENT (RESTORE) REG A CONTENTS
        AND     22H             ;ISOLATE BITS 5 AND 1; THEY REPRESENT
                                ;PW STROBE AND CH RDY
;SIMULATE GATE 37
        JR      Z,CLR0          ;IF NEITHER BIT 1 NOR 5=1, CLR IS 0
        BIT     6,B             ;TEST COMPLEMENT OF RESET
        JR      Z,CLR0          ;IF RESULT IS 0, CLR IS 0
        SCF                     ;CLR IS 1 SO STORE 1 IN CARRY STATUS
        JR      FFAW+2
CLR0:   AND     A               ;CLR IS 0 SO STORE 0 IN CARRY STATUS
;SIMULATE FLIP-FLOP FFAW
FFAW:   JR      NC,FFA0         ;IF CLR=0, SET PORT A1, BIT 0 TO 1
        BIT     5,A             ;CLR IS NOT 0. TEST PW STROBE. IF
                                ;PW STROBE IS 0, CLOCK HAS NOT PULSED.
        JR      Z,FFA0          ;SET BIT 0 OF I/O PORT A1 TO 1
        BIT     1,A             ;PW STROBE IS 1. TEST CH RDY
        JR      Z,FFA0          ;IF CH RDY=0, SET BIT 0 OF PORT A1 TO 1
        IN      A,(4)           ;LOAD I/O PORT A1 INTO REG A
        RES     0,A             ;BIT 0 MUST BE RESET TO 0, SINCE FFA IS
                                ;"ON"
        OUT     (4),A
        JR      FFB             ;JUMP TO FLIP-FLOP B SIMULATION
FFA0:   IN      A,(4)           ;LOAD I/O PORT A1 INTO THE ACCUMULATOR
        SET     0,A             ;BIT 0 MUST BE SET TO 1 SINCE FFA IS "OFF"
        OUT     (4),A
;FLIP-FLOP FFB SIMULATION FOLLOWS
```

It is very important that you understand how instructions fit together to make a program. Read no further until you understand completely how the instruction sequence given above simulates the logic of $FFA_W$ and its three associated gates.

**Let us look at the above simulations.**

The RESET signal, you will recall, has been tied to bit 6 of Z80 PIO I/O Port B0; this port is addressed as Port 2 based on the way in which we have elected to wire the Z80 PIO into our microcomputer system. In order to invert this signal, we input the contents of I/O Port B0 to the Accumulator and complement the contents of the Accumulator:

**INVERTER SIMULATION**

```
                          from I/O Port B0
      IN    A,(2)         X X X X X X X X    to Accumulator B
      CPL                 X̄ X̄ X̄ X̄ X̄ X̄ X̄ X̄    Complement
          Bit 6 ─────────────┘
```

The complement of RESET, and of all the other bits of Port B0, is saved in Accumulator B. **The simulation of gate 27 is complete.**

**The simulation of gate 26 is not quite as straightforward.** We are seeking the OR of PW STROBE and CH RDY. These two signals are represented by bits 5 and 1, respectively, of I/O Port B. Now what we do is restore the contents of I/O Port B0 in the Accumulator by complementing its contents again:

**OR GATE SIMULATION**

**STATUS FLAGS USED TO REPRESENT LOGIC**

```
            XXXXXXXX    Accumulator contents
      CPL   XXXXXXXX    Complement
```

We then execute an AND instruction which sets all bits to 0, except bits 5 and 1. But we do not actually OR these two remaining bits. Why? The reason is because when the AND instruction is executed, it sets the Zero status to the complement of (PW STROBE) OR (CH RDY):

| A5 OR A1 | Accumulator A Contents | | | | | | | | HEX VALUE | ZERO STATUS |
|---|---|---|---|---|---|---|---|---|---|---|
| | A7 | A6 | A5 | A4 | A3 | A2 | A1 | A0 | | |
| 0 | 0 | 0 | 0 | 0 | 0 | 0 | 0 | 0 | 00 | 1 |
| 1 | 0 | 0 | 0 | 0 | 0 | 0 | 1 | 0 | 02 | 0 |
| 1 | 0 | 0 | 1 | 0 | 0 | 0 | 0 | 0 | 20 | 0 |
| 1 | 0 | 0 | 1 | 0 | 0 | 0 | 1 | 0 | 22 | 0 |

PW STROBE ──┘            CH RDY ──┘         Following AND instruction execution, Zero status is complement of (PW STROBE) OR (CH RDY).

**We can therefore move on to gate 37.**

The purpose of gate 37 is to generate the $FFA_W$ CLR input. We are going to simulate CLR using the Carry status. Now we come right out of the gate 26 simulation into the gate 37 simulation; at this time the Zero status will be 0 if the OR of PW STROBE and CH RDY is 1; Zero status will be 1 otherwise. (Recall that Zero statuses always represent the inverse of the 0 condition. In other words, a 0 condition causes the Zero status to be set to 1; a non-zero condition causes the Zero status to be set to 0.)

**ZERO STATUS**

The first instruction of the gate 37 simulation takes advantage of the fact that we have the OR of PW STROBE and CH RDY recorded in the Zero status. If the Zero status is 1, CLR must be 0, so the first JR Z instruction branches to logic that will set the Carry status to 0. The next instruction in the gate 37 simulation tests the complement of RESET as stored in Register B, using a BIT instruction. The BIT instruction will not change the contents of Register B, but it will set the Zero status to reflect the contents of bit 6. If the complement of RESET is 0, then the JR Z instruction which follows will branch to program logic which sets the Carry status to 0. If the complement of RESET is not 0, then all conditions have been met for gate 37 to output a non-zero result — and this condition is simulated by the SCF instruction, which sets the Carry status to 1.

**Flip-flop FFA is simulated next.** The state of this flip-flop may be defined as follows:

If CLR is 0 then $\overline{Q}$ is 1.
If PW STROBE is 0 then $\overline{Q}$ is 1.
If CLR is 1 and PW STROBE is 1 and CH RDY is 0 then $\overline{Q}$ is 1.
If CLR is 1 and PW STROBE is 1 and CH RDY is 1 then $\overline{Q}$ is 0.

CLR is simulated by the Carry status. PW STROBE is simulated by bit 5 of the Accumulator. CH RDY is simulated by bit 1 of the Accumulator.

**The simulation of flip-flop FFA begins with the instruction labeled FFAW.**

First we test the status of CLR using the JR NC instruction. This instruction causes a jump to FFA0 if the Carry status is 0 — which means that CLR is 0. FFA0 is the label for the first instruction in the sequence which sets $\overline{Q}$ to 1.

> **CARRY STATUS**

**Observe that we have some unnecessary steps at this point in the program.** Here is our logic:

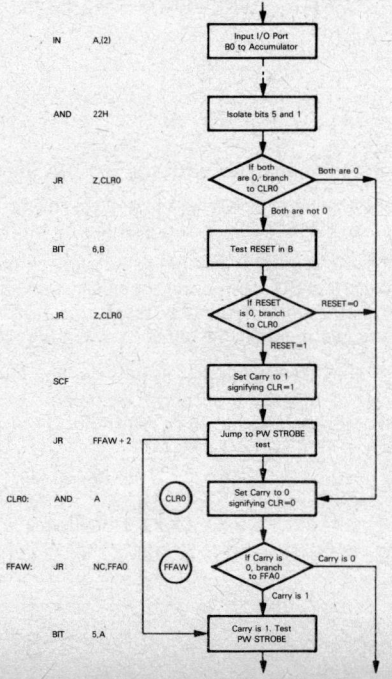

Each rectangular box represents a data movement or manipulation operation.

Each diamond represents logic which tests the condition of a status flag.

The logic sequence illustrated above maintains an orderly instruction flow which conforms with the flip-flop FFA$_W$ and its three preceding gates. But if you look at the instructions labeled CLR0 and FFAW, you will see that they are redundant. The instruction labeled CLR0 sets the Carry status to 0. The instruction labeled FFAW tests the Carry status, and upon detecting 0 branches to the later instruction labeled FFA0. But since we have just set the Carry status to 0, the instruction labeled FFAW must detect a 0 Carry status; therefore, the only allowed logic path following a branch to CLR0 is another branch to FFA0. We can therefore replace the two instructions which branch to CLR0 with instructions that branch directly to FFA0; then we can eliminate instructions labeled CLR0 and FFAW. This also eliminates the instruction which jumps to FFAW+2, since FFAW+2 addresses a BIT instruction which becomes the next sequential instruction. We can also remove the SCF instruction. Since Carry=0 conditions have been accounted for by branches to FFA0, the default is Carry=1, which no longer needs to be identified. Thus, our new instruction sequence may be illustrated as follows:

```
        Old Sequence                        New Sequence
        IN      A,(2)                       IN      A,(2)
        -                                   -
        -                                   -
        -                                   -
        AND     22H                         AND     22H
        JR      Z,CLR0                      JR      Z,FFA0
        BIT     6,B                         BIT     6,B
        JR      Z,CLR0                      JR      Z,FFA0
        SCF
        JR      FFAW+2      } unnecessary
CLR0:   AND     A             instructions
FFAW:   JR      NC,FFA0
        BIT     5,A                         BIT     5,A
        -                                   -
        -                                   -
        -                                   -
```

**Let us continue our program analysis with the BIT 5,A instruction.**

Presuming that CLR has a value of 1, we next test PW STROBE. Again, we use a BIT instruction for this purpose. PW STROBE is represented by bit 5 of the Accumulator.

Assuming that PW STROBE is 1, all that remains is to check the condition of CH RDY. To do this we again execute a BIT instruction; however, this time we test the contents of bit 1. Since the BIT instruction affects only the Zero status flag, we can execute as many BIT instructions as we need on the same byte without changing it.

Assuming that all conditions have been met to turn flip-flop FFA on, we must set bit 0 of I/O Port A1 to 0. This is done by inputting the contents of I/O Port A0 to the Accumulator, resetting the appropriate bit, then returning the result:

```
                    7 6 5 4 3 2 1 0  ← Bit No.
    IN    A,(4)     X X X X X X X Y     Accumulator contents
    RES   0,A       X X X X X X X 0  → Result to Port A1
    OUT   (4),A
```

The last three instructions of the flip-flop FFA simulation are the three instructions which set bit 0 to I/O Port A1 to 1 (reflecting the fact that flip-flop FFA is "off"). These three instructions load the contents of I/O Port A1 into the Accumulator, set the appropriate bit, then return the result:

**SWITCHING A BIT ON**

```
            7 6 5 4 3 2 1 0   ◄── Bit No.
IN   A,(4)  X X X X X X X Y     Accumulator contents
SET  0,A    X X X X X X X 1   ──► Result to Port A1
OUT  (4),A
```

**Now in all honesty, the program sequence we have just described is a ridiculous way of simulating flip-flop FFA and its three associated gates.**

**It is ridiculous because we simulated each gate as an independent transfer function. Instead, let us consider the flip-flop, with its three gates, as a single transfer function. We can represent the transfer function with the following state definition:**

**Set $\overline{Q}$ to 0 if RESET=0, CH RDY=1 and PW STROBE goes from 0 to 1. Set $\overline{Q}$ to 1 otherwise.**

How are we going to test for the transition of PW STROBE from 0 to 1?

Using interrupts, the test would be very simple; but we are not going to use interrupts until Chapter 5.

**Without using interrupts, there is only one way to check for a PW STROBE 0 to 1 transition.** We must input the contents of I/O Port B0 to the Accumulator, test bit 5, save the result, input the contents of I/O Port B0 to the Accumulator again, test bit 5 again, then compare the two bits for an old value of 0 and a new value of 1. But this scheme is risky; it will only catch signal transitions which are lucky enough to occur in between the two instructions which load I/O Port B0 contents to the Accumulator:

**SIGNAL LEVEL CHANGES SENSED WITHOUT INTERRUPTS**

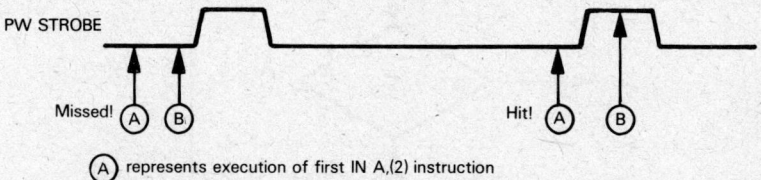

(A) represents execution of first IN A,(2) instruction

(B) represents execution of second IN A,(2) instruction

Within the logic of a microcomputer program, however, we have no need to rely on signal transitions. Event sequences are determined by instruction execution sequence. The whole concept of timing on the leading or trailing edge of a signal pulse has no meaning. Instead of using PW STROBE signal transitions, therefore, we will use PW STROBE signal levels. Flip-flop FFA can now be described with the following state definition:

**EVENT TIMING IN MICROCOMPUTER SYSTEM**

**Set $\overline{Q}$ to 0 if RESET equals 0, CH RDY equals 1 and PW STROBE equals 1. Set $\overline{Q}$ to 1 otherwise.**

If you are a logic designer, you may be deeply troubled by the blithe way in which we simply replace edge triggering with level triggering. We can do this within a microcomputer system because microcomputer programming gives us an extra degree of freedom, as compared with digital logic design: the order in which you stuff logic components into a PC card has nothing to do with the sequence in which logical events occur. Logic sequence is going to be controlled by edge and level triggering. But the order in which you write assembly language instructions is the order in which the instructions will be executed.

**TIMING AND LOGIC SEQUENCE**

To drive this point home, look at the following flowchart which represents the state definition for flip-flop FFA:

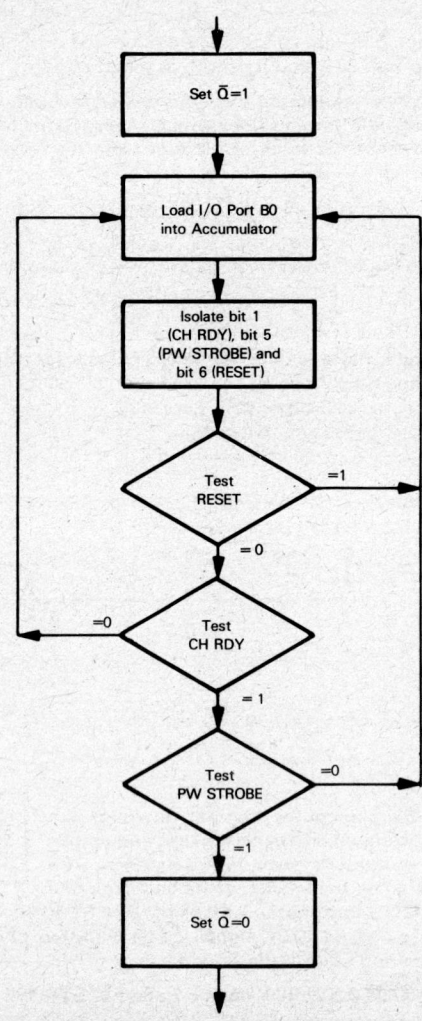

Again each rectangular box represents a data movement or manipulation operation, and each diamond represents logic which tests the condition of a status flag.

The order in which you write down instructions is the order in which instructions will be executed. With regard to the flowchart above, this execution sequence is represented by the continuous line of downward pointing arrows. Special Jump-On-Condition instructions allow the normal sequence to be modified, as represented by the horizontal arrows emanating from the sides of the diamonds. You can follow the arrows to the point where the Jump-On-Condition instruction takes you.

**We will now rewrite the flip-flop FFA simulation treating the flip-flop and the three CLR logic gates as a single transfer function.**

Since RESET, CH RDY and PW STROBE are all connected to pins of I/O Port B0, we load the contents of I/O Port B0 into the Accumulator and isolate all three bits. Now there is only one combination of values that these three bits can have if a new print cycle is to begin. RESET must equal 0, while CH RDY and PW STROBE both equal 1. We will therefore redraw our program flowchart as follows:

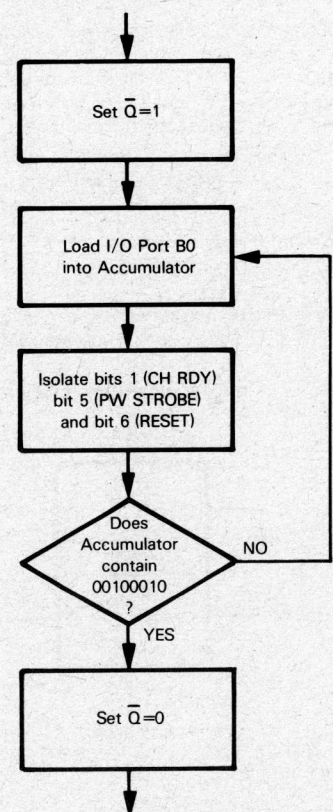

**Our instruction sequence condenses to the following few instructions:**

```
;SIMULATION OF FFAW AND ASSOCIATED LOGIC
         IN      A,(4)           ;INITIALLY SET BIT 0 OF I/O PORT A1 TO 1
         SET     0,A
         OUT     (4),A
;LOAD I/O PORT B0 CONTENTS INTO ACCUMULATOR AND ISOLATE BITS
;1, 5 AND 6 FOR CH RDY, PW STROBE AND RESET, RESPECTIVELY
FFAW:    IN      A,(2)           ;INPUT I/O PORT B0 TO ACCUMULATOR
         AND     62H             ;ISOLATE BITS 6, 5 AND 1. IF RESET=0
         CP      22H             ;CH RDY=1 AND PW STROBE=1, NEW PRINT
                                 ;CYCLE STARTS
         JR      NZ,FFAW         ;OTHERWISE RETURN TO FFAW
         IN      A,(4)           ;START NEW PRINT CYCLE BY SETTING I/O
         RES     0,A             ;PORT A1, BIT 0 TO 0
         OUT     (4),A
;NEW PRINT CYCLE INSTRUCTION SEQUENCE STARTS HERE
```

The first three instructions in the above sequences simply set bit 0 of I/O Port A1 to 1. This is in anticipation of a new print cycle not beginning. Four instructions, beginning with the instruction labeled FFAW, are all that are needed to check for conditions which trigger the start of a new print cycle. These four instructions execute in 36 clock cycles which, assuming a 500 nanosecond clock, means that PW STROBE must pulse high for at least 18 microseconds.

Providing RESET equals 0 while CH RDY and PW STROBE equal 1, a new print cycle must begin, so the last three instructions set bit 0 of I/O Port A1 to 0.

**Our simulation of flip-flop FFA is complete.**

## FLIP-FLOP FFB$_W$

The next device in our logic sequence is another 7474 flip-flop, marked FFB$_W$ in Figure 3-1; it is just to the right of FFA$_W$. This flip-flop may be illustrated as follows:

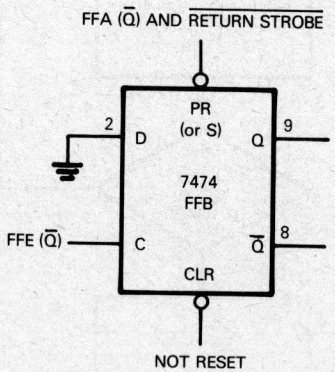

The following function table describes FFB, as wired above, with its D input tied to 0:

| FFA ($\bar{Q}$) | RETURN STROBE | PRESET | NOT RESET (CLR) | FFE ($\bar{Q}$) =CLOCK | Q | $\bar{Q}$ |
|---|---|---|---|---|---|---|
| 0 | 0 | 0 | 1 | X | 1 | 0 |
| 0 | 1 | 0 | 0 | X | unstable | |
| 1 | 0 | 0 | | | | |
| 1 | 1 | 1 | 0 | X | 0 | 1 |
| | | 1 | 1 | 0→1 | 0 | 1 |

Chapter 2 provides the standard 7474 flip-flop function table; all we have done is remove the D column, and the rows that show D=1. We can also remove the CLR column, and all rows that show CLR=0, since CLR is tied to NOT RESET. NOT RESET will always be 1 within a print cycle, since FFA will not turn on if NOT RESET is 0.

**The following simplified function table can now be used for FFB, assuming that CLR (NOT RESET) will always be 1 and D will always be 0:**

| FFA ($\bar{Q}$) AND RETURN STROBE =PRESET | FFE ($\bar{Q}$) =CLOCK | Q | $\bar{Q}$ |
|---|---|---|---|
| 0 | 0 or 1 | 1 | 0 |
| 1 | 0→1 | 0 | 1 |

Let us take a look at the FFB PRESET input: it is FFA ($\bar{Q}$) AND RETURN STROBE.

RETURN STROBE, recall, is a signal input by external logic to initiate a special print cycle which moves the printwheel back to its position of visibility, but does not fire the printhammer or print a character. We call this a "Printwheel Repositioning"

**PRINTWHEEL REPOSITIONING PRINT CYCLE**

print cycle. In between print cycles, therefore, RETURN STROBE must be input high.

**Since RETURN STROBE is input low as an alternative method of initiating a print cycle, when simulating FFB, we must consider RETURN STROBE in two ways:**

**1) As a contributor to the PRESET input.**

**2) As a signal which can initiate a print cycle, bypassing flip-flop FFA.**

**But first, let us define the condition of flip-flop FFB in between print cycles.**

As we have just seen in our simulation of flip-flop FFA, the FFA ($\bar{Q}$) output is high until the beginning of a print cycle, when $\bar{Q}$ goes low; the FFA ($\bar{Q}$) output is therefore high in between print cycles. By definition, RETURN STROBE is high in between print cycles, since RETURN STROBE low is used to initiate a printwheel repositioning print cycle.
**Therefore, the FFB PRESET input will be high in between print cycles:**

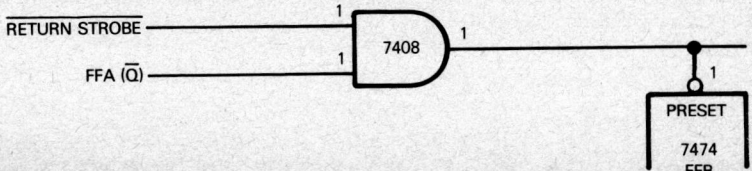

Since PRESET is input high in between print cycles, we are going to assume that at the beginning of a print cycle FFB is off; that is, Q is output low and $\overline{Q}$ is output high. This also assumes that at some recent time PRESET was input high when the $\overline{Q}$ output of flip-flop FFE went from 0 to 1. As you will see later on, this is indeed what happens at the end of every print cycle.

**Coming into a new print cycle, therefore, FFB has a high PRESET input, with a high Q output and a low $\overline{Q}$ output. This flip-flop now acts as a switch: it is turned on by PRESET being input low; it is subsequently turned off by a clock 0 to 1 transition occurring after PRESET has again gone high:**

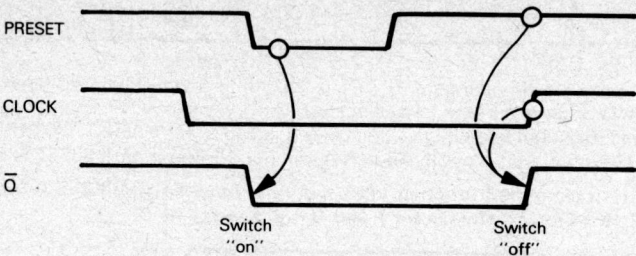

The switch "on" illustrated above occurs under two circumstances:

1) Immediately after the onset of a new print cycle, when FFA outputs $\overline{Q}$ low, thus forcing PRESET low.
2) When RETURN STROBE is input low signalling a printwheel repositioning print cycle.

The switch "off" occurs when the FFE ($\overline{Q}$) output makes a low-to-high transition while PRESET is being input high; this occurs at the end of every print cycle.

## SIMULATING FLIP-FLOP FFB

**Bit 1 of I/O Port A1 has been assigned to the $\overline{Q}$ output of flip-flop FFB. The switch "on" illustrated above is therefore simulated by the following three instructions:**

**SWITCHING BITS ON**

```
IN      A,(4)       ;LOAD FLIP-FLOP DATA BYTE
RES     1,A         ;RESET BIT 1 TO 0
OUT     (4),A       ;RESTORE FLIP-FLOP DATA BYTE
```

Subsequently the switch "off" will be simulated as follows:

**SWITCHING BITS OFF**

```
IN      A,(4)       ;LOAD FLIP-FLOP DATA BYTE
SET     1,A         ;SET BIT 1 TO 1
OUT     (4),A       ;RESTORE FLIP-FLOP DATA BYTE
```

**We now encounter a situation where, with every best intention, we are not going to be able to directly simulate our digital logic.**

It is easy enough to draw one 7474 flip-flop in a logic diagram and connect its pins to suitable signals. Having done that, you no longer need to worry about when a signal does or does not change state. Unfortunately, an assembly language instruction sequence has no pins or signals; **assembly language will simulate events that are occurring at one instant in time only. For flip-flop FFB, this may be illustrated as follows:**

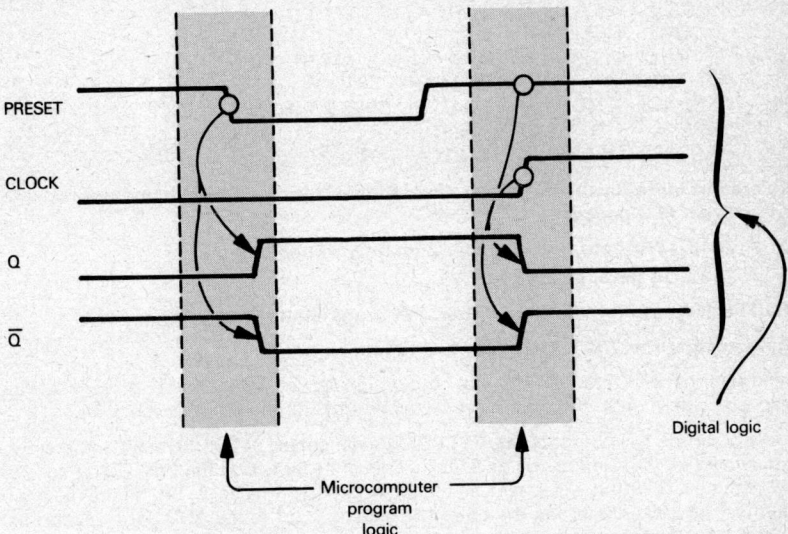

Immediately after flip-flop FFA turns on to usher in a new print cycle, it outputs $\overline{Q}$ low, which in turn switches flip-flop FFB on. FFB will not switch off until some point much later in the print cycle, when FFE outputs $\overline{Q}$ high. **We must therefore divide our simulation of FFB into two parts:**

1) At the beginning of our program we will simulate FFB switching on, since chronologically it is the next event within the print cycle.
2) Later on in the program, when we simulate FFE setting $\overline{Q}$ high, we must remember to simulate FFB switching off.

But that is not all there is to the FFB simulation. **We must also modify the instruction sequence that executes in between print cycles, so that RETURN STROBE input low can be simulated initiating a printwheel repositioning print cycle.**

**With modified or new instructions shaded, this is how our program now looks:**

```
;IN BETWEEN PRINT CYCLES PROGRAM EXECUTION
;INITIALLY SET I/O PORT A1 BITS 1 AND 0 TO 1
        IN      A,(4)           ;INPUT I/O PORT A1 TO ACCUMULATOR
        OR      3               ;SET BITS 1 AND 0
        OUT     (4),A           ;RETURN RESULT
;TEST FOR RETURN STROBE LOW
STBHI:  IN      A,(2)           ;INPUT I/O PORT B0 TO ACCUMULATOR
        BIT     4,A             ;TEST RETURN STROBE BIT
        JR      Z,FFB           ;IF IT IS 0, JUMP TO FFB SIMULATION
;SIMULATION OF FFAW AND ASSOCIATED LOGIC
;LOAD I/O PORT B0 CONTENTS INTO ACCUMULATOR AND ISOLATE BITS
;1, 5 AND 6 FOR CH RDY, PW STROBE AND RESET, RESPECTIVELY
        IN      A,(2)           ;INPUT I/O PORT B0 TO ACCUMULATOR
        AND     62H             ;ISOLATE BITS 6, 5 AND 1. IF RESET=0,
```

```
            CP      22H         ;CH RDY=1 AND PW STROBE=1, NEW
                                ;PRINT CYCLE STARTS
            JR      NZ,STBHI    ;OTHERWISE RETURN TO STBHI
            IN      A,(4)       ;START NEW PRINT CYCLE BY SETTING I/O
            RES     0,A         ;PORT A1, BIT 0 TO 0
            OUT     (4),A
;NEW PRINT CYCLE INSTRUCTION SEQUENCE STARTS HERE
;SIMULATE FLIP-FLOP FFB SWITCHING ON
FFB:        IN      A,(4)       ;LOAD I/O PORT A1 INTO ACCUMULATOR
            RES     1,A         ;RESET BIT 1 TO 0
            OUT     (4),A       ;RESTORE RESULT
```

We are not quite finished with our simulation of flip-flop FFB. Observe that the $\overline{Q}$ output from FFB goes to:

1) **A 7411 AND gate, located approximately at coordinate B5.**
2) **A 7432 OR gate, located at C7.**

The FFB ($\overline{Q}$) output is not idle either, but we will look into it later.

**First consider the 7411 AND gate located at B5.**

If you refer back to the description of output signals, you will notice that CH RDY was declared to be high in between print cycles but low during a print cycle.

In reality, CH RDY is output by the 7411 AND gate located at B5; therefore, in between print cycles, all three inputs to this AND gate must be high. Our analysis of flip-flop FFB shows that its $\overline{Q}$ output will indeed be high in between print cycles, but for the moment you must take it on faith that the other two signals input to the AND gate will also be high in between print cycles.

**In any event, as soon as flip-flop FFB switches on, its $\overline{Q}$ output goes low, which means that no matter what the other two inputs to the 7411 AND gate do, CH RDY will also be driven low. This change in the status of CH RDY is simulated by adding the following instructions to our program:**

```
;TEST FOR RETURN STROBE LOW
STBHI:      IN      A,(2)       ;INPUT I/O PORT B0 TO ACCUMULATOR
            BIT     4,A         ;TEST RETURN STROBE BIT
            JR      Z,FFB       ;IF IT IS 0, JUMP TO FFB SIMULATION
;SIMULATION OF FFAW AND ASSOCIATED LOGIC
;LOAD I/O PORT B0 CONTENTS INTO ACCUMULATOR AND ISOLATE BITS
;1,5 AND 6 FOR CH RDY, PW STROBE AND RESET, RESPECTIVELY
            IN      A,(2)       ;INPUT I/O PORT B0 TO ACCUMULATOR
            AND     62H         ;ISOLATE BITS 6, 5 AND 1. IF RESET=0,
            CP      22H         ;CH RDY=1 AND PW STROBE=1, NEW
                                ;PRINT CYCLE STARTS
            JR      NZ,STBHI    ;OTHERWISE RETURN TO STBHI
            IN      A,(4)       ;START NEW PRINT CYCLE BY SETTING I/O
            RES     0,A         ;PORT A1, BIT 0 TO 0
            OUT     (4),A
;NEW PRINT CYCLE INSTRUCTION SEQUENCE STARTS HERE
;SIMULATE FLIP-FLOP FFB SWITCHING ON
FFB:        IN      A,(4)       ;LOAD I/O PORT A1 INTO ACCUMULATOR
            RES     1,A         ;RESET BIT 1 TO 0
            OUT     (4),A       ;RESTORE RESULT
;SIMULATE 7411 AND GATE SWITCHING CH RDY LOW
            IN      A,(2)       ;INPUT I/O PORT B0 TO ACCUMULATOR
            RES     1,A         ;RESET BIT 1 TO 0
            OUT     (2),A       ;RESTORE RESULT
```

We are now faced with an interesting problem. CH RDY becomes the D input to flip-flop FFA and it contributes to the CLR input of FFA. **What happens when CH RDY goes low in response to FFB switching on?**

Notice that PW STROBE only pulses high, therefore the OR gate located at coordinate B2 relies on CH RDY being high in order to provide a high input to the following AND gate.

This AND gate, in turn, provides a high CLR input to flip-flop FFA. In other words, by the time flip-flop FFB turns "on" and switches CH RDY low, PW STROBE will have already gone low; thus inputs PW STROBE and CH RDY will both be low. **If you look back at flip-flop FFA's CLR truth table, you will find that when CH RDY and PW STROBE are both 0, CLR will always be 0.**

**Therefore, flip-flop FFA will switch off:**

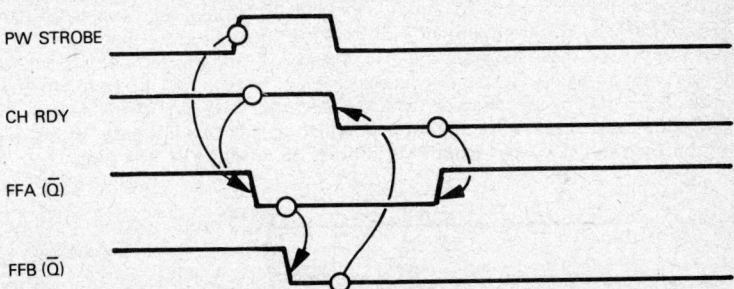

What does this mean? Our conclusion is that flip-flop FFA switches itself "on" at the beginning of a print cycle, but only stays on long enough to switch flip-flop FFB "on". When FFB turns "on" it sets CH RDY low, and that turns flip-flop FFA "off".

But here is the rub: if you look again at Figure 3-1, you will find that **flip-flop FFA helps generate the J input to flip-flop FFC, in addition to switching to flip-flop FFB.**

**TIMING AND LOGIC SEQUENCE**

Now that events are serialized in time, we can go ahead and simulate flip-flop FFA being turned "off", so long as we remember, when simulating flip-flop FFC, that it receives $\overline{Q}$ low from flip-flop FFA. Bearing this precaution in mind, we will extend our program as follows:

```
;TEST FOR RETURN STROBE LOW
STBHI:  IN      A,(2)       ;INPUT I/O PORT B0 TO ACCUMULATOR
        BIT     4,A         ;TEST RETURN STROBE BIT
        JR      Z,FFB       ;IF IT IS 0, JUMP TO FFB SIMULATION
;SIMULATION OF FFAW AND ASSOCIATED LOGIC
;LOAD I/O PORT B0 CONTENTS INTO ACCUMULATOR AND ISOLATE BITS
;1, 5 AND 6 FOR CH RDY, PW STROBE AND RESET, RESPECTIVELY
        IN      A,(2)       ;INPUT I/O PORT B0 TO ACCUMULATOR
        AND     62H         ;ISOLATE BITS 6, 5 AND 1. IF RESET=0,
        CP      22H         ;CH RDY=1 AND PW STROBE=1, NEW
                            ;PRINT CYCLE STARTS
        JR      NZ,STBHI    ;OTHERWISE RETURN TO STBHI
        IN      A,(4)       ;START NEW PRINT CYCLE BY SETTING I/O
        RES     0,A         ;PORT A1, BIT 0 TO 0
        OUT     (4),A
;NEW PRINT CYCLE INSTRUCTION SEQUENCE STARTS HERE
```

```
;SIMULATE FLIP-FLOP FFB SWITCHING ON
FFB:    IN      A,(4)       ;LOAD I/O PORT A1 INTO ACCUMULATOR
        RES     1,A         ;RESET BIT 1 TO 0
        OUT     (4),A       ;RESTORE RESULT
;SIMULATE 7411 AND GATE SWITCHING CH RDY LOW
        IN      A,(2)       ;INPUT I/O PORT B0 TO ACCUMULATOR
        RES     1,A         ;RESET BIT 1 TO 0
        OUT     (2),A       ;RESTORE RESULT
;CH RDY LOW TURNS FFA OFF. SET BIT 0 OF I/O PORT A1 TO 1
        IN      A,(4)       ;LOAD I/O PORT A1 TO ACCUMULATOR
        SET     0,A         ;SET BIT 0 TO 1
        OUT     (4),A       ;RESTORE RESULT
```

**Now look at the OR gate located at coordinate C7.** This gate receives the FFB $\overline{Q}$ output as one of its inputs in order to generate PW REL. The other input to this OR gate is the AND of the Q output from flip-flop FFF, plus the $\overline{Q}$ output of flip-flop FFD. You will find out shortly that these flip-flops are also turned "off" in between print cycles; they are turned on sequentially during the course of the print cycle. At the point where FFB switches on, FFF will be switched off, which means that its Q output will be low; thus, the AND gate located at C6 will output low, which means that **OR gate 26 has been relying on the high $\overline{Q}$ output from FFB in order to output PW REL high:**

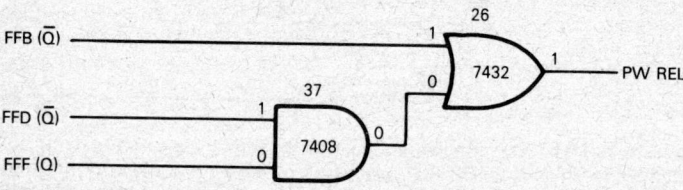

**Now, when FFB switches "on" and outputs $\overline{Q}$ low, PW REL will also output low. We must therefore modify our program to output bits 0 and 1 of I/O Port B0 low, since both PW REL and CH RDY are going to be driven low. This is how our program now looks:**

```
;TEST FOR RETURN STROBE LOW
STBHI:  IN      A,(2)       ;INPUT I/O PORT B0 TO ACCUMULATOR
        BIT     4,A         ;TEST RETURN STROBE BIT
        JR      Z,FFB       ;IF IT IS 0, JUMP TO FFB SIMULATION
;SIMULATION OF FFAW AND ASSOCIATED LOGIC
;LOAD I/O PORT B0 CONTENTS INTO ACCUMULATOR AND ISOLATE BITS
;1, 5 AND 6 FOR CH RDY, PW STROBE AND RESET, RESPECTIVELY
        IN      A,(2)       ;INPUT I/O PORT B0 TO ACCUMULATOR
        AND     62H         ;ISOLATE BITS 6,5 AND 1. IF RESET=0, CH RDY=1
        CP      22H         ;AND PW STROBE=1, NEW PRINT CYCLE STARTS
        JR      NZ,STBHI    ;OTHERWISE RETURN TO STBHI
        IN      A,(4)       ;START NEW PRINT CYCLE BY SETTING I/O PORT
        RES     0,A         ;A1, BIT 0 TO 0
        OUT     (4),A
;NEW PRINT CYCLE INSTRUCTION SEQUENCE STARTS HERE
;SIMULATE FLIP-FLOP FFB SWITCHING ON
FFB:    IN      A,(4)       ;LOAD I/O PORT A1 INTO ACCUMULATOR
        RES     1,A         ;RESET BIT 1 TO 0
        OUT     (4),A       ;RESTORE RESULT
```

```
;SIMULATE 7411 AND GATE SWITCHING CH RDY LOW. ALSO
;7432 OR GATE SWITCHES PW REL LOW
        IN      A,(2)           ;INPUT I/O PORT B0 TO ACCUMULATOR
        AND     0FCH            ;RESET BITS 0 AND 1 TO 0
        OUT     (2),A           ;RESTORE RESULT
;CH RDY LOW TURNS FFA OFF. SET BIT 0 OF I/O PORT A1 TO 1
        IN      A,(4)           ;LOAD I/O PORT A1 TO ACCUMULATOR
        SET     0,A             ;SET BIT 0 TO 1
        OUT     (4),A           ;RESTORE RESULT
```

**Do we have to do anything about the Q output from flip-flop FFB? If you look at this output you will see that it ties directly to the RESET inputs of flip-flops FFC, FFD, and FFE. It also becomes one of the inputs to the 555 multivibrator.**

In fact, the FFB Q output is a clamping signal; when low, it shuts the four connected devices off, and when high, these four devices are switched on.

**The FFB Q output will be taken into account when we simulate the four devices connected to this signal. Therefore, our simulation of flip-flop FFB is done.**

## FLIP-FLOP FFC

This is the 74107 flip-flop at coordinate C2 in Figure 3-1. Since we are going to simulate four 74107 flip-flops, you should refer back to Chapter 2 if you cannot immediately recall the characteristics of this device.

**Let us isolate flip-flop FFC to see how it works:**

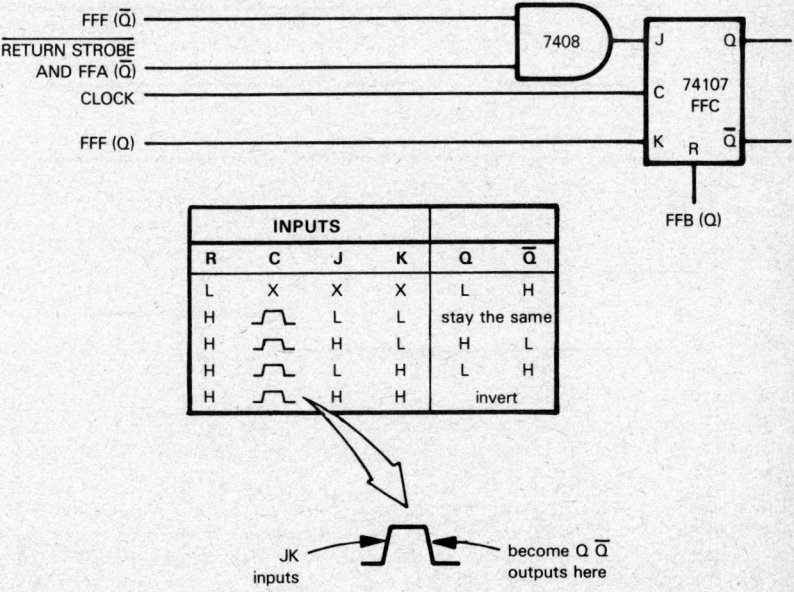

**In between print cycles, the Q output to FFB, being low, switches flip-flop FFC off. FFC, therefore, outputs Q low and $\overline{Q}$ high.**

What happens when FFB is switched on depends on the J and K inputs arriving at FFC.

In between print cycles flip-flop FFF is switched off, therefore its Q output will be low. FFC receives its K input from the FFF Q output, therefore when FFC switches on, its K input will be 0.

The J input to FFC is generated as follows:

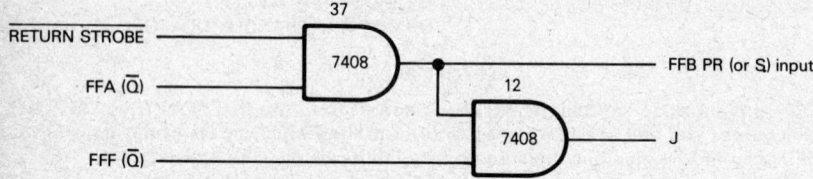

FFF ($\overline{Q}$) will be high, since FFF is switched off. The FFC J input will therefore be identical to the FFB PR input, which we have already described.

**In summary, this is the signal sequence which turns FFC on:**

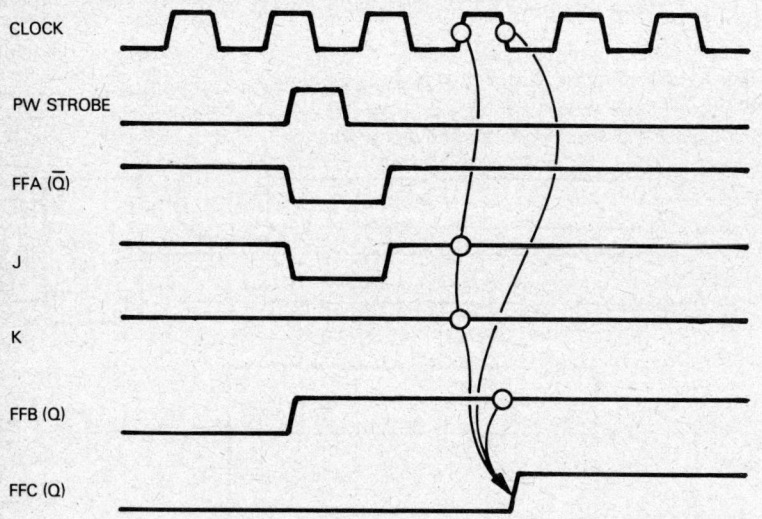

When the FFB Q output goes high, unclamping FFC, FFC waits until the FFA $\overline{Q}$ output goes high again; then FFC will receive a high input at J and a low input at K. On the trailing edge of the clock pulse input to FFC, Q will be output high and $\overline{Q}$ will be output low.

FFC waits for the FFA $\overline{Q}$ output to go high again, because while FFA is switched on, FFA $\overline{Q}$ is output low. While FFA ($\overline{Q}$) (or RETURN STROBE) is pulsed low, FFC receives a low J input. So long as FFC is receiving low J and K inputs, its outputs will not change — that is one of the properties of a 74107 flip-flop.

**Flip-flop FFC will remain in its "on" state until some later point in the print cycle when flip-flop FFF switches on. At that time, flip-flop FFC will receive a high input at K and a low input at J; and that will cause FFC to switch off.**

## SIMULATING FLIP-FLOP FFC
The simulation of flip-flop FFC is indeed straightforward; it involves these three steps:
1) We must adjust our initialization instructions to ensure that flip-flop FFC is reported as "off" in between print cycles.
2) The flip-flop FFB simulation must be followed immediately by instructions which simulate flip-flop FFC turning on.
3) We must remember to simulate FFC turning off — but that will not happen until some later point in the program.

Now the following modifications to the beginning of our program ensure that flip-flop FFC is simulated "off" in between print cycles:

```
;IN BETWEEN PRINT CYCLES PROGRAM EXECUTION
;INITIALLY SET I/O PORT A1 BITS 1 AND 0 TO 1, BIT 2 TO 0
        IN      A,(4)       ;INPUT I/O PORT A1 TO ACCUMULATOR
        OR      3           ;SET BITS 1 AND 0
        RES     2,A         ;RESET BIT 2
        OUT     (4),A       ;RETURN RESULT
;TEST FOR RETURN STROBE LOW
STBHI:  IN      A,(2)       ;INPUT I/O PORT B0 TO ACCUMULATOR
        BIT     4,A         ;TEST RETURN STROBE BIT
        JR      Z,FFB       ;IF IT IS 0, JUMP TO FFB SIMULATION
```

All we have done is add the RES instruction to reset I/O Port A1 bit 2 to 0:

```
                   Accumulator A
                     Contents
                   7 6 5 4 3 2 1 0  ◄──── Bit No.
       IN   A,(4)  X X X X X X X X
       OR   3      0 0 0 0 0 0 1 1
                   X X X X X X 1 1
       RES  2,A    X X X X X 0 1 1
```

Recall that I/O Port A1 bit 2 has been assigned to flip-flop FFC.

**What about the time delay that separates flip-flops B and C switching on?** Recall that flip-flop FFC will not switch on until after flip-flop FFB has switched flip-flop FFA off. If this is a printwheel repositioning print cycle, then FFC will not switch on until RETURN STROBE is input high again.

> TIMING AND LOGIC SEQUENCE

**The simplicity or complexity of our timing problem depends entirely on logic beyond Figure 3-1.** There is nothing within the logic of Figure 3-1 that demands a time delay of fixed duration or, for that matter, any time delay separating FFB and FFC switching on. We will therefore pay no attention to the timing considerations associated with FFC switching on; rather, we will simply add simulation to the end of our program as follows:

```
;NEW PRINT CYCLE INSTRUCTION SEQUENCE STARTS HERE
;SIMULATE FLIP-FLOP FFB SWITCHING ON
FFB:    IN      A,(4)       ;LOAD I/O PORT A1 INTO ACCUMULATOR
        RES     1,A         ;RESET BIT 1 TO 0
        OUT     (4),A       ;RESTORE RESULT
;SIMULATE 7411 AND GATE SWITCHING CH RDY LOW ALSO
```

```
;7432 OR GATE SWITCHES PW REL LOW
          IN      A,(2)    ;INPUT I/O PORT B0 TO ACCUMULATOR
          AND     0FCH     ;RESET BITS 0 AND 1 TO 0
          OUT     (2),A    ;RESTORE RESULT
;CH RDY LOW TURNS FFA OFF. SET BIT 0 OF I/O PORT A1 TO 1
   (A)    IN      A,(4)    ;LOAD I/O PORT A1 TO ACCUMULATOR
          SET     0,A      ;SET BIT 0 TO 1
          OUT     (4),A    ;RESTORE RESULT
;SIMULATE 74107 FLIP-FLOP FFC SWITCHING ON. SET BIT 2 OF
;I/O PORT A1 TO 1
          IN      A,(4)    ;LOAD I/O PORT A1 INTO ACCUMULATOR
          SET     2,A      ;SET BIT 2 TO 1
          OUT     (4),A    ;RESTORE RESULT
```

If you are beginning to think like a programmer, you will detect an opportunity for economy in the simulation of flip-flop FFC switching on. **Observe that the three instructions directly above (A) are also setting a bit of I/O Port A1 to 1.** This generates the following sequence of events:

**PROGRAMS MADE SHORTER**

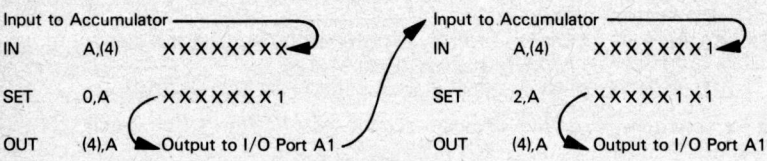

We can combine the two operations as follows:

```
          IN      A,(4)    XXXXXXXX
          OR      5        00000101
                           XXXXX1X1
```

The instructions marked (A) now disappear, and are replaced by these modifications, marked (B):

```
;NEW PRINT CYCLE INSTRUCTION SEQUENCE STARTS HERE
;SIMULATE FLIP-FLOP FFB SWITCHING ON
FFB:      IN      A,(4)    ;LOAD I/O PORT A1 INTO ACCUMULATOR
          RES     1,A      ;RESET BIT 1 TO 0
          OUT     (4),A    ;RESTORE RESULT
;SIMULATE 7411 AND GATE SWITCHING CH RDY LOW. ALSO
;7432 OR GATE SWITCHES PW REL LOW
          IN      A,(2)    ;INPUT I/O PORT B0 TO ACCUMULATOR
          AND     0FCH     ;RESET BITS 0 AND 1 TO 0
          OUT     (2),A    ;RESTORE RESULT
;CH RDY LOW TURNS FFA OFF. SET BIT 0 OF I/O PORT A1 TO 1.
;ALSO SIMULATE FFC TURNING ON. SET BIT 2 OF I/O PORT A1 TO 1
          IN      A,(4)    ;LOAD I/O PORT A1 TO ACCUMULATOR
   (B)    OR      5        ;SET BITS 2 AND 0 TO 1
          OUT     (4),A    ;RESTORE RESULT
```

# START RIBBON MOTION PULSE SIMULATION

**Recall that early in a print cycle the START RIBBON MOTION output signal is pulsed high to trigger external logic which advances the ribbon; thus, when the printhammer fires, fresh ribbon is in front of the character being printed. The**

START RIBBON MOTION signal is generated by a 7411 AND gate (number 7) located at coordinate D6 in Figure 3-1. This AND gate has three inputs:
1) HAMMER ENABLE FF. This is a signal input to identify a printwheel repositioning print cycle.
2) The Q output from flip-flop FFC.
3) The $\overline{Q}$ output from flip-flop FFD.

HAMMER ENABLE FF will be high unless a printwheel repositioning print cycle is in progress, in which case the ribbon does not have to be moved. This signal, therefore, suppresses the START RIBBON MOTION pulse.

In between print cycles, flip-flops FFC and FFD are both switched off; therefore, FFC (Q) is low and FFD ($\overline{Q}$) is high. **The FFC (Q) output holds the START RIBBON MOTION signal low.**

**When FFC switches on during a normal print cycle all inputs to AND gate 7 will be high, so START RIBBON MOTION will pulse high; it will stay high until flip-flop FFD switches on, at which time FFD will output $\overline{Q}$ low;** that will drop START RIBBON MOTION pulse low. **Timing may be illustrated as follows:**

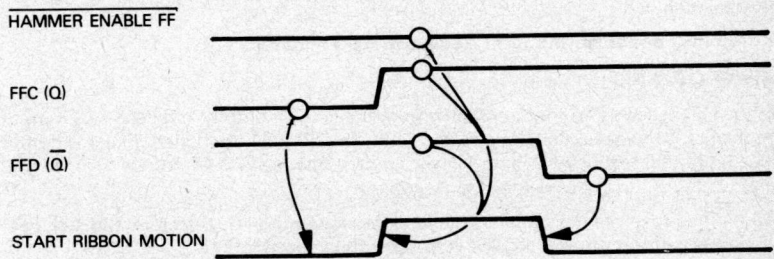

If you look at the timing diagram illustrated in Figure 3-2, you will see that the START RIBBON MOTION output pulse is extremely short. Therefore, instead of using flip-flop FFD to time the end of the START RIBBON MOTION HIGH pulse, we will simply execute instructions to turn bit 3 of I/O Port B0 on, then immediately turn it off, as follows:

```
;NEW PRINT CYCLE INSTRUCTION SEQUENCE STARTS HERE
;SIMULATE FLIP-FLOP FFB SWITCHING ON
FFB:    IN      A,(4)       ;LOAD I/O PORT A1 INTO ACCUMULATOR
        RES     1,A         ;RESET BIT 1 TO 0
        OUT     (4),A       ;RESTORE RESULT
;SIMULATE 7411 AND GATE SWITCHING CH RDY LOW. ALSO
;7432 OR GATE SWITCHES PW REL LOW
        IN      A,(2)       ;INPUT I/O PORT B0 TO ACCUMULATOR
        AND     0FCH        ;RESET BITS 0 AND 1 TO 0
        OUT     (2),A       ;RESTORE RESULT
;CH RDY LOW TURNS FFA OFF. SET BIT 0 OF I/O PORT A1 TO 1
;ALSO SIMULATE FFC TURNING ON. SET BIT 2 OF I/O PORT A1 TO 1
        IN      A,(4)       ;LOAD I/O PORT A1 TO ACCUMULATOR
        OR      5           ;SET BITS 2 AND 0 TO 1
        OUT     (4),A       ;RESTORE RESULT
;PULSE START RIBBON MOTION HIGH
        IN      A,(2)       ;INPUT TO ACCUMULATOR FROM I/O PORT B0
        SET     3,A         ;SET BIT 3 HIGH
        OUT     (2),A       ;OUTPUT TO I/O PORT B0
        RES     3,A         ;SET BIT 3 LOW
        OUT     (2),A       ;OUTPUT TO I/O PORT B0
```

We can calculate the START RIB MOTION pulse width by adding the instruction execution times between pin 3 of I/O Port B being set high, then being reset low:

**PULSE WIDTH CALCULATION**

| Cycles | Instruction | | |
|---|---|---|---|
| 11 | OUT | (2),A | ;OUTPUT TO I/O PORT B0 |
| 8 | RES | 3,A | ;SET BIT 3 LOW |
| 11 | OUT | (2),A | ;OUTPUT TO I/O PORT B0 |

Pulse width = 19 cycles, or 9.5 microseconds using a 500-nanosecond clock.

**What happens next? Our logic sequence may take us to flip-flop FFD, to the right of FFC, or we may drop down to the 74121 one-shot number 36,** just below FFC.

**One-shot 36** has its two A inputs tied to ground, which means that they will both input low. If you look at the 74121 function table given in Chapter 2, you will find that, in this configuration, a one-shot output is triggered by a low-to-high transition at B. FFC ($\overline{Q}$) provides this trigger. Any other B input will keep this one-shot turned off — which means that **Q and $\overline{Q}$ will output low and high, respectively, until much later in the print cycle, when FFC switches off**; that is, when the FFC $\overline{Q}$ output makes a low-to-high transition.

**Flip-flop FFD becomes the next device to be simulated.**

## FLIP-FLOP FFD

Flip-flop FFD receives its J input directly from the FFC (Q) output; it receives its K input from the FFC ($\overline{Q}$) output. Remember, since one-shot 36 is still switched off, its $\overline{Q}$ output will be high; that means AND gate 12 will simply allow the FFC ($\overline{Q}$) output to propagate straight through, to become the FFD (K) input.

Now, flip-flop receives the same reset and clock signals as FFC; therefore, **flip-flop FFD will simply switch on one clock cycle later than flip-flop FFC.**

## SIMULATING FLIP-FLOP FFD

**The simulation of flip-flop FFD is almost identical to the simulation of flip-flop FFC;** the principal difference is that bit 3 of I/O Port A1 has been assigned to flip-flop FFD. Once again, we are going to limit ourselves to switching flip-flop FFD on and ensuring that its setting in between print cycles is correct.

Flip-flop FFD is switched off later in the print cycle; we must therefore remember to switch it off later in the program.

**Here are the necessary program modifications and additions:**

```
;IN BETWEEN PRINT CYCLES PROGRAM EXECUTION
;INITIALLY SET I/O PORT A1 BITS 1 AND 0 TO 1, BITS 3 AND 2 TO 0
        IN      A,(4)       ;INPUT I/O PORT A1 TO ACCUMULATOR
        OR      3           ;SET BITS 1 AND 0
        AND     0F3H        ;RESET BITS 3 AND 2
        OUT     (4),A       ;RETURN RESULT
;TEST FOR RETURN STROBE LOW
STBHI:  IN      A,(2)       ;INPUT I/O PORT B0 TO ACCUMULATOR
        BIT     4,A         ;TEST RETURN STROBE BIT
        JR      Z,FFB       ;IF IT IS 0, JUMP TO FFB SIMULATION
        -
        -
        -
;CH RDY LOW TURNS FFA OFF. SET BIT 0 OF I/O PORT A1 TO 1
```

```
;ALSO SIMULATE FFC TURNING ON. SET BIT 2 OF I/O PORT A1 TO 1
        IN      A,(4)     ;LOAD I/O PORT A1 TO ACCUMULATOR
        OR      5         ;SET BITS 2 AND 0 TO 1
        OUT     (4),A     ;RESTORE RESULT
;PULSE START RIBBON MOTION HIGH
        IN      A,(2)     ;INPUT TO ACCUMULATOR FROM I/O PORT B0
        SET     3,A       ;SET BIT 3 HIGH
        OUT     (2),A     ;OUTPUT TO I/O PORT B0                    C
        RES     3,A       ;SET BIT 3 LOW
        OUT     (2),A     ;OUTPUT TO I/O PORT B0
;SIMULATE FFD TURNING ON. SET BIT 3 OF I/O PORT A1 TO 1
        IN      A,(4)     ;INPUT PORT A1 TO ACCUMULATOR
        SET     3,A       ;SET BIT 3 TO 1
        OUT     (4),A     ;RESTORE RESULT
```

If the program modifications and additions illustrated above are not immediately obvious, compare them to the flip-flop C simulation. Do not go on if you do not understand the flip-flop FFD program changes.

Just as the simulation of FFC switching on (A) was absorbed into the FFB simulation ( (B) ), so **the simulation of FFD switching on ( (C) ) can be absorbed as follows:**

**PROGRAMS MADE SHORTER**

```
;NEW PRINT CYCLE INSTRUCTION SEQUENCE STARTS HERE
;SIMULATE FLIP-FLOP FFB SWITCHING ON
FFB:    IN      A,(4)     ;LOAD I/O PORT A1 INTO ACCUMULATOR
        RES     1,A       ;RESET BIT 1 TO 0
        OUT     (4),A     ;RESTORE RESULT
;SIMULATE 7411 AND GATE SWITCHING CH RDY LOW. ALSO
;7432 OR GATE SWITCHES PW REL LOW
        IN      A,(2)     ;INPUT I/O PORT B0 TO ACCUMULATOR
        AND     0FCH      ;RESET BITS 0 AND 1 TO 0
        OUT     (2),A     ;RESTORE RESULT
;CH RDY LOW TURNS FFA OFF. SET BIT 0 OF I/O PORT A1 TO 1
;ALSO SIMULATE FFC AND FFD TURNING ON. SET BITS 2 AND 3 OF I/O PORT A1
        IN      A,(4)     ;LOAD I/O PORT A1 TO ACCUMULATOR
        OR      0DH       ;SET BITS 3, 2 AND 0 TO 1
        OUT     (4),A     ;RESTORE RESULT                              D
;PULSE START RIBBON MOTION HIGH
        IN      A,(2)     ;INPUT TO ACCUMULATOR FROM I/O PORT B0
        SET     3,A       ;SET BIT 3 HIGH
        OUT     (2),A     ;OUTPUT TO I/O PORT B0
        RES     3,A       ;SET BIT 3 LOW
        OUT     (2),A     ;OUTPUT TO I/O PORT B0
```

**If the simulations are combined ( (D) ), flip-flops FFC and FFD will switch on at exactly the same instant.**

**The logic in Figure 3-1 shows FFD switching on one clock pulse after FFC. If the clock period is two microseconds, then there will be a two-microsecond delay between flip-flops FFD and FFC switching on. Both our simulations are wrong.**

**Does this matter? We honestly cannot tell with the information at hand.** We do not know how external logic uses the FFC and FFD outputs. **If the switching time interval between these two flip-flops must be very close to two microseconds, then our simulation is not going to work.** Either the two flip-flops must become part of "external logic", or some other means of simulating the eventual function must be found.

**TIMING AND LIMITS OF SIMULATION**

If external logic demands some switching time delay but is not fussy about the length of the time delay, then our simulation of flip-flop FFD ( Ⓒ ) is adequate.

It is quite possible that the logic in Figure 3-1 shows a switching time delay between flip-flops FFC and FFD only to define the leading and trailing edges of the START RIBBON MOTION pulse; but we have taken care of this high pulse by sequentially executing instructions that output 1, then 0 to bit 3 of I/O Port B0. So far as logic internal to Figure 3-1 is concerned, therefore, the need for a switching time delay between flip-flops FFC and FFD disappears. This being the case, **we will assume that external logic has no need for a switching time delay between flip-flops FFC and FFD; and we will adopt the shorter, combined simulation identified by** Ⓓ .

## FLIP-FLOP FFE

**The next device in our logic sequence is flip-flop FFE.** The circuitry surrounding this flip-flop is almost identical to FFD.

The FFE (K) input is tied to the FFD ($\overline{Q}$) output, switched by another component of AND gate 12. The other input to this AND gate is the $\overline{Q}$ output of one-shot 49. One-shot 49 is wired in the same way as one-shot 36, which we have just described.

The transition of flip-flop FFD's $\overline{Q}$ output from 0 to 1 will occur when FFD is switched off; this is the transition which will trigger one-shot 49. Therefore, **one-shot 49 will output $\overline{Q}$ high until flip-flop FFD is switched off, which means that when FFD switches on, its $\overline{Q}$ output will propagate straight through the AND gate connecting it to the FFE (K) input:**

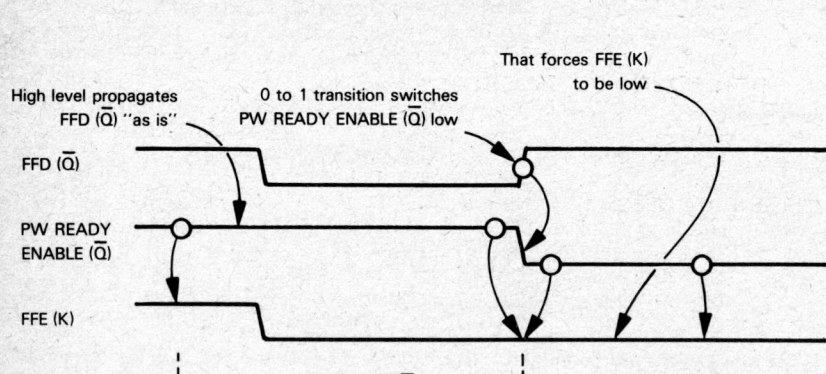

**The unique feature of flip-flop FFE is the way in which its J input is generated.** This input is the AND of the FFD ($\overline{Q}$) output and input signal FFI. Now, the Q output of FFD will go high as soon as FFD switches on; but **FFI is input low from the beginning of**

**the print cycle until the printwheel has correctly positioned itself.** (We described the function of this input signal earlier in the chapter.) **The timing associated with FFI may be illustrated as follows:**

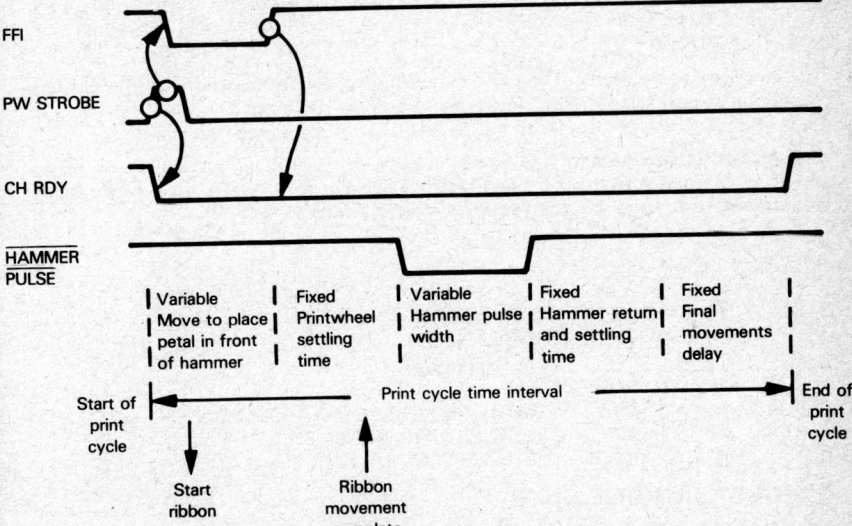

So long as FFI is low, flip-flop FFE will receive a low J input; low J and K inputs, you will recall, hold the Q outputs of a 74107 flip-flop in their prior condition. Thus, **input signal FFI has been used to create the first time delay of the print cycle: a variable time delay needed to move the required printwheel petal in front of the printhammer.** Simulating this time delay is simple enough; it may be illustrated as follows:

```
;PULSE START RIBBON MOTION HIGH
          IN      A,(2)       ;INPUT TO ACCUMULATOR FROM I/O PORT B0
          SET     3,A         ;SET BIT 3 HIGH
          OUT     (2),A       ;OUTPUT TO I/O PORT B0
          RES     3,A         ;SET BIT 3 LOW
          OUT     (2),A       ;OUTPUT TO I/O PORT B0
;TEST VELOCITY DECODE INPUT TO CREATE PRINTWHEEL MOVE DELAY
VLDC:     IN      A,(0)       ;INPUT I/O PORT A0 TO ACCUMULATOR
          RLA                 ;SHIFT BIT 7 INTO CARRY
          JR      NC,VLDC     ;STAY IN LOOP IF CARRY IS ZERO
;AT END OF DELAY SIMULATE FFE SWITCHING ON
          IN      A,(4)       ;INPUT TO ACCUMULATOR FROM I/O PORT A1
          RES     5,A         ;RESET BIT 5
          SET     4,A         ;SET BIT 4
          OUT     (4),A       ;OUTPUT THE RESULT
```

In order to generate the initial time delay, we simply execute a continuous program loop which inputs the contents of I/O Port A0 to the Accumulator. Bit 7 of I/O Port A0 has been assigned to input signal FFI. We test this bit by shifting it into the Carry status. If the Carry status then has a 0 content, FFI must still be low; so we stay within the loop. As soon as a 1 is shifted into the Carry status,

**TIME DELAY OF VARIABLE LENGTH**

**JUMP ON NO CARRY**

the JR NC instruction will create a "false" result; the next sequential instruction executes and we are out of the time delay loop:

```
        VLDC:  IN   A,(0)
C=0            RLA
               JR   NC,VLDC
C=1     →      IN   A,(4)
```

Jump on No Carry means jump if Carry is 0. "Jump" means "do not go on to the next sequential instruction", instead go to VLDC.

**The last four instructions of the FFE simulation show both outputs of this flip-flop becoming output signals. This meets requirements of Figure 3-1.** We therefore reset bit 5 (it represents the $\overline{Q}$ output) and we set bit 4 (it represents the Q output).

**The instruction sequence executed in between print cycles will have to be modified to ensure that bit 5 has initially been set to 1, while bit 4 has initially been reset to 0. Here are the required modifications:**

```
;IN BETWEEN PRINT CYCLES PROGRAM EXECUTION
;INITIALLY SET I/O PORT A1 BITS 5, 1 AND 0 TO 1, BITS 4, 3 AND 2 TO 0
        IN      A,(4)       ;INPUT I/O PORT A1 TO ACCUMULATOR
        OR      23H         ;SET BITS 5, 1 AND 0 TO 1
        AND     0E3H        ;RESET BITS 4, 3 AND 2 TO 0
        OUT     (4),A       ;RETURN RESULT
;TEST FOR RETURN STROBE LOW
STBHI:  IN      A,(2)       ;INPUT I/O PORT B0 TO ACCUMULATOR
        BIT     4,A         ;TEST RETURN STROBE BIT
        JR      Z,FFB       ;IF IT IS 0, JUMP TO FFB SIMULATION
```

## PW SETTLING ONE-SHOT

**The PW SETTLING one-shot is the 74121 device at coordinate B6 in Figure 3-1.** We have described this device in Chapter 2. With its two A inputs tied to ground, **this one-shot is triggered by a low-to-high transition at its B input. Since the B input is tied to the FFE Q output, this transition occurs as soon as flip-flop FFE switches on.**

**The PW SETTLING one-shot has a two millisecond delay.** This delay results from the external capacitor/resistor combination marked C1 and R1. Therefore, as soon as FFE switches on, the PW SETTLING one-shot outputs $\overline{Q}$ low for two milliseconds:

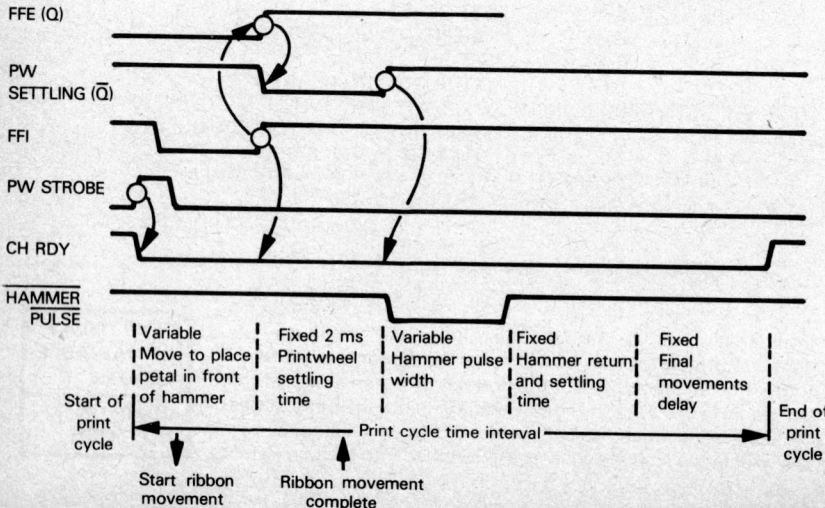

## SIMULATING THE PW SETTLING ONE-SHOT

Simulating the one-shot time delay is simple enough, and may be illustrated as follows:

**ONE-SHOT TIME DELAY SIMULATION**

```
;PULSE START RIBBON MOTION HIGH
        IN      A,(2)           ;INPUT TO ACCUMULATOR
                                 FROM I/O PORT B0
        SET     3,A             ;SET BIT 3 HIGH
        OUT     (2),A           ;OUTPUT TO I/O PORT B0
        RES     3,A             ;SET BIT 3 LOW
        OUT     (2),A           ;OUTPUT TO I/O PORT B0
;TEST VELOCITY DECODE INPUT TO CREATE PRINTWHEEL MOVE DELAY
VLDC:   IN      A,(0)           ;INPUT I/O PORT A0 TO AC-
                                 CUMULATOR
        RLA                     ;SHIFT BIT 7 INTO CARRY
        JR      NC,VLDC         ;STAY IN LOOP IF CARRY IS
                                 ZERO
;AT END OF DELAY SIMULATE FFE SWITCHING ON
        IN      A,(4)           ;INPUT TO ACCUMULATOR
                                 FROM I/O PORT A1
        RES     5,A             ;RESET BIT 5
        SET     4,A             ;SET BIT 4
        OUT     (4),A           ;OUTPUT THE RESULT
;SIMULATE 2 MS PW SETTLING TIME DELAY
        LD      A,0FAH          ;LOAD INITIAL TIME DELAY
                                 CONSTANT
PWS:    DEC     A               ;DECREMENT ACCUMULATOR
        JR      NZ,PWS          ;REDECREMENT IF RESULT IS
                                 NOT ZERO
```

There are two instructions in the time delay loop: DEC A and JR NZ; thus, the total time delay can be computed as follows:

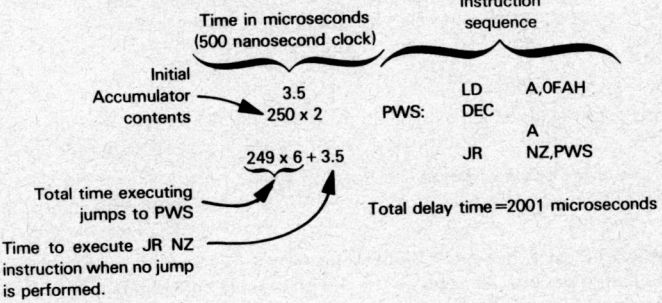

## FLIP-FLOP FFF

Once the PW SETTLING one-shot has timed out, we are ready to fire the printhammer. The 555 multivibrator is actually going to generate the printhammer firing pulse, but it is most important to ensure that the printhammer does not fire while any part of the print or carriage mechanisms is moving. The 555 one-shot is therefore triggered by flip-flop FFF, which, in turn, is switched on by a J input that is the AND of many safeguard signals. Let us isolate flip-flop FFF and examine its inputs.

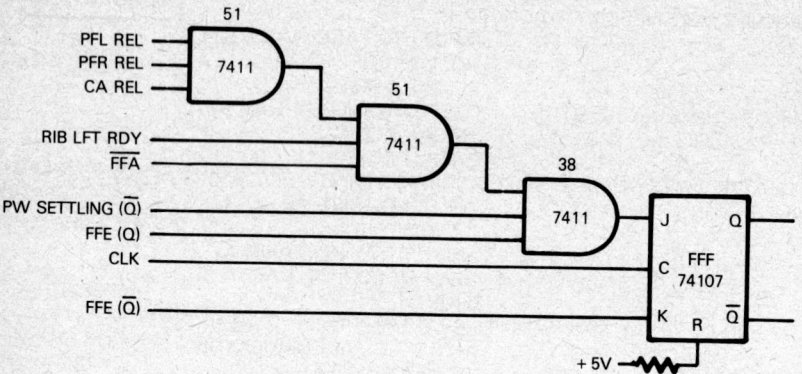

With its Clear (R) input tied to +5V, flip-flop FFF has the following function table:

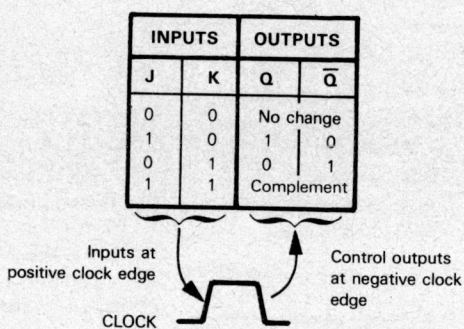

In between print cycles, FFE is "off", so the K input to FFF is high. The flip-flop J input will be low since the FFE (Q) output will be low, and FFE (Q) is one contributor to FFF (J).

**In between print cycles, therefore, flip-flop FFF is "off",** since a low J input and a high K input generate steady outputs of $Q=0$, $\overline{Q}=1$; this is characteristic of a flip-flop in its "off" condition.

Now when FFE switches on, it inputs a low K to FFF. So long as the J input is also low, no change occurs. As soon as the seven signals contributing to FFF (J) are all high, flip-flop FFF will receive a high J input; this will switch flip-flop FFF on — Q is then output high and $\overline{Q}$ is output low.

## SIMULATING FLIP-FLOP FFF

Coming out of the simulation of FFE, we know that FFE (Q) and FFE ($\overline{Q}$) have correct levels for FFF to switch on.

**Coming out of the simulation of the PW SETTLING one-shot, the one-shot $\overline{Q}$ output must be high:**

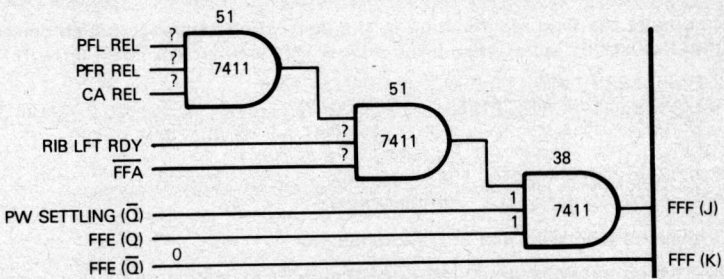

**All that is needed is to test the five remaining interlock signals; as soon as they are all high, we simulate flip-flop FFF switching on. This is the instruction sequence:**

```
;TEST VELOCITY DECODE INPUT TO CREATE PRINTWHEEL MOVE DELAY
VLDC:   IN      A,(0)       ;INPUT I/O PORT A0 TO ACCUMULATOR
        RLA                 ;SHIFT BIT 7 INTO CARRY
        JR      NC,VLDC     ;STAY IN LOOP IF CARRY IS ZERO
;AT END OF DELAY SIMULATE FFE SWITCHING ON
        IN      A,(4)       ;INPUT TO ACCUMULATOR FROM I/O PORT A1
        RES     5,A         ;RESET BIT 5
        SET     4,A         ;SET BIT 4
        OUT     (4),A       ;OUTPUT THE RESULT
;SIMULATE 2 MS PW SETTLING TIME DELAY
        LD      A,0FAH      ;LOAD INITIAL TIME DELAY CONSTANT
PWS:    DEC     A           ;DECREMENT ACCUMULATOR
        JR      NZ,PWS      ;REDECREMENT IF RESULT IS NOT ZERO
;SIMULATE FLIP-FLOP FFF SWITCHING ON
FFF:    IN      A,(0)       ;INPUT I/O PORT A0 CONTENTS TO ACCUMULATOR
        CPL                 ;COMPLEMENT TO TEST FOR 1 BITS
        AND     1FH         ;ISOLATE BITS 0 THROUGH 4
        JR      NZ,FFF      ;IF ANY BITS ARE 1, STAY IN LOOP
        IN      A,(4)       ;SET BIT 6 OF I/O PORT A1 TO 1
        SET     6,B
        OUT     (4),A
```

By now, you should be able to understand instructions as they are added to the program.

The first four instructions simply load the contents of I/O Port A0 into the Accumulator and test for 1s in the low-order five bits. Until such time as all five bits are 1, the program will remain in the four-instruction loop that begins with IN A,(0) and ends with JR NZ,FFF.

When bits 0 through 4 all equal 1, the CPL instruction changes all these bits to 0:

|  |  |  | Accumulator contents |  |
|---|---|---|---|---|
| FFF: | IN | A,(0) | X X X 1 1 1 1 1 |  |
|  | CPL |  | $\overline{X}\,\overline{X}\,\overline{X}$ 0 0 0 0 0 |  |
|  | AND | 1FH | 0 0 0 1 1 1 1 1 |  |
|  |  |  | 0 0 0 0 0 0 0 0 | Zero Status = 1 |
|  | JR | NZ,FFF | Return to FFF only if Zero status = 0 |  |
|  | IN | A,(4) | Continue here if Zero status is 1 |  |

The JR NZ instruction no longer deflects program execution back to FFF; rather, it allows the next sequential instruction to be executed.

**We can make the final modification to the instruction sequence which correctly sets flip-flop status in between print cycles. This is what we finish up with:**

```
;IN BETWEEN PRINT CYCLES PROGRAM EXECUTION
;INITIALLY SET I/O PORT A1 BITS 5, 1 AND 0 TO 1, BITS 6, 4, 3 AND 2 TO 0
    IN      A,(4)       ;INPUT I/O PORT A1 TO ACCUMULATOR
    OR      23H         ;SET BITS 5, 1 AND 0 TO 1
    AND     0A3H        ;RESET BITS 6, 4, 3 AND 2 TO 0
    OUT     (4),A       ;RETURN RESULT
```

**What happens when flip-flop FFF switches on?**

**The FFF (Q) output goes up to pin 9 of AND gate 37 at coordinate C6.** This is part of the logic which contributes to the PW REL signal. However, **the transition of the FFF (Q) output from low-to-high is not significant,** since the other input to AND gate 37 is the FFD (Q̄) output which is currently low. The FFF (Q) output is connected to AND gate 37 to hold PW REL low early in the print cycle when FFD (Q̄) is high.

**The FFF Q and Q̄ outputs contribute to the FFC J and K inputs.** FFF (Q̄) is one contributor to AND gate 12, the output of which becomes the FFC (J) input. The other contributor to this AND gate is the output of AND gate 37 at coordinate A4, which is constantly high by this time in the print cycle; therefore, when the FFF (Q̄) output goes low, the FFC (J) input also goes low. The K input to FFC is the FFF (Q) output. **FFC will therefore switch off when K goes high, and that will not happen until FFF switches on.**

In our <u>simulation, however</u>, we are going to postpone FFC switching off until the end of HAMMER PULSE. This is because the purpose of FFC switching off is to trigger the **PW RELEASE ENABLE** one-shot, which creates the time delay needed by the printhammer to settle back. Thus, instead of using parallel delays:

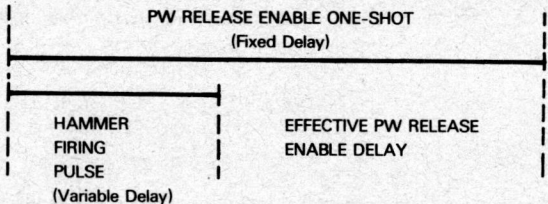

we will implement serial delays, which more immediately meet logic needs:

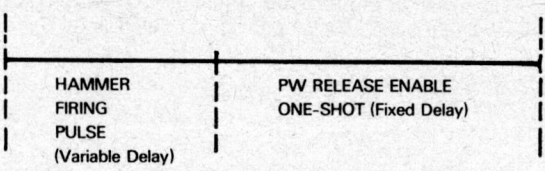

The hammer firing pulse is generated by the 555 multivibrator. Therefore, the 555 multivibrator provides the next event in our chronological sequence; it is triggered by a high-to-low transition at pin 2. This pin's input is created as follows:

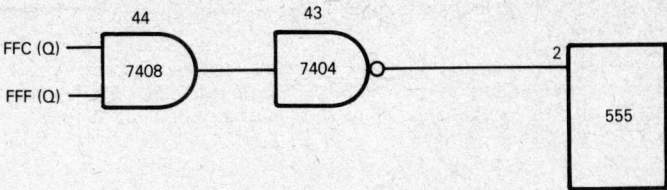

This is the sequence of events that must be simulated:

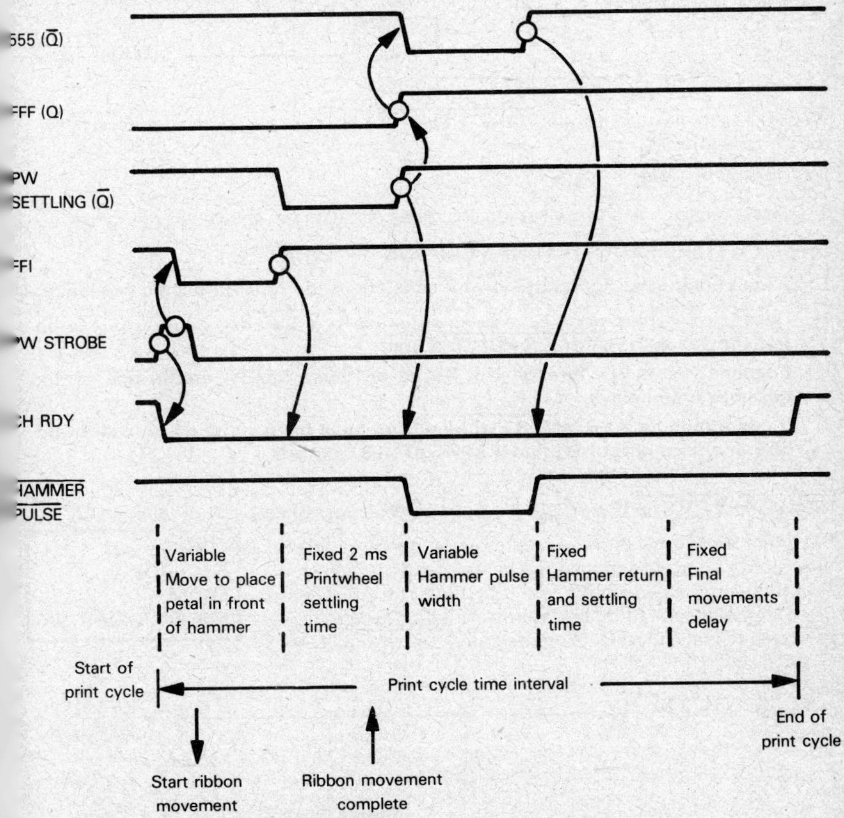

```
555 (Q̄)

FFF (Q)

PW
SETTLING (Q̄)

FFI

PW STROBE

CH RDY

HAMMER
PULSE
```

| Variable Move to place petal in front of hammer | Fixed 2 ms Printwheel settling time | Variable Hammer pulse width | Fixed Hammer return and settling time | Fixed Final movements delay |

Start of print cycle ← Print cycle time interval → End of print cycle

↓ Start ribbon movement
↑ Ribbon movement complete

## THE 555 MULTIVIBRATOR

Compare the way in which the 555 multivibrator has been wired in Figure 3-1 with the description of the multivibrator as given in Chapter 2; you will see that **flip-flop FFB switches the multivibrator "off" in between print cycles** by inputting a low reset at pin 4. **The flip-flop FFF (Q) output triggers the multivibrator,** as we have just described.

**The duration of the one-shot output pulse is controlled by inputs H1 through H6.** One of these six inputs will be true while the other five will be false; thus, the multivibrator, once triggered, will output a one-shot which can have a "high" pulse with one of six possible durations.

> ONE-SHOT VARIABLE PULSE

The 555 multivibrator one-shot output is eventually inverted to become a $\overline{\text{HAMMER PULSE}}$ output; however, for the $\overline{\text{HAMMER PULSE}}$ output to occur, additional inputs to AND gates 37 and 38, located at coordinate C7, must also be high. We may represent the $\overline{\text{HAMMER PULSE}}$ logic as follows:

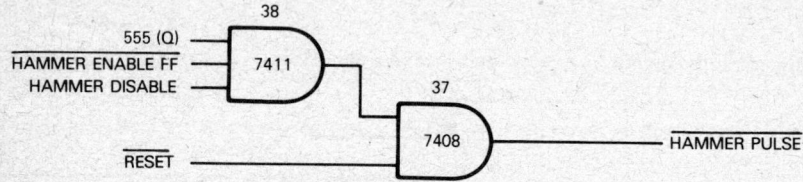

We will simply have to test the $\overline{\text{HAMMER ENABLE FF}}$ input before generating a $\overline{\text{HAMMER PULSE}}$ output.

The HAMMER DISABLE switch must be simulated.

$\overline{\text{RESET}}$ we can ignore, since reset logic is being simulated in between print cycles.

## SIMULATING MULTIVIBRATOR 555

The simulation of the 555 multivibrator consists of the following logic sequence:

1) **Determine if conditions have been satisfied for a 555 one-shot output to be transmitted as a $\overline{\text{HAMMER PULSE}}$ output.**
2) **Examine inputs H1 through H6. Based on these inputs, create one of six possible time delays.**
3) **If conditions for a $\overline{\text{HAMMER PULSE}}$ output have been satisfied, translate the 555 one-shot output into a $\overline{\text{HAMMER PULSE}}$ output.**

Let us first look at the $\overline{\text{HAMMER PULSE}}$ output enabling logic. Testing the condition of HAMMER ENABLE FF is simple enough; it has been assigned pin 6 of I/O Port A0.

But there are no switches in assembly language programs; how are we going to simulate the hammer disable? We could assign the one remaining pin — pin 5 of I/O Port A0 — to an input signal generated by an external switch. It would be just as simple to place this switch in the path of HAMMER ENABLE FF as follows:

> LOGIC EXCLUDED FROM MICROCOMPUTER

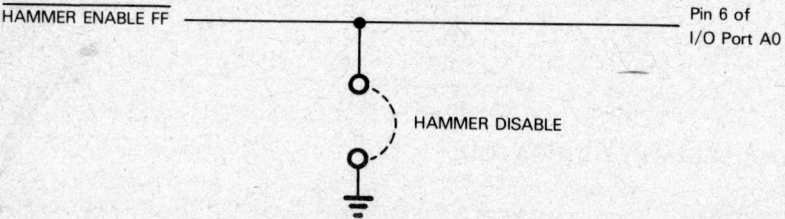

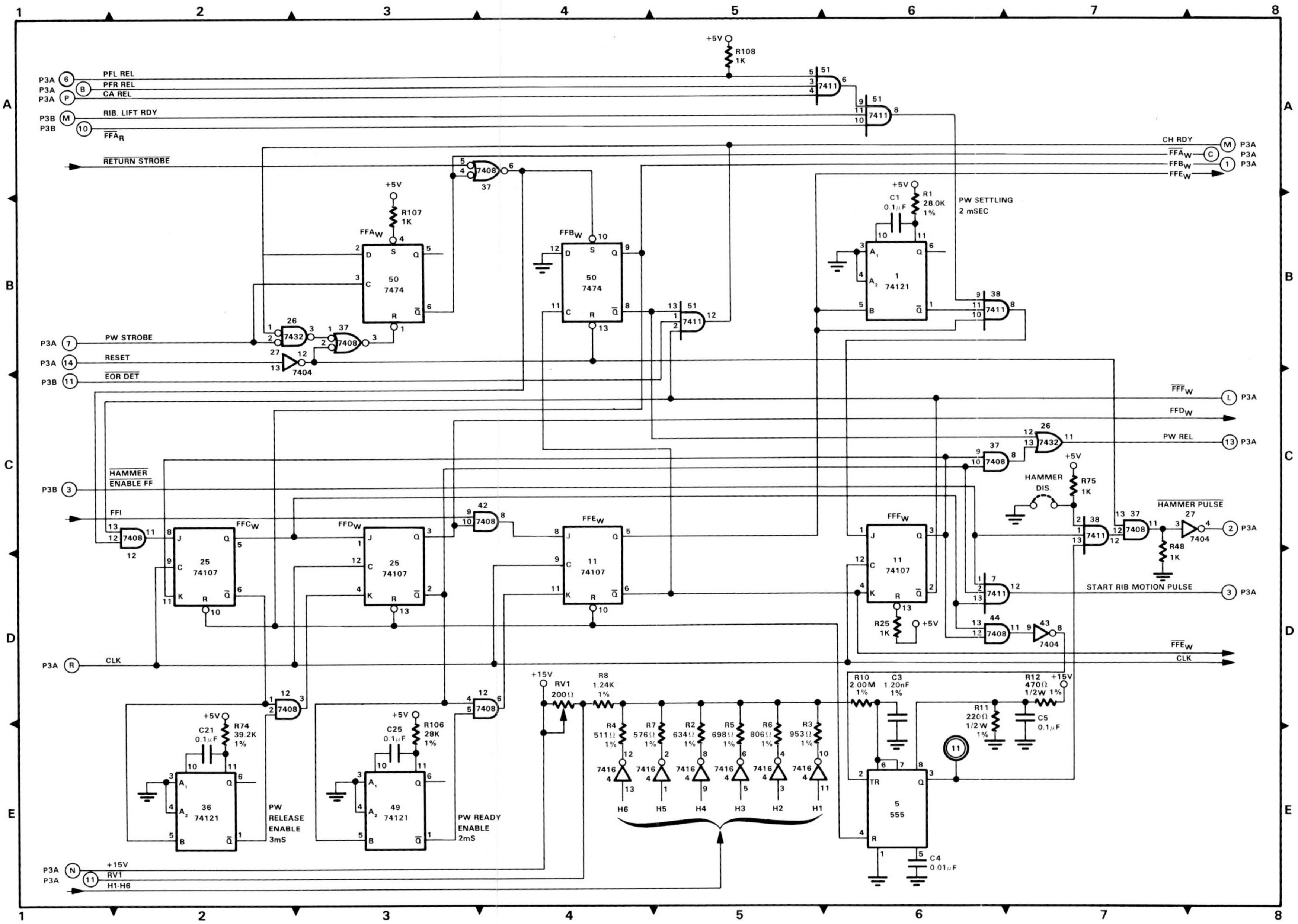

Figure 3-1. Printwheel Control Logic

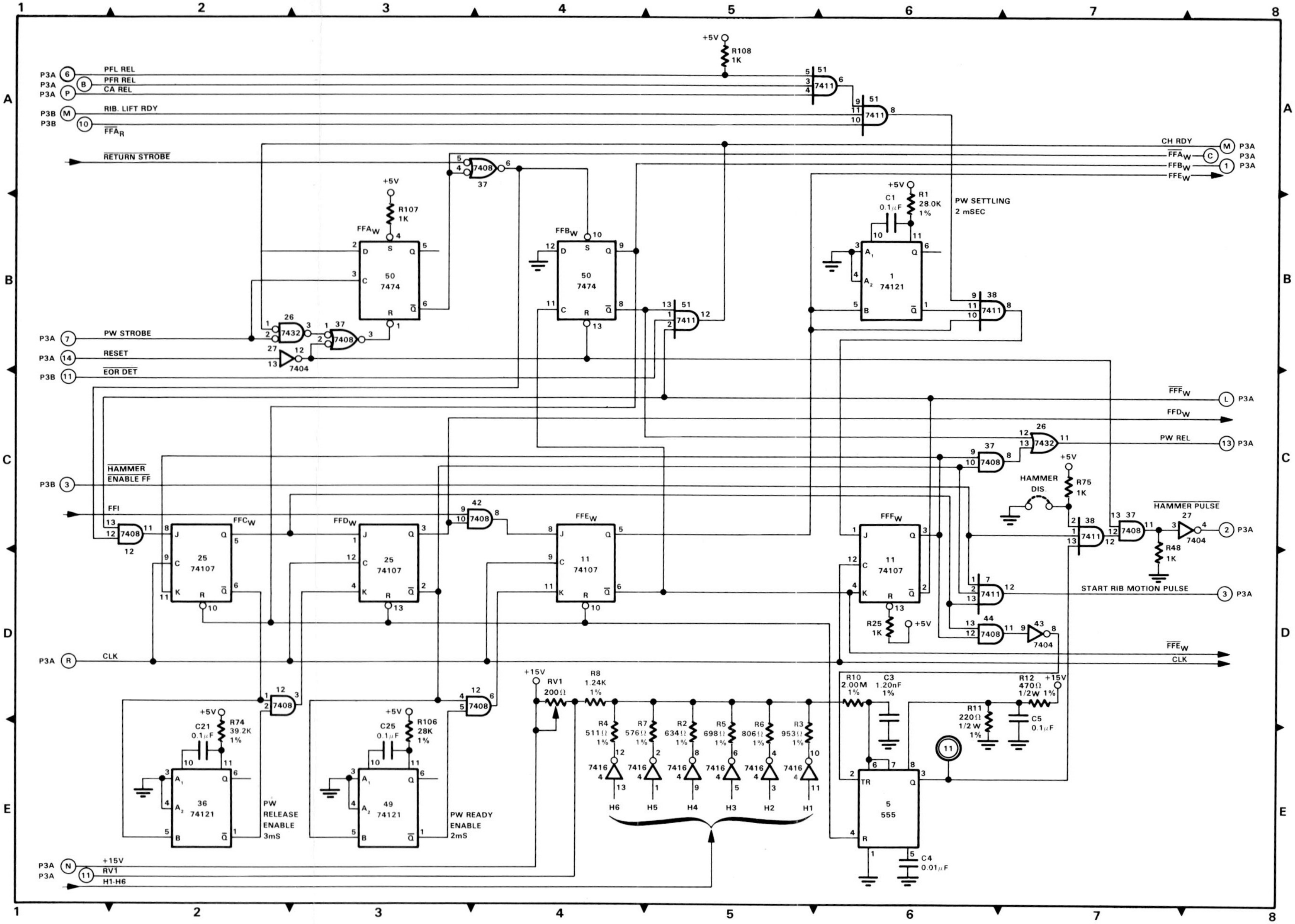

Figure 3-1. Printwheel Control Logic

We will therefore ignore the hammer disable switch and enable a HAMMER PULSE output, providing the HAMMER ENABLE FF input is high.

**What about the six possible durations for the 555 multivibrator output?** We described in Chapter 2 how a time delay can be created by loading a 16-bit value into a register pair, then decrementing this register pair within a program loop, remaining in the program loop until a decrement to zero occurs. **Selecting one of six possible time delays is as simple as selecting one of six possible initial time constants. We can now simulate our 555 multivibrator as follows:**

```
FFF:    IN      A,(0)       ;INPUT I/O PORT A0 CONTENTS TO ACCUMULATOR
        CPL                 ;COMPLEMENT TO TEST FOR 1 BITS
        AND     1FH         ;ISOLATE BITS 0 THROUGH 4
        JR      NZ,FFF      ;IF ANY BITS ARE 1, STAY IN LOOP
        IN      A,(4)       ;SET BIT 6 OF I/O PORT A1 TO 1
        SET     6,B
        OUT     (4),A
;TEST HAMMER ENABLE FF
        IN      A,(0)       ;INPUT I/O PORT A0 TO ACCUMULATOR
        BIT     6,A         ;TEST BIT 6
        JR      Z,HP0       ;IF ZERO, BYPASS SETTING HAMMER PULSE LOW
;HAMMER ENABLE FF IS HIGH, SO HAMMER PULSE MUST BE OUTPUT LOW
;THEREFORE SET BIT 2 OF I/O PORT B0 TO 0
        IN      A,(2)       ;INPUT I/O PORT B0 TO ACCUMULATOR
        RES     2,A         ;SET BIT 2 TO 0
        OUT     (2),A       ;OUTPUT RESULT
;COMPUTE TIME DELAY
HP0:    LD      HL,DELY     ;LOAD DELAY BASE ADDRESS INTO HL PAIR
        IN      A,(6)       ;INPUT SELECTOR (PORT B1) TO ACCUMULATOR
HP1:    RRA                 ;ROTATE ACCUMULATOR RIGHT THROUGH CARRY
        INC     HL          ;INCREMENT HL CONTENTS BY 2
        INC     HL
        JR      NC,HP1      ;IF NO CARRY, ROTATE AND INCREMENT AGAIN
        LD      E,(HL)      ;LOAD 16-BIT TIME DELAY CONSTANT INTO DE
        INC     HL
        LD      D,(HL)
TDLY:   DEC     DE          ;EXECUTE TIME DELAY LOOP
        LD      A,D
        OR      E
        JR      NZ,TDLY
;OUTPUT HAMMER PULSE HIGH AGAIN
        IN      A,(2)       ;INPUT I/O PORT B0 TO ACCUMULATOR
        SET     2,A         ;SET BIT 2 TO 1
        OUT     (2),A       ;OUTPUT RESULT
```

Compared to the other devices we have simulated thus far, the 555 multivibrator requires a lot of simulation instructions. While it may look as though there is a lot to understand, the logic is, in fact, quite simple; so let us take it one piece at a time.

Initially we test **HAMMER ENABLE FF.** HAMMER PULSE will be output low only if HAMMER ENABLE FF is high. The three instructions which test the status of HAMMER ENABLE FF are:

**SIGNAL ENABLE**

```
        IN      A,(0)       ;INPUT I/O PORT A0 TO ACCUMULATOR
        BIT     6,A         ;TEST BIT 6
        JR      Z,HP0       ;IF ZERO, BYPASS SETTING HAMMER
                            ;PULSE LOW
```

There are two aspects of these three instructions which need to be explained. First, there is the logic being implemented. We are determining if conditions have been met for HAMMER PULSE to be output low. If conditions have been met, the JR Z,HPO instruction branches around the instruction sequence that outputs HAMMER PULSE low.

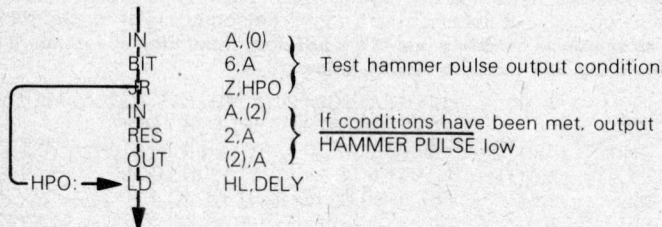

**We output HAMMER PULSE low before starting to compute the duration of the time pulse;** why is this? The reason is to save time. Instructions which compute the length of the time delay can be executed at the beginning of the time delay.

**EVENT SEQUENCE**

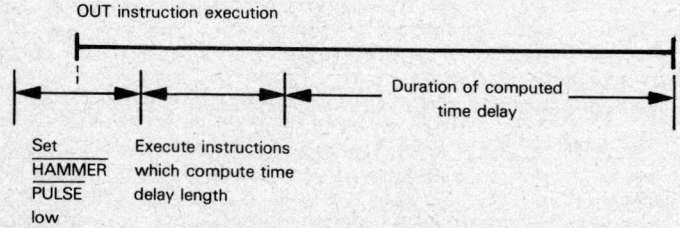

We could just as easily have computed the time delay, then set HAMMER PULSE low, and then executed the time delay; events would have occurred chronologically as follows:

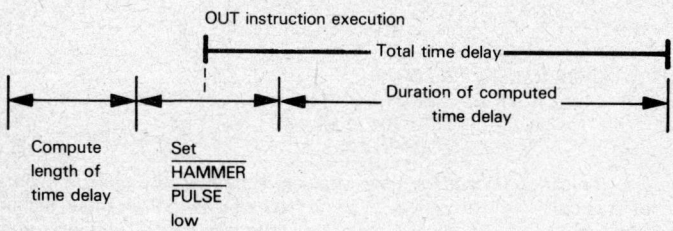

**Overlapping events in time makes a lot more sense.**

The actual method used to compute the time delay needs a little explanation. **At the end of our program, there will be 12 bytes of memory in which six 16-bit constants are stored.** This is how the source program will look:

```
;OUTPUT HAMMER PULSE HIGH AGAIN
        IN      A,(2)       ;INPUT I[O PORT B0 TO ACCUMULATOR
        SET     2,A         ;SET BIT 2 TO 1
        OUT     (2),A       ;OUTPUT RESULT
        —
        —
        —
        ORG     DELY+2
        DEFW    PPQQH       ;H1 TIME DELAY
        DEFW    RRSSH       ;H2 TIME DELAY
        DEFW    TTUUH       ;H3 TIME DELAY
        DEFW    VVWWH       ;H4 TIME DELAY
        DEFW    XXYYH       ;H5 TIME DELAY
        DEFW    ZZOOH       ;H6 TIME DELAY
```

The letters O through Z have been used to represent hexadecimal values. The six time delays can be represented by any numeric values, ranging from $0000_{16}$ through $FFFF_{16}$.

**The address of the first memory byte in which the first time delay is stored is given by the expression DELY+2. Suppose this memory location happened to be 2138:**

| Data Memory | Arbitrary Memory Address |
|---|---|
| QQ | 2138 |
| PP | 2139 |
| SS | 213A |
| RR | 213B |
| UU | 213C |
| TT | 213D |
| WW | 213E |
| VV | 213F |
| YY | 2140 |
| XX | 2141 |
| OO | 2142 |
| ZZ | 2143 |

Each 16-bit value will occupy two memory locations. The Z80 assembler will place the least significant byte in the location with the lower address. This is consistent with the object code representation of addresses and 16-bit immediate data values, as we mentioned in Chapter 2.

DELY is a label to which the value 2136 must be assigned. This assignment is made using an Equate directive, which would appear at the beginning of the program, as follows:

```
        DELY    EQU     2136H
```

**Now we begin our computation of the time delay by loading the address DELY into the HL register pair.** Assume that the label DELY has the value 2136, as illustrated above. After the LD HL,DELY instruction has been executed, this is the situation:

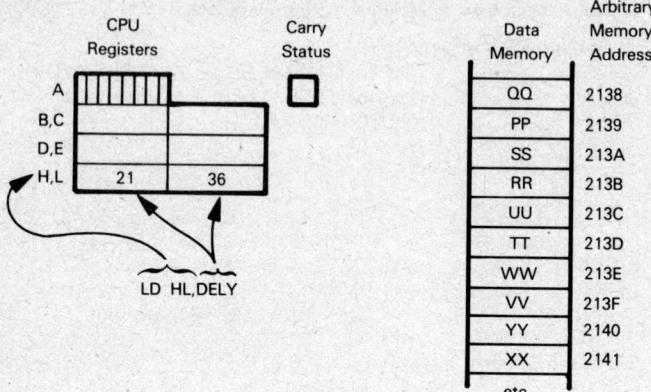

The next instruction, IN A,(6), loads the contents of I/O Port B1 into the Accumulator. From our discussion of input signals, recall that, of the six inputs H1 through H6, one signal will be high while the other five signals are low.

Therefore, **after the IN instruction has executed, the Accumulator will contain a 1 in one of the six low-order bits:**

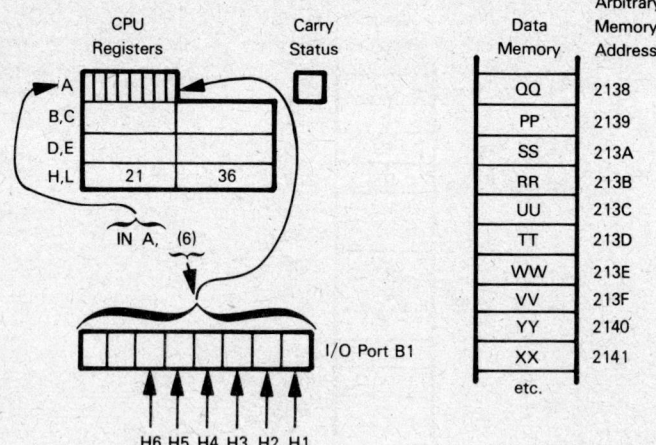

We can compute the address of the required time delay by adding 2 to the contents of the HL register pair a number of times given by the position of the Accumulator 1 bit. This may be illustrated as follows:

**DATA MEMORY ADDRESS COMPUTATION**

① Rotate Accumulator contents right one bit, through Carry:

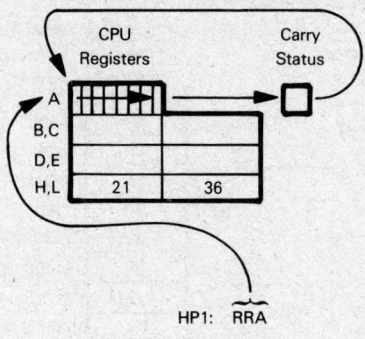

HP1: RRA

② Add 2 to the register pair HL:

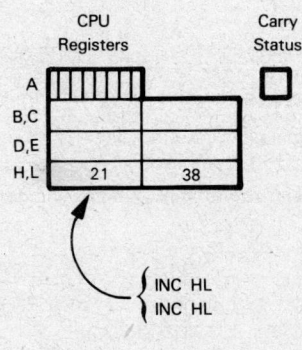

{ INC HL
{ INC HL

③ If Carry status is not 1, go back to ① ; otherwise, HL contains the correct address.

The logic to make the required address addition is provided by these four instructions:

```
HP1:    RRA             ;ROTATE ACCUMULATOR RIGHT THROUGH CARRY
        INC     HL      ;INCREMENT HL CONTENTS BY 2
        INC     HL
        JR      NC,HP1  ;IF NO CARRY, ROTATE AND INCREMENT AGAIN
```

When the JR NC instruction causes program execution to continue with the next sequential instruction rather than branching back to HP1, HL will contain the address of the initial time delay constant's first byte. **Now that the correct time delay is ad-**

**dressed by the H and L registers, we load the appropriate 16-bit delay constant into D and E.** Suppose H2 was the high input signal; this is the result:

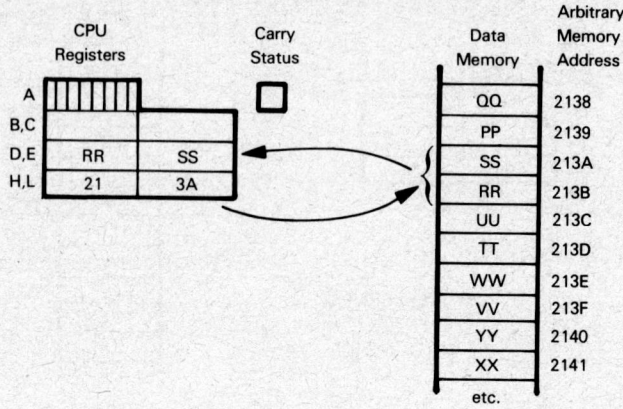

The selected delay constant RRSS is moved to the D and E registers by these three instructions:

```
LD      E,(HL)    ;MOVE CONTENTS OF BYTE 213A TO REGISTER E
INC     HL        ;ADDRESS BYTE 213B
LD      D,(HL)    ;MOVE CONTENTS OF BYTE 213B TO REGISTER D
```

Note that we load the low-order register, Register E, first, since the low-order byte is at the lower address.

**The actual time delay is created by this instruction loop, which was described in Chapter 2:**

```
TDLY:   DEC     DE         ;DECREMENT DELAY COUNTER
        LD      A,D        ;TEST FOR 0 IN DE BY ORING D WITH
        OR      E          ;E IN THE ACCUMULATOR
        JR      NZ,TDLY    ;RETURN IF RESULT IS NOT ZERO
```

**The last three instructions output HAMMER PULSE high,** without making any test for whether HAMMER PULSE was low. This logic will work since outputting HAMMER PULSE high, if it was already high, will have no discernible effect. Under these circumstances, the time required to execute the last three instructions is simply wasted. Since it would take three instructions to test if HAMMER PULSE had been set low, the waste is justified.

**Let us now give a little thought to the time it will take to compute the time delay.** Execution times for relevant instructions are listed as follows:

**TIME DELAY COMPUTATION**

| Cycles | | Instruction | |
|---|---|---|---|
| | | IN | A,(2) |
| | | RES | 2,A |
| | | OUT | (2),A  ◄── HAMMER PULSE low starts here |
| 10 | HP0: | LD | HL,DELY |
| 10 | | IN | A,(6) |
| 4 | HP1: | RRA | |
| 6 | | INC | HL |
| 6 | | INC | HL |
| 7/12 | | JR | NC,HP1 |
| 7 | | LD | E,(HL) |
| 6 | | INC | HL |
| 7 | | LD | D,(HL) |
| 6 | TDLY: | DEC | DE |
| 4 | | LD | A,D |
| 4 | | OR | E |
| 7/12 | | JR | NZ,TDLY |
| 10 | | IN | A,(2) |
| 8 | | SET | 2,A |
| 11 | | OUT | (2),A  ◄── HAMMER PULSE low ends here |
| 113 | | | |

The four instructions INC HL, INC HL, JR NC,HP1 (grouped with RRA) will be executed between one and six times. 28 cycles are in the loop.

The four instructions DEC DE, LD A,D, OR E, JR NZ,TDLY constitute the time delay. 26 cycles are in this loop.

Assuming a 500-nanosecond clock, the time taken to initiate and terminate the HAMMER PULSE signal is given by:

$$56.5 - 13 - 14 + 14N \text{ microseconds}$$

where N is a number between 1 and 6, representing the bit position of the Accumulator that is set to 1. **Thus, initiation and termination time will vary between 43.5 microseconds and 113.5 microseconds.** The shortest time applies to N = 1 (H1), whereas the longest time applies to N = 6 (H6).

**These times must be subtracted from the delays subsequently generated.** For example, suppose H1 high requires the 555 to output a one-shot signal which is high for 1.65 milliseconds (approximately); then a delay of 1.6 milliseconds added to a set-up time of 43.5 microseconds will suffice.

## THE PW RELEASE ENABLE FLIP-FLOP

**As soon as the 555 one-shot output becomes high again, flip-flop FFC is simulated switching off. When FFC switches off, its $\overline{Q}$ output makes a low-to-high transition and this triggers the PW RELEASE ENABLE one-shot.** This is a 74121 one-shot identified by the 36 at coordinate E2. The purpose of this one-shot is to allow the printhammer time to settle back before any attempt is made to reposition the printwheel. **This was illustrated as the fixed hammer return and settling time delay.**

## SIMULATING THE PW RELEASE ENABLE FLIP-FLOP

**This is really a two-part simulation; first we must simulate flip-flop FFC switching off, then we must execute an appropriate time delay. A three-millisecond time delay is sufficient.** Instructions which turn flip-flop FFC off will execute within the three-millisecond time

**TIME DELAY**

delay. The computed time delay will therefore be a little less than three-milliseconds. Here is the appropriate instruction sequence:

```
;OUTPUT HAMMER PULSE HIGH AGAIN
        IN      A,(2)       ;INPUT I/O PORT B0 TO ACCUMULATOR
        SET     2,A         ;SET BIT 2 TO 1
        OUT     (2),A       ;OUTPUT RESULT
;SWITCH FLIP-FLOP FFC OFF
        IN      A,(4)       ;SET BIT 2 OF I/O PORT A1 TO 0
        RES     2,A
        OUT     (4),A
;EXECUTE A 3 MILLISECOND TIME DELAY
        LD      DE,230      ;LOAD TIME CONSTANT INTO D,E
PWRL:   DEC     DE          ;DECREMENT REGISTER PAIR
        LD      A,D         ;TEST FOR ZERO
        OR      E
        JR      NZ,PWRL     ;REDECREMENT IF NOT ZERO
```

Notice that the initial time constant has been identified as a decimal number, 230. The time constant could be specified as a hexadecimal number thus:

```
        LD      DE,0E6H
```

Assuming a 500-nanosecond clock, the three instructions which precede the time delay loop (RES, OUT and LD) execute in 14.5 microseconds, and the four instructions in the delay loop execute in 13 microseconds. Therefore, the total delay time is given by the equation:

$$229 \times 13 + 10.5 + 14.5 = 3002 \text{ microseconds}$$

To be honest, the three-millisecond time delay is not a critical number; 2.5 or 3.5 milliseconds would probably do just as well, so our worrying about 10 microseconds is not meaningful in this instance. Nevertheless, in your next application the duration of a time delay may be very critical; then the timing considerations discussed above will be very meaningful.

**In order to determine what happens at the conclusion of the PW RELEASE time delay, we must look at the FFC Q and $\overline{Q}$ outputs.** The Q output connects to the START RIBBON MOTION PULSE AND gate and to the 555 one-shot trigger logic; in neither case does the Q high-to-low transition have any effect. The START RIBBON MOTION pulse signal is already low, and the 555 one-shot is triggered by a high-to-low Q transition. The low-to-high transition simply raises the trigger signal to a high level which requires no simulation:

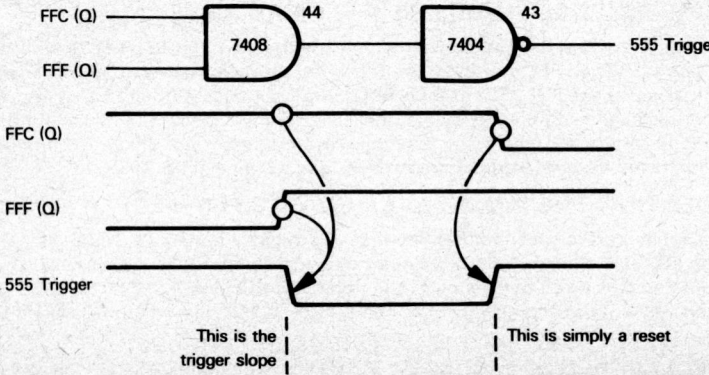

The FFC ($\overline{Q}$) output is ANDed with the PW RELEASE ENABLE $\overline{Q}$ one-shot in order to generate the FFD (K) input. The FFD (J) input comes directly from FFC (Q), therefore **as soon as the PW RELEASE ENABLE one-shot goes high again, FFD will receive a low J input and a high K input:**

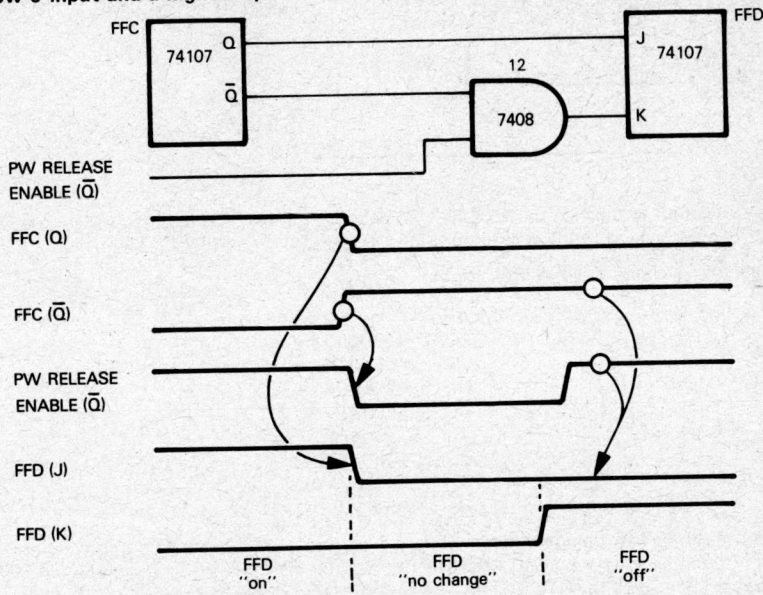

**A low J and high K input to flip-flop FFD switches this flip-flop off, and that triggers the PW READY ENABLE one-shot.**

## SIMULATING THE PW READY ENABLE ONE-SHOT

Logic associated with this one-shot is almost identical to the PW RELEASE ENABLE one-shot. FFD switching off causes a low-to-high $\overline{Q}$ output, which triggers the PW READY ENABLE one-shot.

**We must now simulate a two-millisecond time delay;** otherwise the next instruction sequence is almost identical to the PW RELEASE ENABLE one-shot simulation and may be illustrated as follows:

```
;EXECUTE A 3 MILLISECOND TIME DELAY
        LD      DE,230      ;LOAD TIME CONSTANT INTO D,E
PWRL:   DEC     DE          ;DECREMENT REGISTER PAIR
        LD      A,D         ;TEST FOR ZERO
        OR      E
        JR      NZ,PWRL     ;REDECREMENT IF NOT ZERO
;SWITCH FLIP-FLOP FFD OFF
        IN      A,(4)       ;SET BIT 3 OF I/O PORT A1 TO 0
        RES     3,A
        OUT     (4),A
;EXECUTE A 2 MILLISECOND TIME DELAY
        LD      A,250       ;LOAD INITIAL TIME CONSTANT INTO
                            ;ACCUMULATOR
PWRD:   DEC     A           ;DECREMENT ACCUMULATOR
        JR      NZ,PWRD     ;REDECREMENT IF NOT ZERO
```

**When FFD switches off, the PW REL output goes high again.** Here is the PW REL creation logic:

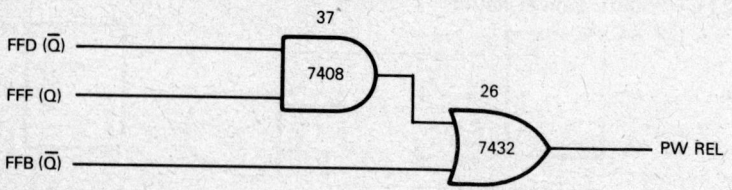

FFB ($\overline{Q}$) is still low at this time. But FFD ($\overline{Q}$) and FFF (Q) are both high, so AND gate 37 outputs a high level which passes through OR gate 26 to set PW REL high.

**These instructions set PW REL high:**

```
;EXECUTE A 2 MILLISECOND TIME DELAY
        LD      A,250           ;LOAD INITIAL TIME CONSTANT INTO
                                ;ACCUMULATOR
PWRD:   DEC     A               ;DECREMENT ACCUMULATOR
        JR      NZ,PWRD         ;REDECREMENT IF NOT ZERO
;SET PW REL HIGH
        IN      A,(2)           ;INPUT I/O PORT B0 TO ACCUMULATOR
        SET     0,A             ;SET BIT 0 TO 1
        OUT     (2),A           ;RETURN RESULT
```

**Now the whole print cycle ends in a hurry.** The flip-flop FFD Q and $\overline{Q}$ outputs become the FFE J and K inputs. Q is first ANDed with FFI, which, at this time, is constantly high; therefore, the moment FFD switches off, FFE receives a low J input.

The FFE (K) input does not go high until the end of the PW READY ENABLE one-shot, since the PW READY ENABLE $\overline{Q}$ output is ANDed with $\overline{Q}$ from FFD in order to generate FFE (K).

**FFE switching off is our next chronological event.**

**FFE switching off, in turn, causes FFB and FFF to switch off.** FFB is switched off by the low-to-high transition of FFE ($\overline{Q}$), which becomes the FFB clock input. FFF switches off because its J and K inputs are tied directly to the Q and $\overline{Q}$ outputs of FFE.

**Once FFB and FFF have switched off, all conditions have been met for CH RDY to go high again, providing $\overline{EOR\ DET}$ is not signaling the end of ribbon:**

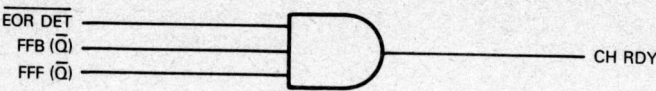

```
;EXECUTE A 2 MILLISECOND TIME DELAY
        LD      A,250           ;LOAD INITIAL TIME CONSTANT INTO ACCUMULATOR
PWRD:   DEC     A               ;DECREMENT ACCUMULATOR
        JR      NZ,PWRD         ;REDECREMENT IF NOT ZERO
;SET PW REL HIGH
        IN      A,(2)           ;INPUT I/O PORT B0 TO ACCUMULATOR
        SET     0,A             ;SET BIT 0 TO 1
        OUT     (2),A           ;RETURN RESULT
;TURN OFF FLIP-FLOPS FFB, FFE AND FFF
        IN      A,(4)           ;INPUT PORT A1 TO ACCUMULATOR
        AND     0AFH            ;RESET BITS 4 AND 6 TO 0
        OR      22              ;SET BITS 5 AND 1 TO 1
        OUT     (4),A           ;OUTPUT RESULT
;SET CH RDY HIGH
        IN      A,(2)           ;INPUT I/O PORT B0 TO ACCUMULATOR
        SET     1,A             ;SET BIT 1 TO 1
        OUT     (2),A           ;OUTPUT RESULT
;BRANCH TO TEST FOR VALID END OF PRINT CYCLE
        JP      VALND
```

## SIMULATION SUMMARY

The complete simulation program developed in this chapter is given in Figure 3-3. The circled numbers correspond to the numbers on Figure 3-2.

**We can conclude that an absolutely exact, one-for-one simulation of digital logic using assembly language instructions within a microcomputer system is not feasible; but then, it is not particularly desirable.**

If you are not a digital logic designer, you will probably be very confused by the various signal combinations required within the logic of Figure 3-1. A great deal of what is going on has nothing to do with the ultimate requirements of the Qume printer; rather, it reflects one logic designer's internal logic implementation, aimed at ensuring appropriate external signal sequences under all conceivable circumstances.

If you are a logic designer, chances are you would have implemented the specific requirements of the Qume printer interface in a totally different way; you may even be grumbling at this implementation.

**The important point to bear in mind is that digital logic contains innumerable subtleties which are specific to discrete logic devices. These subtleties are not tied to the requirements of the overall implementation.**

**Now, assembly language has its own set of subtleties, which also have nothing to do with the ultimate implementation;** rather, they are aimed at making most effective use of individual instructions or instruction sequences.

It should therefore come as no surprise that an exact duplication of digital logic, using assembly language, is neither feasible nor desirable. So, we will move away from digital logic and start treating a problem from a programming viewpoint.

**The principal difference between digital logic and assembly language is that assembly language treats events chronologically, while digital logic segregates logic into functional nodes.** Thus, one logic device may be responsible for a number of events occurring at different times during any logic cycle; when translated into an assembly language program, each event becomes an isolated instruction sequence.

| ASSEMBLY LANGUAGE VERSUS DIGITAL LOGIC |

In Figure 3-1, for example, the print cycle began with a cascade of flip-flops switching on and ended with the same flip-flops switching off. In many cases a flip-flop switching on triggered one event, while the same flip-flop switching off triggered an entirely different event. Within an assembly language program, the two events will have nothing in common. Each event will be represented by a completely independent instruction sequence occurring at substantially different parts of the program.

**The major difference between digital logic and assembly language is the concept of timing.** Within synchronous digital logic, as illustrated in Figure 3-1, timing is bound to clock signals and the need for clean signal interactions. Within an assembly language program, timing results strictly from the sequence in which instructions are executed. Moreover, whereas components in a digital logic circuit may switch and operate in parallel, within an assembly language program everything must occur serially.

**The key concept to grasp from this chapter is that there is nothing innately correct about digital logic as a means of implementing anything.** The fact that we have been unable to exactly duplicate digital logic using assembly language instructions does not mean that assembly language is in any way inferior; it simply means that assembly language is going to do the job in a different way.

Having spent our time in Chapter 3 drawing direct parallels between assembly language and digital logic, we will now abandon any attempt to favor digital logic. Moving on to Chapter 4, the logic illustrated in Figure 3-1 will be resimulated — but from the programmer's point of view.

```
;ASSIGN LOCATION TO BEGINNING OF DELAY COUNT TABLE
DELY   EQU     NNNNH
;TEST FOR VALID END OF PRINT CYCLE
VALND:
       IN      A,(2)        ;INPUT I/O PORT B0 TO ACCUMULATOR
       RLA                  ;SHIFT BIT 7 INTO CARRY
       JR      NC,VALND     ;IF ZERO IN CARRY, STAY IN PRINT CYCLE
;IN BETWEEN PRINT CYCLES PROGRAM EXECUTION
;INITIALLY SET I/O PORT A1 BITS 5, 1 AND 0 TO 1, BITS 6, 4, 3 AND 2 TO 0
       IN      A,(4)        ;INPUT I/O PORT A1 TO ACCUMULATOR
       OR      23H          ;SET BITS 5, 1 AND 0 TO 1
       AND     0A3H         ;RESET BITS 6, 4, 3 AND 2 TO 0
       OUT     (4),A        ;RETURN RESULT
;TEST FOR RETURN STROBE LOW
STBHI: IN      A,(2)        ;INPUT I/O PORT B0 TO ACCUMULATOR
       BIT     4,A          ;TEST RETURN STROBE BIT
       JR      Z,FFB        ;IF IT IS 0, JUMP TO FFB SIMULATION
;SIMULATION OF FFAW AND ASSOCIATED LOGIC
;LOAD I/O PORT B0 CONTENTS INTO ACCUMULATOR AND ISOLATE BITS
;1, 5 AND 6 FOR CH RDY, PW STROBE AND RESET, RESPECTIVELY
① IN    A,(2)        ;INPUT I/O PORT B0 TO ACCUMULATOR
   AND   62H          ;ISOLATE BITS 6, 5 AND 1. IF RESET=0,
   CP    22H          ;CH RDY=1 AND PW STROBE=1, NEW PRINT
                      ;CYCLE STARTS
   JR    NZ,STBHI     ;OTHERWISE RETURN TO STBHI
② IN    A,(4)        ;START NEW PRINT CYCLE BY SETTING I/O PORT
   RES   0,A          ;A1, BIT 0 TO 0
   OUT   (4),A
;NEW PRINT CYCLE INSTRUCTION SEQUENCE STARTS HERE
;SIMULATE FLIP-FLOP FFB SWITCHING ON
```

Figure 3-3. The Complete Simulation Program

```
FFB:    IN      A,(4)           ;LOAD I/O PORT A1 INTO ACCUMULATOR
③       RES     1,A             ;RESET BIT 1 TO 0
        OUT     (4),A           ;RESTORE RESULT
;SIMULATE 7411 AND GATE SWITCHING CH RDY LOW. ALSO
;7432 OR GATE SWITCHES PW REL LOW
⑤       IN      A,(2)           ;INPUT I/O PORT B0 TO ACCUMULATOR
        AND     0FCH            ;RESET BITS 0 AND 1 TO 0
        OUT     (2),A           ;RESTORE RESULT
;CH RDY LOW TURNS FFA OFF. SET BIT 0 OF I/O PORT A1 TO 1
;ALSO SIMULATE FFC AND FFD TURNING ON. SET BITS 2 AND 3
;OF I/O PORT A1
⑥       IN      A,(4)           ;LOAD I/O PORT A1 TO ACCUMULATOR
⑧       OR      0DH             ;SET BITS 3, 2 AND 0 TO 1
        OUT     (4),A           ;RESTORE RESULT
;PULSE START RIBBON MOTION HIGH
⑦       IN      A,(2)           ;INPUT TO ACCUMULATOR FROM I/O PORT B0
        SET     3,A             ;SET BIT 3 HIGH
        OUT     (2),A           ;OUTPUT TO I/O PORT B0
        RES     3,A             ;SET BIT 3 LOW
        OUT     (2),A           ;OUTPUT TO I/O PORT B0
;TEST VELOCITY DECODE INPUT TO CREATE PRINTWHEEL MOVE DELAY
VLDC:   IN      A,(0)           ;INPUT I/O PORT A0 TO ACCUMULATOR
        RLA                     ;SHIFT BIT 7 INTO CARRY
        JR      NC,VLDC         ;STAY IN LOOP IF CARRY IS ZERO
;AT END OF DELAY SIMULATE FFE SWITCHING ON
⑨       IN      A,(4)           ;INPUT TO ACCUMULATOR FROM I/O PORT A1
        RES     5,A             ;RESET BIT 5
        SET     4,A             ;SET BIT 4
        OUT     (4),A           ;OUTPUT THE RESULT
;SIMULATE 2 MS PW SETTLING TIME DELAY
        LD      A,0FAH          ;LOAD INITIAL TIME DELAY CONSTANT
PWS:    DEC     A               ;DECREMENT ACCUMULATOR
        JR      NZ,PWS          ;REDECREMENT IF RESULT IS NOT ZERO
;SIMULATE FLIP-FLOP FFF SWITCHING ON
FFF:    IN      A,(0)           ;INPUT I/O PORT A0 CONTENTS TO ACCUMULATOR
        CPL                     ;COMPLEMENT TO TEST FOR 1 BITS
        AND     1FH             ;ISOLATE BITS 0 THROUGH 4
        JR      NZ,FFF          ;IF ANY BITS ARE 1, STAY IN LOOP
        IN      A,(4)           ;SET BIT 6 OF I/O PORT A1 TO 1
⑩       SET     6,A
        OUT     (4),A
;TEST HAMMER ENABLE FF
        IN      A,(0)           ;INPUT I/O PORT A0 TO ACCUMULATOR
        BIT     6,A             ;TEST BIT 6
        JR      Z,HP0           ;IF ZERO, BYPASS SETTING HAMMER PULSE LOW
;HAMMER ENABLE FF IS HIGH, SO HAMMER PULSE MUST BE OUTPUT LOW
;THEREFORE SET BIT 2 OF I/O PORT B0 TO 0
⑪       IN      A,(2)           ;INPUT I/O PORT B0 TO ACCUMULATOR
        RES     2,A             ;SET BIT 2 TO 0
        OUT     (2),A           ;OUTPUT RESULT
;COMPUTE TIME DELAY
HP0:    LD      HL,DELY         ;LOAD DELAY BASE ADDRESS INTO HL PAIR
        IN      A,(6)           ;INPUT SELECTOR (PORT B1) TO ACCUMULATOR
HP1:    RRA                     ;ROTATE ACCUMULATOR RIGHT THROUGH CARRY
        INC     HL              ;INCREMENT HL CONTENTS BY 2
```

Figure 3-3. The Complete Simulation Program (Continued)

```
            INC     HL
            JR      NC,HP1          ;IF NO CARRY, ROTATE AND INCREMENT AGAIN
            LD      E,(HL)          ;LOAD 16-BIT TIME DELAY CONSTANT INTO DE
            INC     HL
            LD      D,(HL)
TDLY:       DEC     DE              ;EXECUTE TIME DELAY LOOP
            LD      A,D
            OR      E
            JR      NZ,TDLY
;OUTPUT HAMMER PULSE HIGH AGAIN
            IN      A,(2)           ;INPUT I/O PORT B0 TO ACCUMULATOR
            SET     2,A             ;SET BIT 2 TO 1
            OUT     (2),A           ;OUTPUT RESULT
;SWITCH FLIP-FLOP FFC OFF
⑫          IN      A,(4)           ;SET BIT 2 OF I/O PORT A1 TO 0
            RES     2,A
            OUT     (4),A
;EXECUTE A 3 MILLISECOND TIME DELAY
            LD      DE,230          ;LOAD TIME CONSTANT INTO D,E
PWRL:       DEC     DE              ;DECREMENT REGISTER PAIR
            LD      A,D
            OR      E               ;TEST FOR ZERO
            JR      NZ,PWRL         ;REDECREMENT IF NOT ZERO
;SWITCH FLIP-FLOP FFD OFF
⑬          IN      A,(4)           ;SET BIT 3 OF I/O PORT A1 TO 0
            RES     3,A
            OUT     (4),A
;EXECUTE A 2 MILLISECOND TIME DELAY
            LD      A,250           ;LOAD INITIAL TIME CONSTANT INTO ACCUMULATOR
PWRD:
            DEC     A               ;DECREMENT ACCUMULATOR
            JR      NZ,PWRD         ;REDECREMENT IF NOT ZERO
;SET PW REL HIGH
⑭          IN      A,(2)           ;INPUT I/O PORT B0 TO ACCUMULATOR
            SET     0,A             ;SET BIT 0 TO 1
            OUT     (2),A           ;RETURN RESULT
;TURN OFF FLIP-FLOPS FFB, FFE AND FFF
⑮          IN      A,(4)           ;INPUT PORT A1 TO ACCUMULATOR
            AND     0AFH            ;RESET BITS 4 AND 6 TO 0
⑯          OR      22              ;SET BITS 5 AND 1 TO 1
            OUT     (4),A           ;OUTPUT RESULT
;SET CH RDY HIGH
⑰          IN      A,(2)           ;INPUT I/O PORT B0 TO ACCUMULATOR
            SET     1,A             ;SET BIT 1 TO 1
            OUT     (2),A           ;OUTPUT RESULT
;BRANCH TO TEST FOR VALID END OF PRINT CYCLE
            JP      VALND
;DELAY COUNT TABLE
            ORG     DELY+2
            DEFW    PPQQH           ;H1 TIME DELAY
            DEFW    RRSSH           ;H2 TIME DELAY
            DEFW    TTUUH           ;H3 TIME DELAY
            DEFW    VVWWH           ;H4 TIME DELAY
            DEFW    XXYYH           ;H5 TIME DELAY
            DEFW    ZZOOH           ;H6 TIME DELAY
```

Figure 3-3. The Complete Simulation Program (Continued)

# Chapter 4
# A SIMPLE PROGRAM

The problems associated with simulating digital logic, as we did in Chapter 3, can be attributed to one fact: we tried to divide logic into a number of isolated transfer functions, each of which corresponded to a digital logic device. We are now going to abandon digital and combinatorial logic, pretend it does not exist, and take another look at Figures 3-1 and 3-2.

## ASSEMBLY LANGUAGE TIMING VERSUS DIGITAL LOGIC TIMING

Returning to Figure 3-1, simply ignore everything that exists between the left and right-hand margins of the figure. What remains is a set of input signals and a set of output signals. **TRANSFER FUNCTION**
The output signals are related to the input signals by a set of transfer functions which have nothing to do with digital logic devices.

The transfer functions for Figure 3-1 are loosely represented by the timing diagram in Figure 3-2. What does "loosely represented" mean? It means that timing which relates to system requirements is mixed indiscriminately with timing that simply reflects the needs of digital logic. We can abandon timing considerations that simply reflect the needs of digital logic. To be specific, the printhammer must still be fired by outputting one of six solenoid pulses; the various movement and settling delays must also be maintained. But we can abandon time delays that separate one signal's change of state from another simply to keep the digital logic clean.

From the programmer's point of view, therefore, the timing diagram illustrated in Figure 4-1 is a perfectly valid substitution for the logic designer's timing diagram illustrated in Figure 3-2.

## INPUT AND OUTPUT SIGNALS

Looking at Figure 4-1 you will see that we have abandoned a lot more than minor timing delays; we have also abandoned most of our signals. But there is a simple criterion for determining whether a signal is really necessary within a microcomputer system. This is the criterion: if the signal is uniquely associated with real time events in logic external to the microcomputer system, then the signal must remain. If the source and destination of the signal are within the microcomputer system "black box", then the signal may be abandoned. Based on this criterion, let us take another look at our input and output signals.

First consider the input signals.

RETURN STROBE and PW STROBE are meaningless signals. **INPUT SIGNALS**
As digital logic, these two signals are print cycle sequence initiators. Within an assembly language program, jumping to the first instruction of a sequence is all the initiation you need. The fact that RETURN STROBE represents a print cycle during which the printhammer is not fired is unimportant, because HAMMER ENABLE is used to actually suppress HAMMER PULSE.
**We will combine the various hammer firing inhibit signals into one hammer status input.** There are five such signals: PFL REL, RIB LIFT RDY, RIBBON ADVANCE, PFR REL

and CA REL. Each of these signals owes its origin to different logic external to Figure 3-1; in the digital logic implementation, these signals are ANDed in order to create a master HAMMER INTERLOCK signal. In our assembly language implementation we will wire-OR all of these external signals to a single pin which becomes a HAMMER INTERLOCK status.

**RESET will remain as a master Reset signal tied to the CPU $\overline{\text{RESET}}$ pin.** $\overline{\text{RESET}}$ can therefore be ignored by the assembly language program; however, recall that once RESET is activated, program execution is going to resume with the instruction stored at memory location 0.

**EOR DET will be retained.** This is the signal which detects end of ribbon and prevents a print cycle from ever ending, thus inhibiting further character printing after the ribbon is exhausted.

**HAMMER ENABLE FF must be retained;** it suppresses the printhammer firing pulse during printwheel repositioning print cycles.

**The function performed by the six hammer pulse length signals, H1 through H6, must remain, but the signals themselves will disappear.** Instead of using six pins of an I/O port to identify hammer pulse width, we are going to create time delays directly from ASCII character codes.

**Let us now turn our attention to the output signals.**

**To begin with, we can eliminate all of the flip-flop outputs.** The boundary of each time interval within the print cycle is already identified by an existing signal changing state. If more than one external logic event must be triggered by a transition from one time interval to the next, there is nothing to stop the appropriate signal from being buffered externally, then used to trigger numerous external logic events. Within the microcomputer program, there is no reason why duplicate signals should be output simply to identify the transition from one print cycle time interval to the next.

**The remaining output signals are maintained.** It is possible that some of these signals would disappear if additional external logic were replaced by more assembly language programs within the microcomputer system; but given the bounds of the problem, as stated, the remaining signals are needed in order to define the print cycle time intervals.

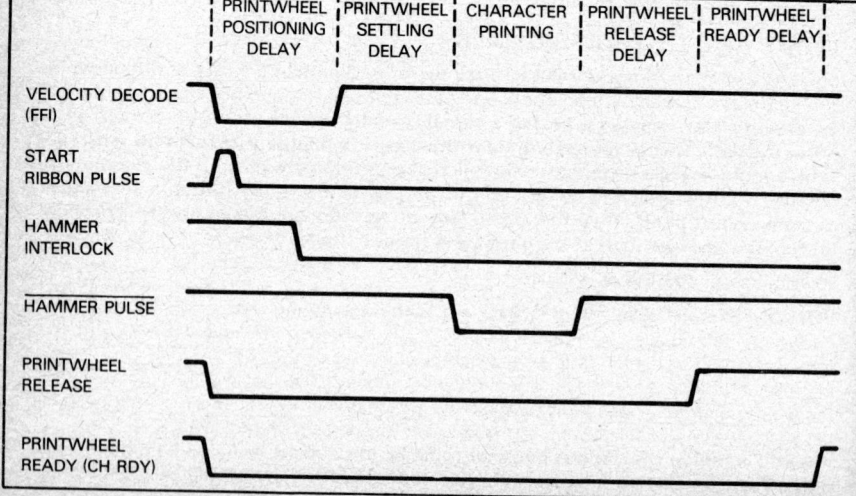

Figure 4-1. Timing for Figure 3-1, from the Programmer's Viewpoint

Given our new, simplified set of signals, we can eliminate one Z80 PIO; for the single remaining Z80 PIO, I/O ports and pins are assigned as follows:

**PIN ASSIGNMENTS**

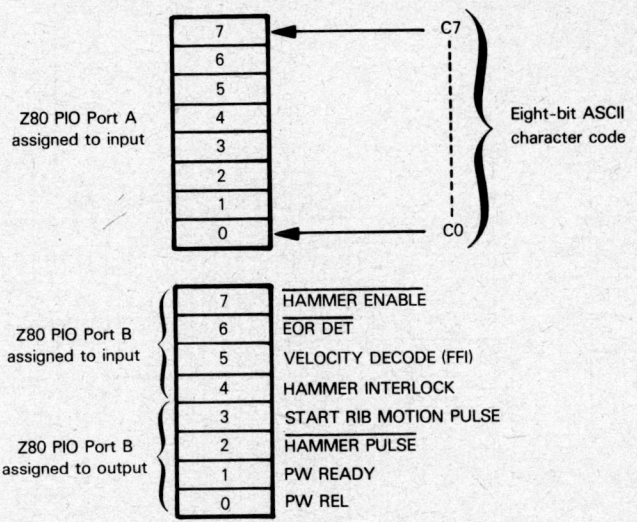

## MICROCOMPUTER DEVICE CONFIGURATION

We are now in a position to select the devices needed for program implementation. The selection is really quite straightforward; in addition to the CPU, we will need one Z80 Parallel Input/Output device, some read-only memory for program storage, and some read/write memory for general data storage. Figure 4-2 illustrates the microcomputer system which results when we combine these devices. Now, if you don't immediately understand Figure 4-2 do not despair; there are only a few aspects of this figure which are consequential to our immediate discussion.

## GENERAL DESIGN CONCEPTS

**This is the most important concept to derive from Figure 4-2: when designing logic by writing assembly language programs within a microcomputer system, the program you write is going to be highly dependent upon the device configuration.** There is nothing unique about the way in which devices have been combined as illustrated in Figure 4-2; alternative configurations would be equally feasible. The assembly language programs created, however, might differ markedly from one microcomputer configuration to the next, and this is a factor you should not lose sight of when writing microcomputer programs. Also, do not be afraid of modifying the selected hardware configuration. **Microcomputer device configuration and assembly language programming interact strongly and should not be separated.** These two steps should be within one iterative loop. During the early stages of writing a microcomputer program, you should assume that in the course of writing the assembly language program you will discover features of the hardware that can be improved; that in turn means the program will have to be rewritten.

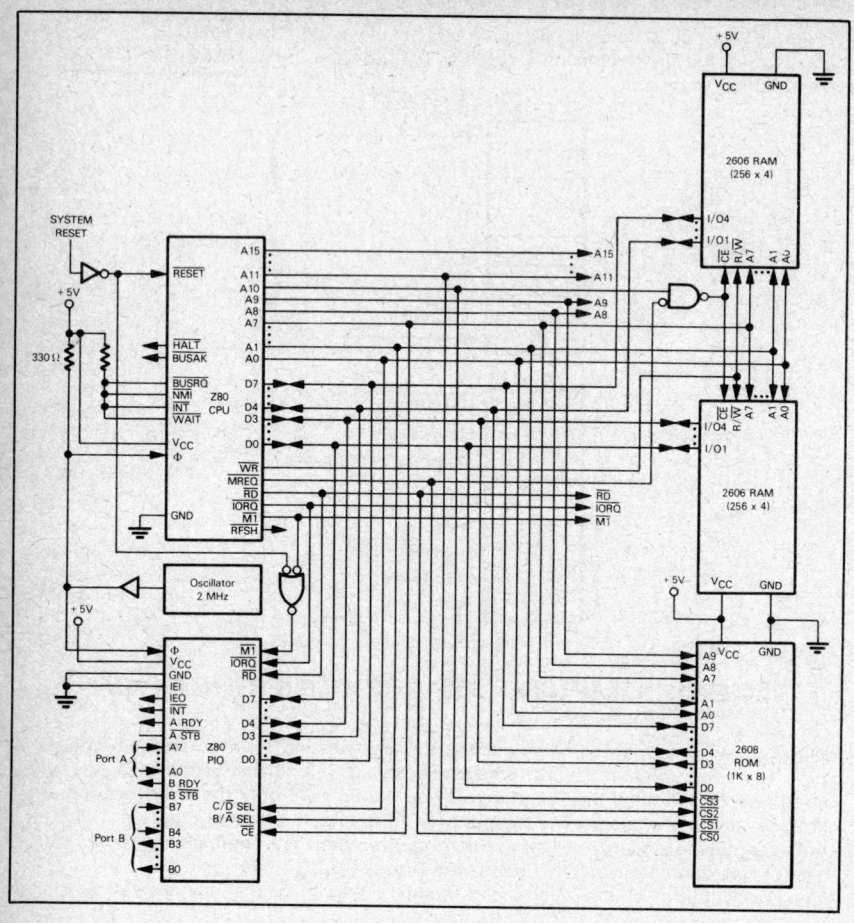

Figure 4-2. Z80 Microcomputer Configuration

This is a good point at which to bring up one of the reasons why higher-level languages are not desirable when you are programming a microcomputer to replace digital logic.

**HIGHER LEVEL LANGUAGES**

Higher-level languages are problem-oriented. For example, it is hard to look at a PL/M program statement and visualize the exact way in which data will be moved around a microcomputer system in response to the statement's execution. It is even harder to relate PL/M programs to exact device configurations. Assembly language, on the other hand, has a one-for-one relationship with your hardware.

## Z80 PARALLEL INPUT/OUTPUT INTERFACE (PIO)

Now let us turn to the specific way in which devices have been incorporated into Figure 4-2.

The Z80 PIO will respond to I/O addresses as follows:

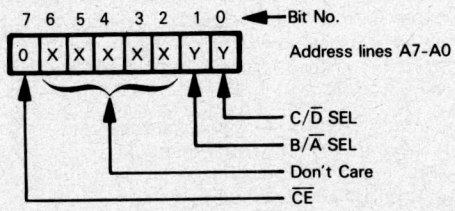

We will assume that all of the don't care address bits are 0; as a result **we will use the four I/O port addresses 0 through 3 to address the single Z80 PIO in Figure 4-2. The four addresses will access Z80 PIO locations as follows:**

**CHIP SELECT IN SIMPLE SYSTEMS**

```
0: Port A data
1: Port A control
2: Port B data
3: Port B control
```

If a microcomputer configuration contains a large number of I/O ports, the chip select logic may become a little more complex. If a Z80 PIO is to respond to exactly four I/O port addresses, excluding all others, then the chip select input must be created by combining eight low-order address lines in some unique way.

**Suppose the Z80 PIO in Figure 4-2 must respond to I/O port addresses 0 through 3 only.** Now all of the don't care signal lines must input to logic which is true only when these signal lines are all low. **This is one way of creating chip select logic:**

**CHIP SELECT IN LARGER SYSTEMS**

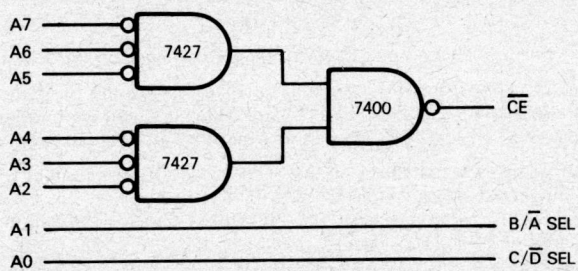

The CE input can be created using two of the three gates of a 7427 Triple 3-Input Positive-NOR gate and one of the four gates in a 7400 Quadruple 2-Input NAND gate.

Given the select logic above, the Z80 PIO will consider itself selected if and only if one of the four specified I/O port addresses is output on the Address Bus.

**The data direction and port utilization illustrated for the Z80 PIO in Figure 4-2 is not a hardware feature. At any time port utilization may be modified by writing the appropriate control words** into the Control registers of the Z80 PIO.

The Reset logic needs comment. Instead of testing for a Reset condition in between print cycles, as we did in Chapter 3, **we are going to use a hardware Reset signal,** but in a microcomputer environment.

| RESET LOGIC |

The signal $\overline{RESET}$ is connected to the $\overline{M1}$ input of the Z80 PIO. When the Z80 PIO receives $\overline{M1}$ low while $\overline{RD}$ and $\overline{IORQ}$ are both high, it is reset; at Reset, both ports are in Mode 1 -- input with handshaking. At some point following a hardware Reset, the CPU program must set the Z80 PIO for our particular purpose by executing instructions which write the appropriate control words.

| Z80 PIO RESET LOGIC |

Activating the $\overline{RESET}$ input to the Z80 CPU will clear the Program Counter; this means that program execution will restart with the instruction stored in the memory byte whose address is 0. We must therefore have post-reset and system initialization program steps beginning at this memory location.

**Memory select logic illustrated in Figure 4-2 will satisfy Reset logic requirements.**

## ROM AND RAM MEMORY

A Signetics 2608 provides our microcomputer system with 1024 bytes of read-only memory. Two of the four select lines, plus ten address lines, create ROM addresses as follows:

| ROM ADDRESSES |

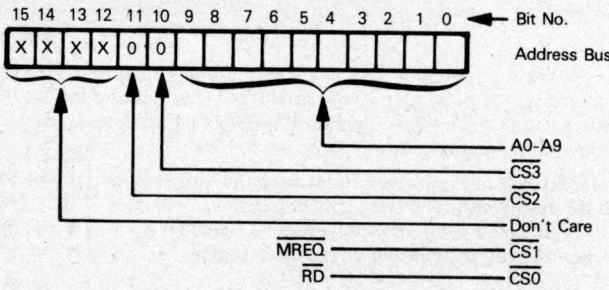

The other two select lines are connected to the control signals $\overline{MREQ}$ and $\overline{RD}$.

If the don't care address bits are assumed equal to 0, then the ROM device will be selected by addresses 0 through $03FF_{16}$. This provides the memory byte that we must have at address 0 when the Z80 CPU begins executing instructions after a Reset.

Notice that under no circumstance will the ROM address space conflict with Z80 PIO addresses: $\overline{MREQ}$ must be active for the 2608 ROM to be selected, while $\overline{IORQ}$ must be active for the Z80 PIO to be selected. The Z80 CPU never activates both $\overline{MREQ}$ and $\overline{IORQ}$ at the same time.

**Once again we are using a primitive ROM chip select on account of the microcomputer system's simplicity.** We define the address range 0 through $03FF_{16}$ for the 1024 bytes of ROM memory; but in fact, a wide variety of other addresses would also access ROM memory -- the address lines A12 through A15 can have any value. Providing A10 and A11 are both 0, ROM memory will be accessed. There is nothing to prevent you from selecting memory in this primitive way, providing yours is a small microcomputer system. There is no reason why you should incur additional expense creating complex chip select codes using all of the high-order six address bus lines. Even from the programming point of view, **you will not have to rewrite programs should you expand your system and include more memory at a later date. Providing you do not now use any of the alternative addresses that would also select the ROM, then at some future time you could take one of these alternative sets of addresses, use it to select another ROM, and in no way affect programs already written.**

| ROM SELECT IN SIMPLE SYSTEMS |
|---|

**By specifying ROM for program storage, we are assuming that the product will be developed in sufficient volume to justify the expense of creating a ROM mask.** If your volume does not justify the expense of creating ROM, then you can use Programmable Read Only Memory (PROM).

**The two Signetics 2606 RAM devices each provide 1024 bits of read/write memory, organized into 256 4-bit units.** Each RAM therefore provides half of a read/write memory byte. The 256 bytes of RAM will have addresses $0400_{16}$ through $04FF_{16}$. This may be illustrated as follows:

| RAM MEMORY ADDRESSES |
|---|

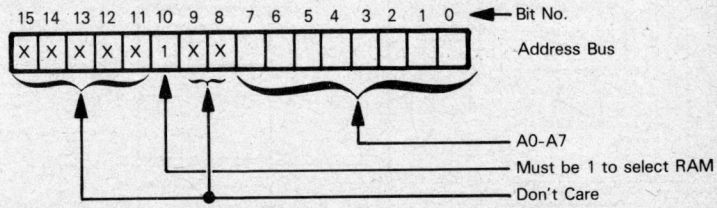

**Even though memory addresses $0400_{16}$ through $04FF_{16}$ have been specified as providing the RAM address space, once again a large number of other addresses would also select RAM. Note, however, that in no case will a RAM address coincide with a ROM address;** Address Bus line A10 must always be 0 to select ROM, while it must always be 1 to select RAM. Address contentions will therefore never arise.

In summary, addresses for the microcomputer system illustrated in Figure 4-2 will be interpreted as follows:

|  | ADDRESSES |  |
|---|---|---|
| I/O PORTS | $00_{16} - 03_{16}$ | Z80 PIO |
| MEMORY | $0000_{16} - 03FF_{16}$ | Read-only memory |
| MEMORY | $0400_{16} - 04FF_{16}$ | Read/Write memory |

## SYSTEM INITIALIZATION

Let us now turn our attention to system operations.

**When the system is initialized, "in between print cycles" conditions must be reestablished immediately. These are the necessary steps:**

1) If the printhammer has been fired, discontinue the firing pulse and allow the printhammer time to retract.

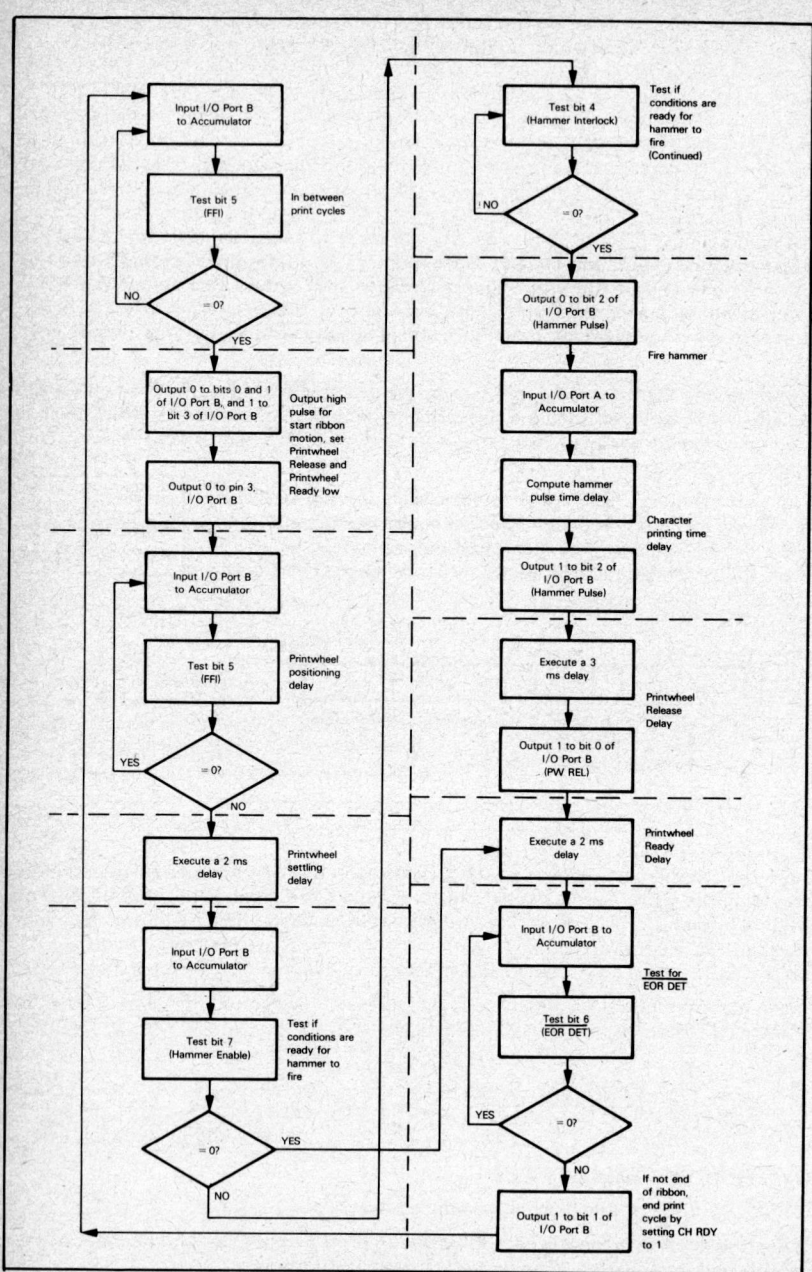

Figure 4-3. First Attempt at Program Flowchart

2) Move the printwheel back to its position of visibility.
3) Ensure that output signals have their "in between print cycles" status.

We now arrive at another fundamental programming concept: **there is a "most efficient" sequence in which you should write assembly language source programs.** We could go ahead and write an initialization program to implement a Reset, but that would require a lot of guessing. How do we know that the printhammer has been fired? How do we move the printwheel back to its position of visibility? Reset is going to abort a print cycle -- therefore the print cycle program must be created before we can know how to abort it.

PROGRAM<br>IMPLEMENTATION<br>SEQUENCE

**Generally stated, you should start writing a program by implementing the most important event in your logic, then you should work away from this beginning, implementing dependent events.**

Specifically, we are going to postpone creating a program to implement the Reset logic until the print cycle program has been created.

## PROGRAM FLOWCHART

**Let us now turn our attention to the functions which must be performed by the microcomputer system. These functions are identified by the flowchart illustrated in Figure 4-3. We will analyze this flowchart, step-by-step.**

**We are going to use the velocity decode input signal (FF) to identify the start of a new print cycle.** In between print cycles, therefore, the program continuously inputs I/O Port B contents to the Accumulator, testing bit 5. So long as this bit equals 1, a new print cycle has not begun. As soon as this bit equals 0, a new print cycle is identified:

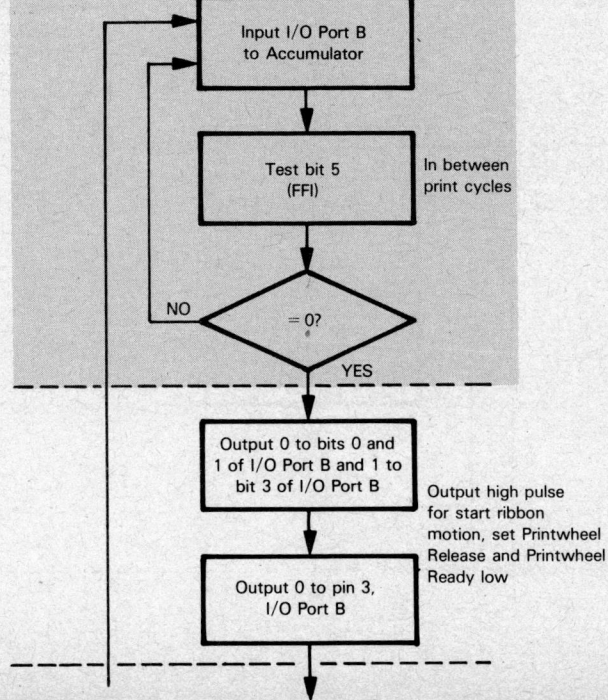

The first thing that happens within the new print cycle is that **a high START RIBBON MOTION pulse is output by sequentially writing a 1, then a 0 to bit 3 of I/O Port B. Also, 0s are output to bits 0 and 1 of I/O Port B,** since PRINTWHEEL RELEASE and PRINTWHEEL READY must both be output low at the start of the cycle:

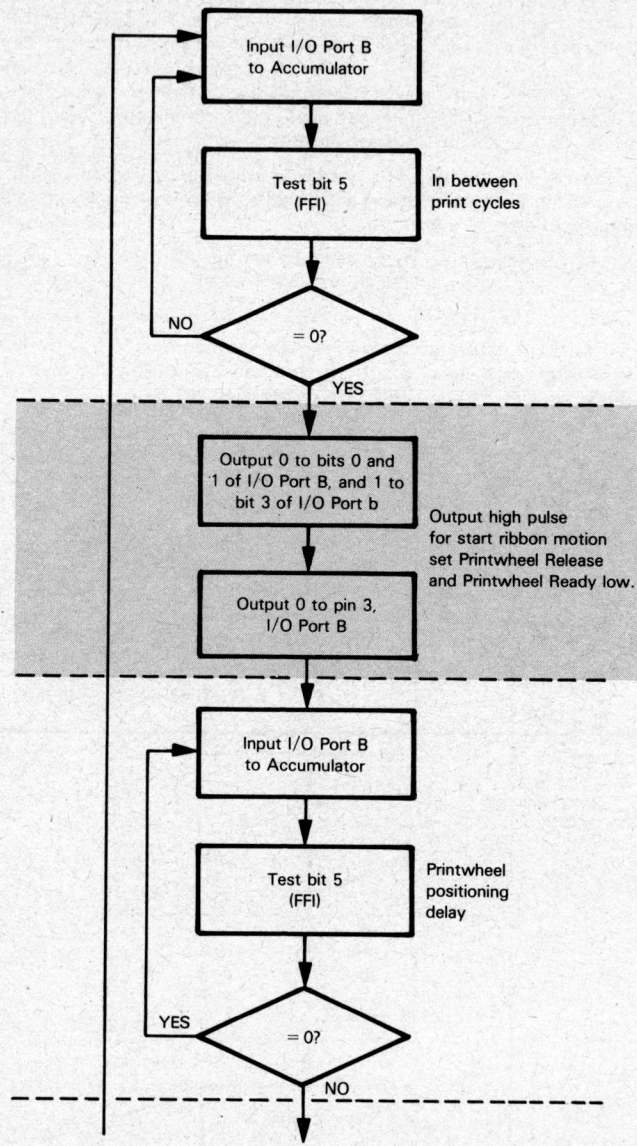

**The printwheel positioning delay is computed by the velocity decode signal FFI.**
So long as this signal is low, the printwheel is still being positioned. We therefore go into a variable delay loop, which in terms of program logic is the inverse of the "in between print cycles" delay loop. Once again, I/O Port B contents are input to the Accumulator and bit 5 is tested; however, we stay in the delay loop until bit 5 is 1. At that time the printwheel positioning delay is over:

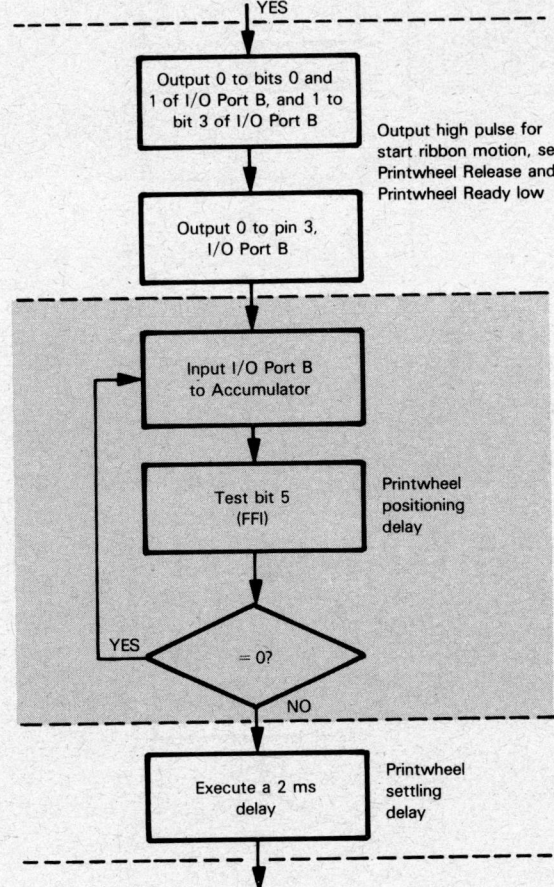

**The printwheel positioning delay must be followed by a two-millisecond printwheel settling delay.** The usual delay loop will be executed here:

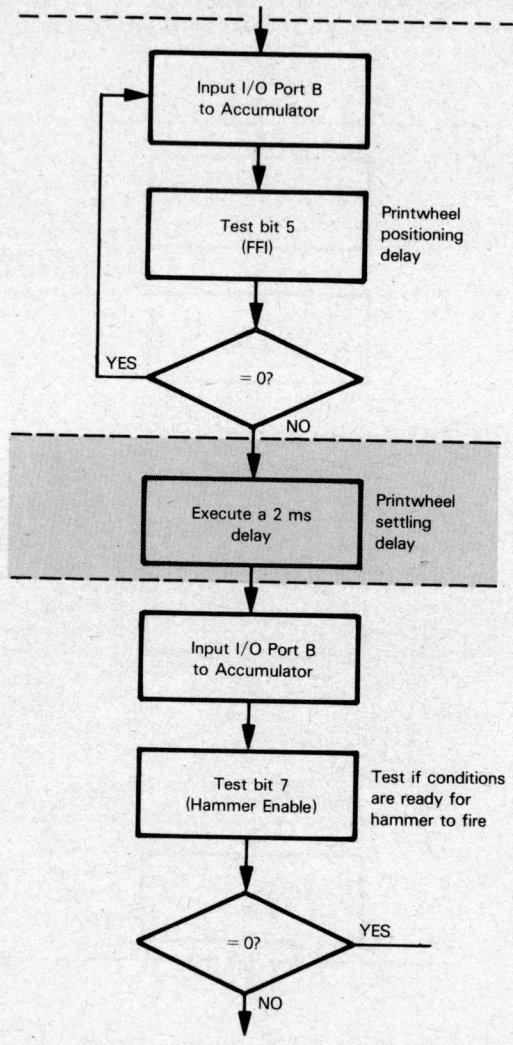

At the end of the printwheel settling delay, **the printhammer is fired, providing the HAMMER INTERLOCK signal is low and HAMMER ENABLE is high.** Recall that HAMMER INTERLOCK is a signal status bit, used by all external conditions that can prevent the hammer from being fired. Any signal inputting a high level to this status pin will suppress printhammer firing.

**A printwheel repositioning print cycle is identified by $\overline{\text{HAMMER ENABLE}}$ being input low.** This condition is detected by testing bit 7 of I/O Port B before testing the condition of HAMMER INTERLOCK. If bit 7 of I/O Port B equals 0, then the entire printhammer firing sequence is skipped and we jump directly to the printwheel ready delay, which is the last time delay of the print cycle:

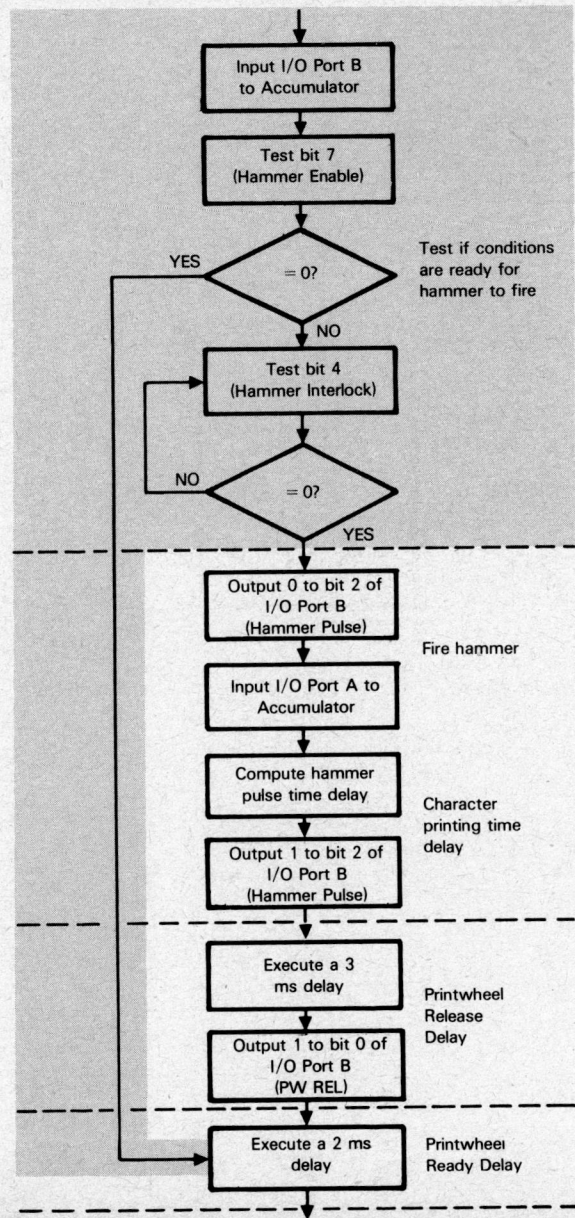

**If HAMMER ENABLE is high, this is a character printing cycle,** so the printhammer will be fired, but only when HAMMER INTERLOCK is 0. So long as any signal wire-ORed to pin 4 of I/O Port B is high, the program will stay in an endless loop, continuously testing the status of this I/O port pin. When finally the I/O port pin equals 0, the program will advance to the printhammer firing instruction sequence:

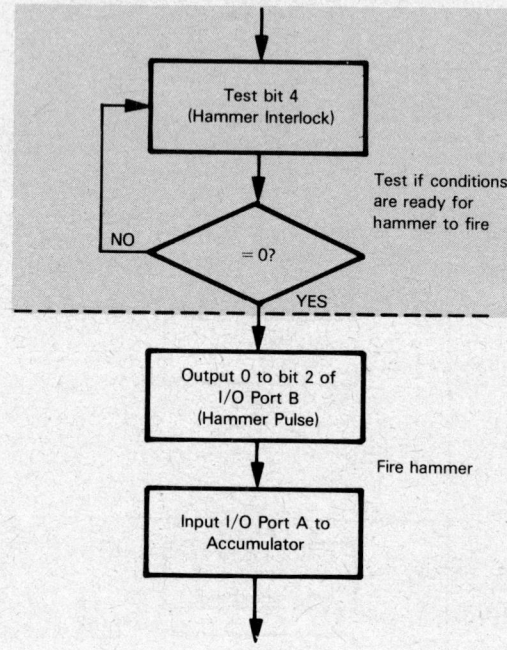

**In order to fire the printhammer, a variable length firing pulse must be output.** To do this a 0 is output to pin 2 of I/O Port B, since this is the pin via which the hammer pulse is output. Next the hammer pulse time delay is computed. We will describe how the hammer pulse width is computed after completing a description of the flowchart. At the end of the printhammer firing time delay, a 1 is output to bit 2 of I/O Port B. This terminates the printhammer firing pulse:

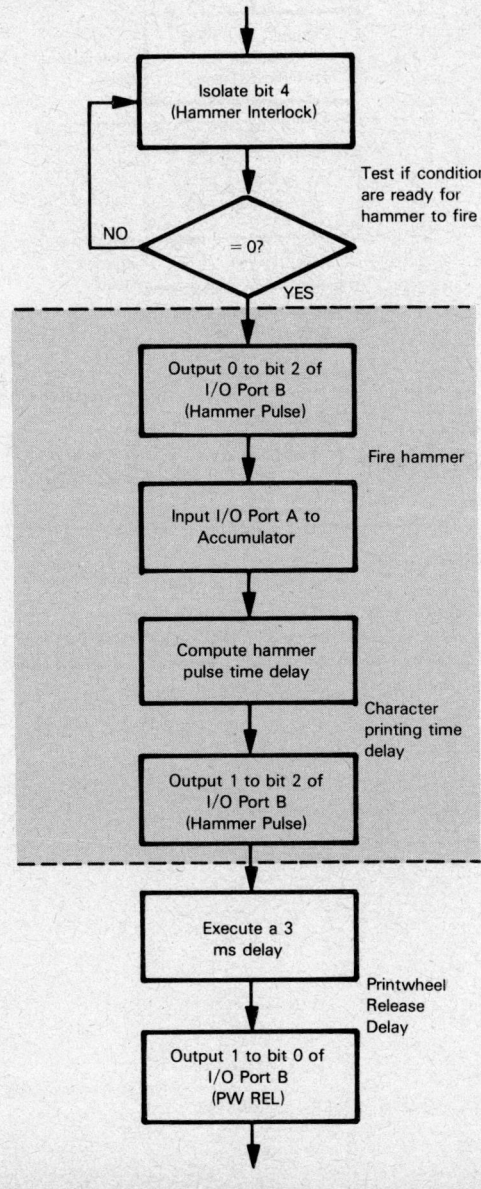

**Now two settling delays follow. First there is a three-millisecond printwheel release delay,** the termination of which is marked by a 1 being output to bit 0 of I/O Port B. This causes PW REL to output high:

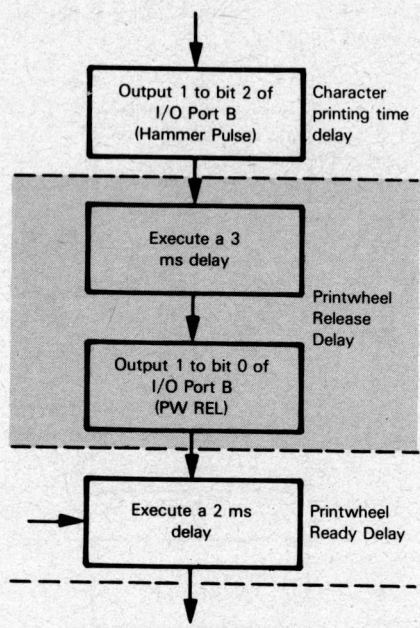

**Next, a two-millisecond printwheel ready delay is executed.** The end of this delay and **the end of the print cycle are marked by a 1 output to bit 1 of I/O Port B;** this sets CH RDY high. **We do not want to do this, however, if there is an end-of-ribbon status.** This status is identified by $\overline{\text{EOR DET}}$ being low.

The program therefore inputs I/O Port B and tests bit 6, via which $\overline{\text{EOR DET}}$ is input to the microcomputer system. If $\overline{\text{EOR DET}}$ equals 0, then the program stays in an endless loop continuously retesting bit 6 of I/O Port B; thus another print cycle cannot begin. Only if $\overline{\text{EOR DET}}$ is detected equal to 1 will the print cycle terminate with CH RDY set to 1:

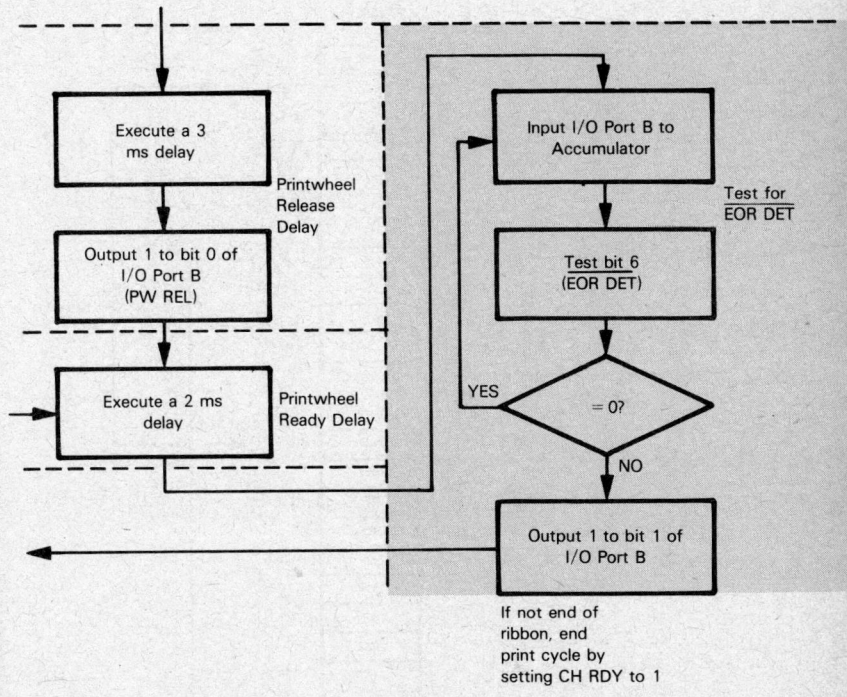

Now let us turn our attention to the method via which the appropriate printhammer firing delay is computed. In Figure 3-1, the appropriate printhammer firing delay was signaled by one of six lines (H1 through H6) being input true. Some external logic had to generate the true line, based on the nature of the character being printed; **this kind of operation is easier to do within a microcomputer program.**

**PRINTHAMMER FIRING DELAY**

**This is the method we will use to compute the appropriate printhammer firing pulse time delay:** every character to be printed is represented by one ASCII code data byte, as illustrated in Appendix A.

If we ignore the high-order parity bit, then 128 possible bit combinations remain. If you look at the ASCII codes given in Appendix A, you will see that only character codes between $20_{16}$ and $7A_{16}$ are significant. Therefore, only $5A_{16}$ (or $90_{10}$) code combinations need to be accounted for. Each of these code combinations will have assigned to it one byte in a 90-byte table; in this byte will be stored a number between 1 and 6.

This number will identify the time delay required by the character. A 12-byte table will contain the six actual time delays associated with the six digits. This scheme may be illustrated as follows:

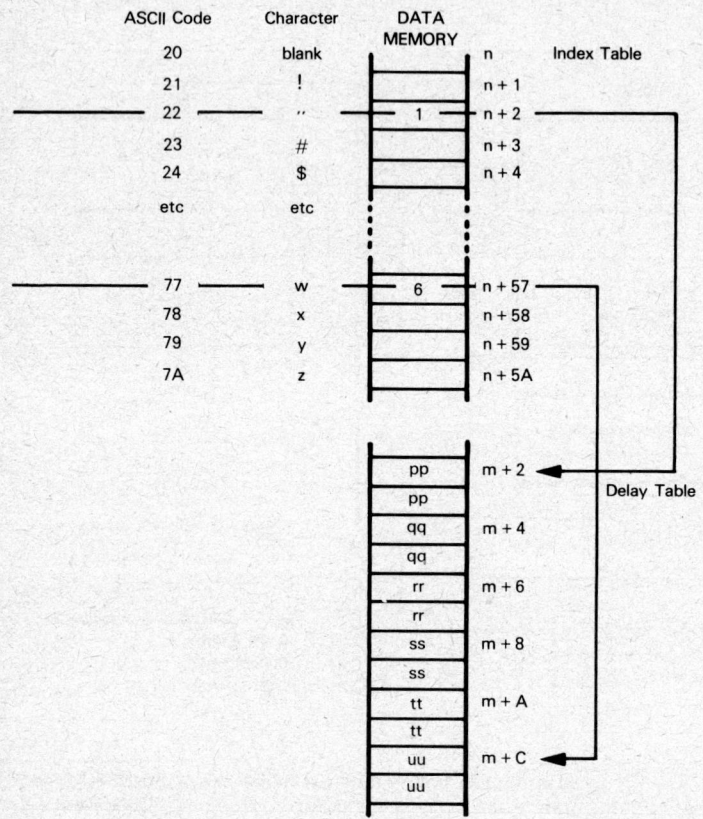

In the above illustration the letters "n" and "m", to the right of the data memory, represent any valid base memory addresses. For example, "n" might represent $0390_{16}$ while "m" represents $03F0_{16}$.

**Consider two examples.**

ASCII code $22_{16}$ signifies the double quotes character ("), which requires the shortest time delay. The data memory byte with address n + 2 corresponds to this ASCII code. 1 is stored in this data memory byte. Therefore, the first time delay, represented by pppp, is the value which must be loaded into the Index register before executing the long time delay loop which creates the printhammer firing pulse for the " character.

ASCII code $77_{16}$ represents "w". The data memory byte with address n + $57_{16}$ corresponds to this ASCII code. Within this data memory byte the value 6 is stored, which means that the longest printhammer firing delay is required for a "w". Therefore, a value represented by uuuu will be loaded into the Index register before executing the long time delay loop which creates the printhammer firing pulse for the w character.

**Figure 4-4 identifies the program steps via which the printhammer firing delay will be computed.**

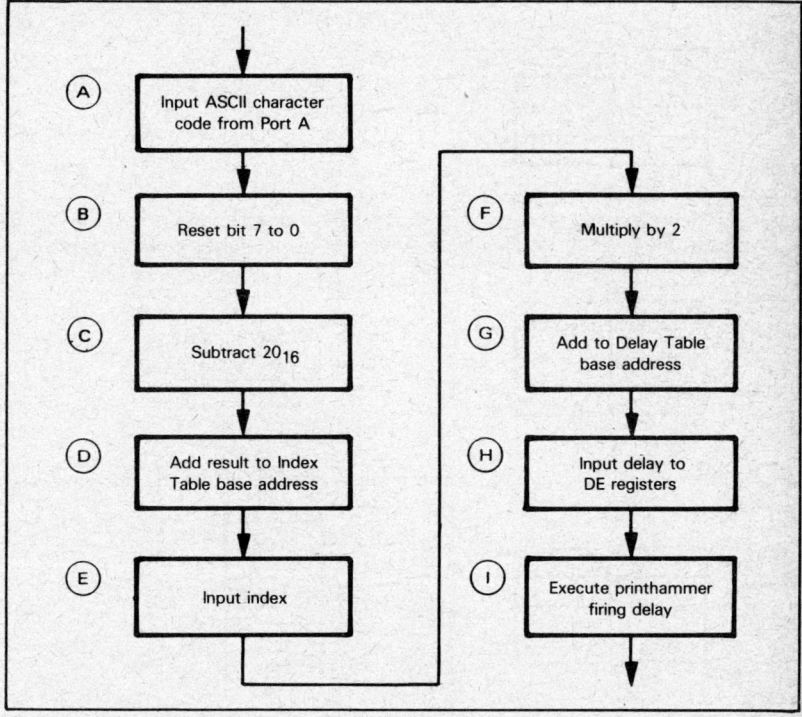

Figure 4-4. Program Flowchart to Compute Printhammer Firing Pulse Length

In order to better understand Figure 4-4, we will go down steps Ⓐ through Ⓘ for the case of "w".

Ⓐ The ASCII representation of lower-case w is input to the Accumulator:

| A | X1110111 | ← From I/O Port A |
|---|---|---|
| B,C | | |
| D,E | | |
| H,L | | |

Ⓑ We then set the parity bit to 0:

0 ⟶

| A | X1110111 |
|---|---|
| B,C | |
| D,E | |
| H,L | |

Ⓒ The index table entry corresponding to lower-case w is computed by adding the ASCII code less $20_{16}$ to the index table base address. We must subtract $20_{16}$ because the first $1F_{16}$ codes have no ASCII equivalent:

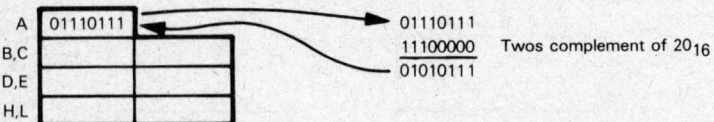

Ⓓ The index table base address is loaded into the H and L registers. We will assume this address is $0390_{16}$. Then the Accumulator contents are added to this 16-bit address:

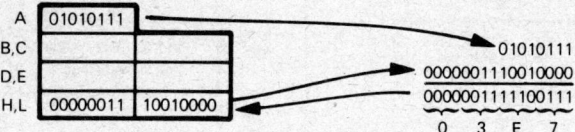

Ⓔ The appropriate index is loaded from the index table into the Accumulator:

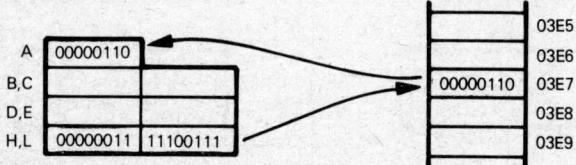

Ⓕ Since the actual delay is two bytes long, we are going to calculate the address of the appropriate delay by adding twice the index to the delay table base address. First we multiply the index by 2:

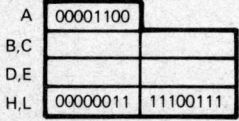

Ⓖ Next we add the index multiplied by 2 to the delay table base address. Assume this base address to be $03F0_{16}$. This base address is again loaded into the H and L registers, after which the Accumulator contents are added to the H and L registers' contents:

(H) The two bytes addressed by H and L are loaded into the D and F registers:

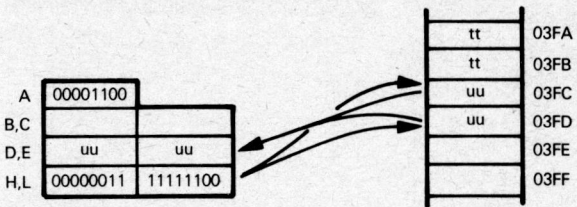

(I) The D and E registers now contain the correct initial value for a long delay to be executed as described in Chapter 2.

```
;PRINT CYCLE PROGRAM
;IN BETWEEN PRINT CYCLES TEST FF1 (BIT 5 OF I/O PORT B)
;FOR A ZERO VALUE
START:  IN      A,(2)        ;INPUT I/O PORT B TO ACCUMULATOR
        BIT     5,A          ;TEST BIT 5
        JR      NZ,START     ;IF NOT ZERO, RETURN TO START
;INITIALIZE PRINT CYCLE. OUTPUT 0 TO BITS 0 AND 1 OF I/O PORT B
;OUTPUT 1 TO BITS 2 AND 3 OF I/O PORT B
        LD      A,0CH        ;LOAD MASK INTO ACCUMULATOR
        OUT     (2),A        ;OUTPUT TO I/O PORT B
;OUTPUT 0 TO BIT 3 OF I/O PORT B, COMPLETING START RIBBON
;MOTION PULSE
        RES     3,A          ;RESET BIT 3 OF MASK IN ACCUMULATOR
        OUT     (2),A        ;OUTPUT TO I/O PORT B
;TEST FOR END OF PRINTWHEEL POSITIONING
;BIT 5 OF I/O PORT B (FFI) WILL BE 1
PWPOS:  IN      A,(2)        ;INPUT I/O PORT B TO ACCUMULATOR
        BIT     5,A          ;TEST BIT 5
        JR      Z,PWPOS      ;IF 0 RETURN TO CHECK AGAIN
;EXECUTE PRINTWHEEL SETTLING 2 MILLISECOND DELAY
        LD      A,0FAH       ;LOAD INITIAL TIME DELAY CONSTANT
PWSET:  DEC     A            ;DECREMENT ACCUMULATOR
        JR      NZ,PWSET     ;RE-DECREMENT IF NOT ZERO
;TEST PRINTHAMMER FIRING CONDITIONS
PHFIR:  IN      A,(2)        ;INPUT I/O PORT B TO ACCUMULATOR
        BIT     7,A          ;TEST BIT 7 (HAMMER ENABLE)
        JP      Z,PWRDY      ;IF IT IS 0, BYPASS PRINTHAMMER FIRING
        BIT     4,A          ;TEST HAMMER INTERLOCK
        JR      Z,PHFIR      ;WAIT FOR NONZERO VALUE BEFORE FIRING
;FIRE PRINTHAMMER
        RES     2,A          ;SET HAMMER PULSE LOW:
        OUT     (2),A        ;OUTPUT 0 TO BIT 2 OF I/O PORT B
        IN      A,(0)        ;INPUT ASCII CHARACTER TO ACCUMULATOR
        RES     7,A          ;RESET HIGH ORDER BIT
        SUB     20H          ;SUBTRACT 20H
        LD      HL,INDX      ;LOAD INDEX TABLE BASE ADDRESS TO HL
        ADD     L            ;ADD ACCUMULATOR CONTENTS TO HL
        LD      L,A
        LD      A,(HL)       ;LOAD INDEX INTO ACCUMULATOR
```

Figure 4-5. A Simple Print Cycle Instruction Sequence without Initialization or Reset

```
                ADD     A               ;MULTIPLY BY 2
                LD      HL,DELY         ;LOAD DELAY TABLE BASE ADDRESS INTO HL
                ADD     L               ;ADD ACCUMULATOR CONTENTS TO HL
                LD      L,A
                LD      E,(HL)          ;LOAD DELAY CONSTANT INTO D,E
                INC     HL
                LD      D,(HL)
        PRDLY:  DEC     DE              ;EXECUTE PRINTING DELAY
                LD      A,D
                OR      E
                JR      NZ,PRDLY
                IN      A,(2)           ;AT END OF DELAY OUTPUT 1 TO BIT 2 OF I/O
                SET     2,A             ;PORT B. THIS SETS HAMMER PULSE HIGH
                OUT     (2),A
        ;EXECUTE A 3 MILLISECOND PRINTWHEEL RELEASE TIME DELAY
                LD      DE,231          ;LOAD INITIAL TIME DELAY CONSTANT
        PWREL:  DEC     DE              ;EXECUTE LONG TIME DELAY
                LD      A,D
                OR      E
                JR      NZ,PWREL
        ;OUTPUT 1 TO BIT 0 OF I/O PORT B. THIS SETS PW REL HIGH
                IN      A,(2)           ;INPUT I/O PORT B TO ACCUMULATOR
                SET     0,A             ;SET BIT 0 TO 1
                OUT     (2),A           ;OUTPUT RESULT
        ;EXECUTE A 2 MILLISECOND PRINTWHEEL READY DELAY
        PWRDY:  LD      A,0FAH          ;LOAD TIME DELAY CONSTANT
        RDYDLY: DEC     A               ;DECREMENT ACCUMULATOR
                JR      NZ,RDYDLY       ;RE-DECREMENT IF NOT ZERO
        ;TEST FOR EOR DET (BIT 6 OF I/O PORT B) EQUAL TO 1 AS A PREREQUISITE
        ;FOR ENDING THE PRINT CYCLE
        EORCHK: IN      A,(2)           ;INPUT I/O PORT B TO ACCUMULATOR
                BIT     6,A             ;TEST BIT 6
                JR      Z,EORCHK        ;RETURN AND RETEST IF 0
        ;AT END OF PRINT CYCLE SET BIT 1 OF I/O PORT B TO 1
        ;THIS SETS CH READY HIGH
                SET     1,A             ;SET BIT 1 OF PORT B (IN ACCUMULATOR)
                OUT     (2),A           ;OUTPUT RESULT
                JP      START           ;JUMP TO NEW PRINT CYCLE TEST
```

Figure 4-5. A Simple Print Cycle Instruction Sequence
without Initialization or Reset (Continued)

**Putting together the program flowcharts illustrated in Figures 4-3 and 4-4, we generate the entire required program, as illustrated in Figure 4-5. This program is now described, section-by-section.**

**In between print cycles, the following three-instruction loop continuously tests the status of I/O Port B, bit 5.** The FFI signal is input to this pin. So long as this signal is input high, a new print cycle cannot start. As soon as this signal is input low, the printwheel is identified as being in motion -- which means that a new print cycle is underway:

```
;PRINT CYCLE PROGRAM
;IN BETWEEN PRINT CYCLES TEST FFI (BIT 5 OF I/O PORT B) FOR A ZERO VALUE
Enter    START:  IN    A,(2)      ;INPUT I/O PORT B TO ACCUMULATOR
Program          BIT   5,A        ;TEST BIT 5
                 JR    NZ,START   ;IF NOT ZERO, RETURN TO START
;INITIALIZE PRINT CYCLE. OUTPUT 0 TO BITS 0 AND 1 OF I/O PORT B
;OUTPUT 1 TO BITS 2 AND 3 OF I/O PORT B
                 LD    A,0CH      ;LOAD MASK INTO ACCUMULATOR
```

**As soon as a new print cycle starts, the PRINTWHEEL RELEASE and PRINTWHEEL READY signals must be output low. Also, a high START RIBBON MOTION pulse must be output** so that when the printhammer fires, fresh ribbon is in front of the character which is to be printed. These initial signal changes may be illustrated as follows:

```
;INITIALIZE PRINT CYCLE. OUTPUT 0 TO BITS 0 AND 1 OF I/O PORT B
;OUTPUT 1 TO BITS 2 AND 3 OF I/O PORT B
                 LD    A,0CH      ;LOAD MASK INTO ACCUMULATOR
                 OUT   (2),A      ;OUTPUT TO I/O PORT B
;OUTPUT 0 TO BIT 3 OF I/O PORT B, COMPLETING START RIBBON
;MOTION PULSE
                 RES   3,A        ;RESET BIT 3 OF MASK IN ACCUMULATOR
                 OUT   (2),A      ;OUTPUT TO I/O PORT B
```

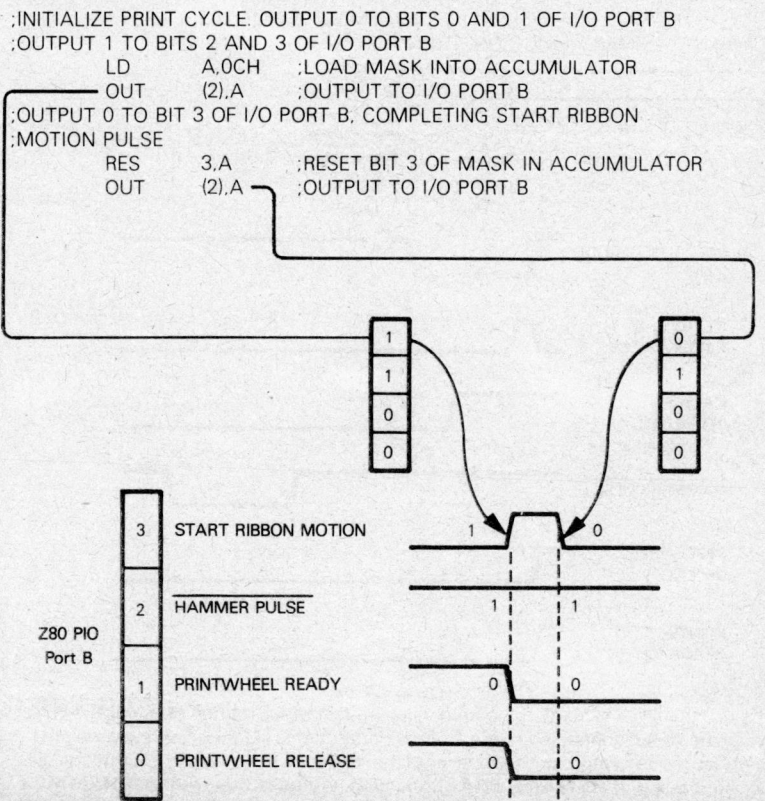

In the above illustration, notice that I/O Port B, pin 2 has been forced to output 1. This is the HAMMER PULSE pin, which goes low only for the duration of the printhammer firing pulse. At this point in the print cycle, this signal is high, so outputting 1 is harmless.

**PROGRAMMED SIGNAL PULSE**

**The program now executes a variable length delay, during which time the printwheel either moves until the appropriate character petal is in front of the printhammer, or the printwheel moves back to its position of visibility.** In either case, external logic inputs signal FFI low for the duration of the printwheel positioning delay. As soon as the printwheel has been positioned, FFI is detected high -- and program **logic advances to the two-millisecond printwheel settling delay.** We have seen this three-instruction delay loop frequently before:

**TIME DELAY OF VARIABLE LENGTH**

```
            ;TEST FOR END OF PRINTWHEEL POSITIONING.
            ;BIT 5 OF I/O PORT B (FFI) WILL BE 1.
    PWPOS:   IN      A,(2)       ;INPUT I/O PORT B TO ACCUMULATOR
             BIT     5,A         ;TEST BIT 5
             JR      Z,PWPOS     ;IF 0 RETURN TO CHECK AGAIN
            ;EXECUTE PRINTWHEEL SETTLING 2 MILLISECOND DELAY
             LD      A,0FAH      ;LOAD INITIAL TIME DELAY CONSTANT
    PWSET:   DEC     A           ;DECREMENT ACCUMULATOR
             JR      NZ,PWSET    ;RE-DECREMENT IF NOT ZERO
            ;TEST PRINTHAMMER FIRING CONDITIONS
```

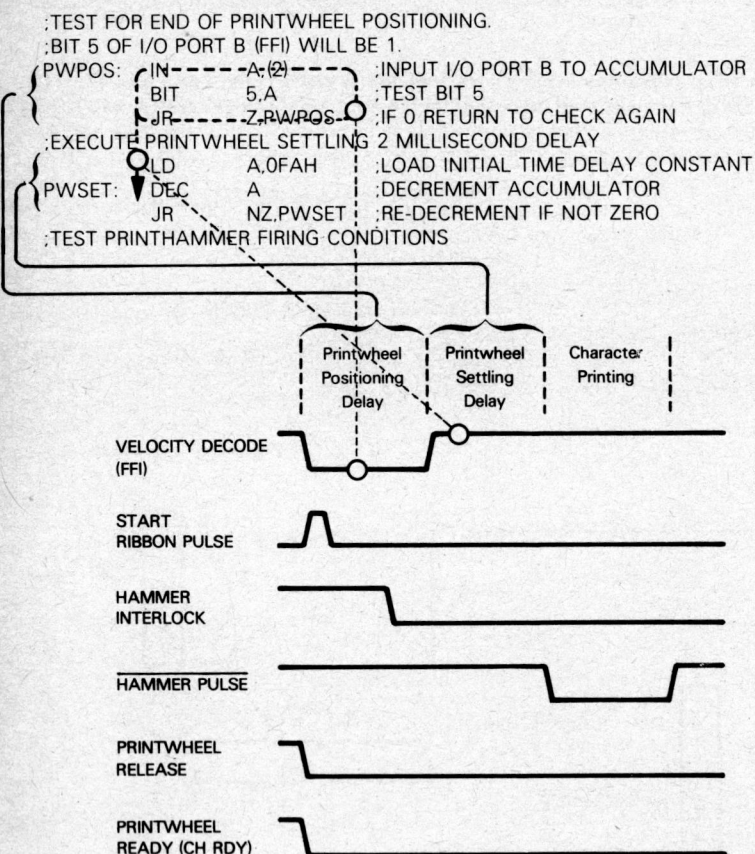

Now the printhammer is ready to be fired. First we test the condition of HAMMER ENABLE, which has been connected to pin 7 of I/O Port B. If this signal is low, then we are in a printwheel repositioning print cycle and the entire hammer firing instruction sequence is bypassed. **If HAMMER ENABLE is high, we pass this test. But HAMMER INTERLOCK must still be tested;** this signal is input to I/O Port B, pin 4. Since the BIT instruction which tests bit 7 leaves the Accumulator's contents intact, we simply execute another BIT instruction to test HAMMER INTERLOCK.

**If HAMMER ENABLE is detected low, execution branches to** the instruction labeled PWRDY. You will find this instruction close to the end of the program, at the beginning of the instruction sequence which executes **a two-millisecond PRINT-WHEEL READY delay.**

Note that the five-instruction sequence illustrated in Figure 4-5 tests for HAMMER ENABLE low within the loop that tests for HAMMER INTERLOCK high. HAMMER ENABLE will be either high or low for the duration of the print cycle; it will not change levels during the print cycle. Therefore, the fact that it is continuously being tested is redundant -- it serves no purpose, but it does no harm.

**Next, the printhammer is fired.** The instruction sequence which causes the printhammer to fire implements steps (A) through (I), which we have already described. In order to make the instruction sequence easier to understand, it is reproduced below with labels (A) through (I) added:

```
;FIRE PRINTHAMMER
            RES     2,A             ;SET HAMMER PULSE LOW;
            OUT     (2),A           ;OUTPUT 0 TO BIT 2 OF I/O PORT B
(A)         IN      A,(0)           ;INPUT ASCII CHARACTER TO ACCUMULATOR
(B)         RES     7,A             ;RESET HIGH ORDER BIT
(C)         SUB     20H             ;SUBTRACT 20H
           ┌LD      HL,INDX         ;LOAD INDEX TABLE BASE ADDRESS TO HL
(D)        ┤ADD     L               ;ADD ACCUMULATOR CONTENTS TO HL
           └LD      L,A
(E)         LD      A,(HL)          ;LOAD INDEX INTO ACCUMULATOR
(F)         ADD     A               ;MULTIPLY BY 2
           ┌LD      HL,DELY         ;LOAD DELAY TABLE BASE ADDRESS INTO HL
(G)        ┤ADD     L               ;ADD ACCUMULATOR CONTENTS TO HL
           └LD      L,A
           ┌LD      E,(HL)          ;LOAD DELAY CONSTANT INTO D,E
(H)        ┤INC     HL
           └LD      D,(HL)
PRDLY:     ┌DEC     DE              ;EXECUTE PRINTING DELAY
(I)        ┤LD      A,D
           │OR      E
           └JR      NZ,PRDLY
            IN      A,(2)           ;AT END OF DELAY OUTPUT 1 TO BIT 2 OF I/O
            SET     2,A             ;PORT B. THIS SETS HAMMER PULSE HIGH.
            OUT     (2),A
;EXECUTE A 3 MILLISECOND PRINTWHEEL RELEASE TIME DELAY
```

Notice that the bit tests of HAMMER ENABLE and HAMMER INTERLOCK left the contents of Port B intact in the Accumulator. We need not input from Port B, therefore, before setting HAMMER PULSE low; we simply reset bit 2 of the Accumulator to 0 and then output the result to I/O Port B.

A three-millisecond **PRINTWHEEL RELEASE** time delay is now executed, and the end of this time delay is marked by the **PRINTWHEEL RELEASE** signal being output high. Next, the two-millisecond **PRINTWHEEL READY** delay is executed:

```
       ;EXECUTE A 3 MILLISECOND PRINTWHEEL RELEASE TIME DELAY
          LD      DE,231      ;LOAD INITIAL TIME DELAY CONSTANT
PWREL:    DEC     DE          ;EXECUTE LONG TIME DELAY
          LD      A,D
          OR      E
          JR      NZ,PWREL
       ;OUTPUT 1 TO BIT 0 OF I/O PORT B. THIS SETS PW REL HIGH.
          IN      A,(2)       ;INPUT I/O PORT B TO ACCUMULATOR
          SET     0,A         ;SET BIT 0 TO 1
          OUT     (2),A       ;OUTPUT RESULT
       ;EXECUTE A 2 MILLISECOND PRINTWHEEL READY DELAY
PWRDY:    LD      A,0FAH      ;LOAD TIME DELAY CONSTANT
RDYDLY:   DEC     A           ;DECREMENT ACCUMULATOR
          JR      NZ,RDYDLY   ;RE-DECREMENT IF NOT ZERO
```

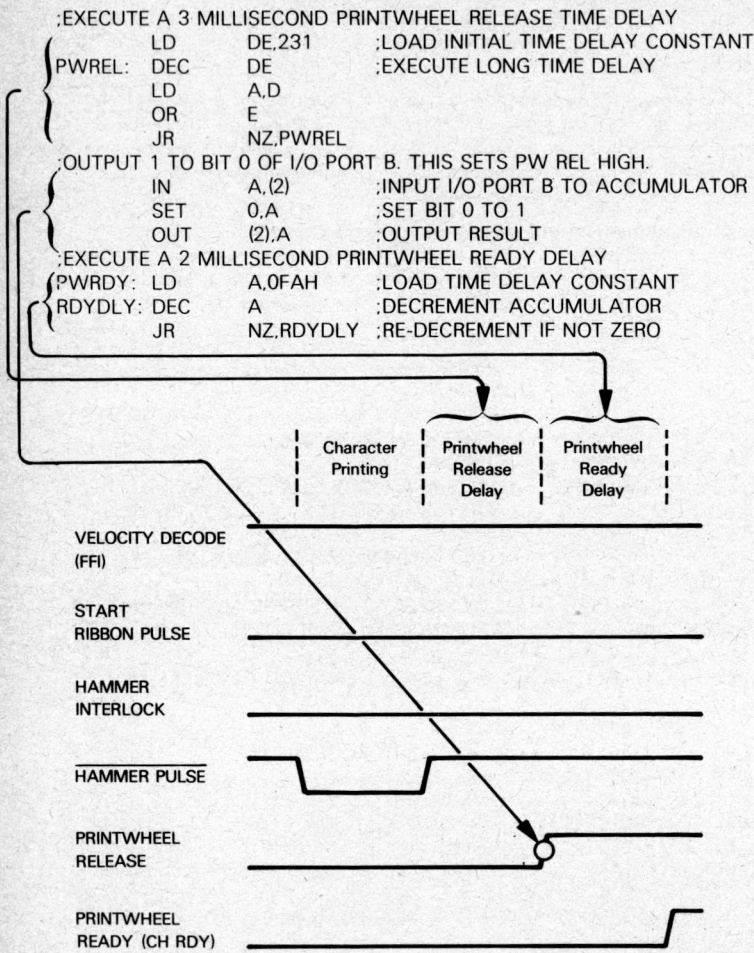

Before terminating the print cycle by outputting **PRINTWHEEL READY (CH RDY) high**, the program must ensure that the end of ribbon has not been reached. If $\overline{\text{EOR DET}}$ is detected low, the program stays in an endless loop until the ribbon has been changed; then $\overline{\text{EOR DET}}$ will be input high by external logic. When $\overline{\text{EOR DET}}$ is detected high, the final instructions of the program set PRINTWHEEL READY high, then return to the beginning of the program and wait for the next print cycle.

## PROGRAM LOGIC ERRORS

The program we have developed in this chapter contains a logic error which could not occur in a digital logic implementation. The error is in the hammer pulse time delay computation.

In a digital logic implementation, the ASCII code for any character would be processed as seven individual signals. These signals would be combined in some way to generate one of the time delay signals H1 through H6. **It does not matter what ASCII code combination is input, one of the time delay signals H1 through H6 will be output high;** if the signal generation logic is unsound, a time delay signal will still be created, although it may be the wrong signal.

Now look at the assembly language program implementation. It is simple enough for us to look up the table in Appendix A and see that valid ASCII codes only cover the range $20_{16}$ through $7A_{16}$. That does not prevent a logic designer from using the microcomputer system we create in a special system that includes unusual characters, represented by codes outside the normal ASCII range. Our program could output some very strange results under these circumstances. Suppose the ASCII code $10_{16}$ had been adopted to represent a special character. Then, our attempt to look up the Index Table would load into the Accumulator whatever happened to be in memory byte $n-10_{16}$.

| LIMIT CHECKING |

There is no telling what could be in this memory byte; in all probability this byte will be used to store an instruction code, perhaps a two-hexadecimal-digit value. Suppose it contained $2A_{16}$; the next program step will double $2A_{16}$, add it to the base address of the Delay Table, and access the initial delay code from memory location $m + 54_{16}$.

Given the microcomputer configuration illustrated in Figure 4-2, this memory location could easily be one of the duplicate addresses which spuriously access some memory byte, because we have used disarmingly simple chip select logic. Had we used more complex chip select logic, chances are we would now be attempting to access a memory byte that did not exist. In the former case, there is no telling what length of hammer pulse would be generated; in the latter case, an extremely long hammer pulse would be generated, since we would retrieve 0 from a non-existent memory location, and this value would be interpreted as the initial delay constant for the long delay program loop. The hammer pulse would be 852 milliseconds long:

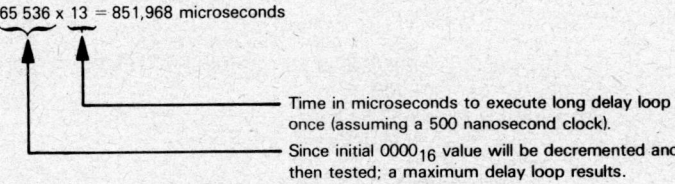

65 536 x 13 = 851,968 microseconds

- Time in microseconds to execute long delay loop once (assuming a 500 nanosecond clock).
- Since initial $0000_{16}$ value will be decremented and then tested; a maximum delay loop results.

Now, in order to avoid this problem we have two options:

1) **Program logic can simply ignore any invalid ASCII code.**
2) **Program logic can generate a default hammer pulse width for invalid ASCII codes.**

If we ignore special characters, the conclusion is obvious: the microcomputer system cannot be used in any application that requires special characters to be printed. Since the special character is ignored, nothing will happen when such a character code is detected on input -- there will be no hammer pulse, no carriage movement, and no positioning.

Providing a default hammer pulse for special characters means that such characters will be printed, but they may create unevenness in the density of the typed text.

You, as the logic designer, would have to specify your preference.

Either instruction sequence may be inserted into the existing program, as follows:

```
                IN      A,(0)       ;INPUT ASCII CHARACTER TO ACCUMULATOR
                RES     7,A         ;RESET HIGH ORDER BIT
Check for       SUB     20H         ;SUBTRACT 20H
valid           LD      HL,INDX     ;LOAD INDEX TABLE BASE ADDRESS TO HL
ASCII           ADD     L           ;ADD ACCUMULATOR CONTENTS TO HL
codes           LD      L,A
inserted        LD      A,(HL)      ;LOAD INDEX INTO ACCUMULATOR
here            ADD     A           ;MULTIPLY BY 2
```

**Here is the instruction sequence which ignores non-standard ASCII codes:**

```
                •
                •
                IN      A,(0)       ;INPUT ASCII CHARACTER TO ACCUMULATOR
                RES     7,A         ;RESET HIGH ORDER BIT
;COMPARE ASCII CODE WITH LOWEST LEGAL VALUE
                CP      20H
                JP      M,PWRDY     ;IF CODE IS 1FH OR LESS, BYPASS
                                    ;HAMMER FIRING
;COMPARE ASCII CODE WITH HIGHEST LEGAL VALUE
                CP      7BH
                JP      P,PWRDY     ;IF CODE IS 7BH OR GREATER, BYPASS
                                    ;HAMMER FIRING
;ASCII CODE IS VALID
                SUB     20H         ;SUBTRACT 20H
                •
                •
```

**The second option, illustrated below, prints unknown characters with a median density, using density code 3:**

```
                •
                •
                IN      A,(0)       ;INPUT ASCII CHARACTER TO ACCUMULATOR
                RES     7,A         ;RESET HIGH ORDER BIT
;COMPARE ASCII CODE WITH LOWEST LEGAL VALUE
                CP      20H
                JP      P,OK        ;IF CODE IS 20H OR HIGHER, TEST FOR HIGH
                                    ;LIMIT
;CODE IS ILLEGAL. ASSUME A DENSITY OF 3
NOK:            LD      A,6         ;LOAD TWICE THE DENSITY
                JP      NEXT
;COMPARE ASCII CODE WITH LARGEST LEGAL VALUE
OK:             CP      7BH         ;IF CODE IS 7BH OR GREATER, ASSUME
                JP      P,NOK       ;A DENSITY OF 3
;ASCII CODE IS VALID
                SUB     20H         ;SUBTRACT 20H
                LD      HL,INDX     ;LOAD INDEX TABLE BASE ADDRESS TO HL
                ADD     L           ;ADD ACCUMULATOR CONTENTS TO HL
                LD      L,A
                LD      A,(HL)      ;LOAD INDEX INTO ACCUMULATOR
                ADD     A           ;MULTIPLY BY 2
NEXT:           LD      HL,DELY     ;LOAD DELAY TABLE BASE ADDRESS INTO HL
                •
                •
```

**Both of the invalid ASCII code instruction sequences are simplistic in their solution to the problem.**

The only new feature introduced is the use of the Compare Immediate (CP) instruction. This instruction subtracts the immediate data in the operand from the contents of the Accumulator. The result of the subtraction is discarded, which means that the Accumulator contents are not altered;

**COMPARE IMMEDIATE**

**BRANCH ON CONDITION**

however, status flags are set to reflect the results of the subtraction. We use a JP M (Jump on Minus) instruction to identify a negative result, which means that the immediate data in the operand was larger than the value in the Accumulator. Similarly, a JP P (Jump on Plus) instruction identifies a value in the immediate operand which is equal to or less than the contents of the Accumulator.

**In the second instruction sequence,** if the value in the immediate operand is less than or equal to the contents of the Accumulator, the JP P instruction causes a branch to a later instruction labeled OK. The actual program execution paths for the second instruction sequence may appear a trifle confusing

**CONDITIONAL INSTRUCTION EXECUTION PATHS**

to you if you are new to programming; **we** therefore **illustrate execution paths as follows:**

```
            IN      A,(0)       ;INPUT ASCII CHARACTER TO ACCUMULATOR
            RES     7,A         ;RESET HIGH ORDER BIT
;COMPARE ASCII CODE WITH LOWEST LEGAL VALUE
   (A)      CP      20H
            JP      P,OK        ;IF CODE IS 20H OR HIGHER, TEST FOR HIGH LIMIT
;CODE IS ILLEGAL. ASSUME A DENSITY OF 3.
 NOK:  (B)  LD      A,6         ;LOAD TWICE THE DENSITY
            JP      NEXT   (E)
;COMPARE ASCII CODE WITH LARGEST LEGAL VALUE
  OK:       CP      7BH         ;IF CODE IS 7BH OR GREATER, ASSUME
   (D)      JP      P,NOK       ;A DENSITY OF 3.
;ASCII CODE IS VALID
            SUB     20H         ;SUBTRACT 20H
   (F)      LD      HL,INDX     ;LOAD INDEX TABLE BASE ADDRESS TO HL
            ADD     L,A         ;ADD ACCUMULATOR CONTENTS TO HL
            LD      L,A
            LD      A,(HL)      ;LOAD INDEX INTO ACCUMULATOR
   (C)      ADD     A           ;MULTIPLY BY 2
  NEXT:     LD      HL,DELY     ;LOAD DELAY TABLE BASE ADDRESS INTO HL
```

Execution paths, illustrated by circled letters above, can be interpreted as follows:

(A) An ASCII code passes the "lowest legal value" test, but now must be tested for the "highest legal value".

(B) The ASCII code failed the "lowest legal value" test. The program loads twice the default density into the Accumulator and branches to the instruction sequence which accesses the delay constant appropriate to this default density. This Jump is illustrated by (C)

(D) A character which has passed the "lowest legal ASCII value" test is next checked for "highest legal ASCII value"; if it fails this test then program execution branches, as shown by (E), to instructions which assume a default density of 3. (E), in fact, meets (B)

**(F)** An ASCII character that passes both the "lowest legal value" test and the "highest legal value" test is processed via instruction path **(F)**. Instructions in this path load the appropriate density index into the Accumulator.

# RESET AND INITIALIZATION

**In order to complete our program, we must create the necessary Reset and Initialization instructions.**

Reset instructions will be executed whenever RESET is input true to the microcomputer system.

Initialization instructions will be executed whenever the system is started up.

**There is no reason why Reset and Initialization instruction sequences should coincide;** in many applications two separate and distinct instruction sequences may be needed. **On the other hand, it is quite common to use Reset in lieu of system initialization.** This means that when you first power up the system, RESET is pulsed true; this starts the entire microcomputer-based logic system.

**In our case, the Reset program is indeed simple.** All we have to do is output Control codes to the Z80 Parallel Input/Output/Interface, then set output signals to the "in between print cycles" condition. **Here is the necessary Initialization instruction sequence:**

```
        ORG     0
;FIRST OUTPUT CONTROL CODES TO I/O PORT A CONTROL REGISTER
        LD      A,0FFH          ;SET MODE 3
        OUT     (1),A
        OUT     (1),A           ;ALL LINES INPUT
;NEXT OUTPUT CONTROL CODES TO I/O PORT B CONTROL REGISTER
        OUT     (3),A           ;SET MODE 3
        LD      A,0F0H          ;SET PINS 0 THROUGH 3 TO OUTPUT AND
        OUT     (3),A           ;PINS 4 THROUGH 7 TO INPUT
;SET HAMMER PULSE, PW READY AND PW REL HIGH
;SET START RIBBON MOTION LOW
        LD      A,7
        OUT     (2),A
```

This is how Control codes for each port of the Z80 PIO are constructed:

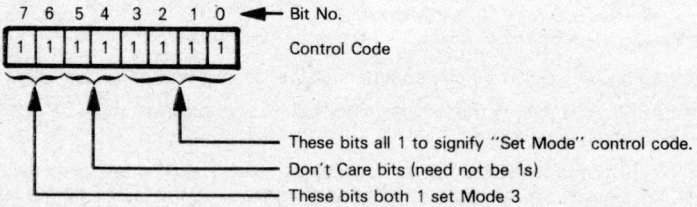

After Mode 3 is set, another byte must be written to the port's Control register; this second byte specifies the direction of each of the port's pins. A bit set to 1 in the direction byte specifies an input line, and a 0 bit specifies an output line.

# A PROGRAM SUMMARY

**First of all, it would be a good idea to put together the entire program, as developed in this chapter. We will include the necessary Assembler directives. This final program is illustrated in Figure 4-6.**

**Now that the program is finished, notice that RAM memory has not been used.** The CPU registers have provided sufficient read/write memory to handle all variable data.

The 1K bytes of ROM program memory are sufficient to contain the entire program, plus the two data tables.

Were you implementing a microcomputer system within the limited confines of the logic included in this chapter, you could now eliminate the two RAM memory chips. In all probability, there would be numerous other logic functions more economically included within the microcomputer system; these would almost certainly require the presence of some RAM memory. There are nine bytes of read/write memory provided by the seven CPU registers and the CPU Stack Pointer; these are usually insufficient for any real application.

**Here is the final program memory map identifying the way in which the program illustrated in Figure 4-6 uses ROM memory:**

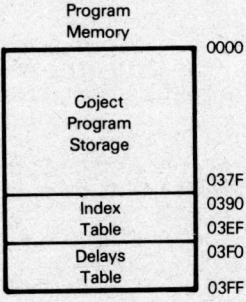

```
INDX      EQU     390H         ;INDEX TABLE BASE ADDRESS
DELY      EQU     3F2H         ;FIRST LOCATION IN DELAY TABLE
          ORG     0
;FIRST OUTPUT CONTROL CODES TO I/O PORT A CONTROL REGISTER
          LD      A,0FFH       ;SET MODE 3
          OUT     (1),A
          OUT     (1),A        ;ALL LINES INPUT
;NEXT OUTPUT CONTROL CODES TO I/O PORT B CONTROL REGISTER
          OUT     (3),A        ;SET MODE 3
          LD      A,0F0H       ;SET PINS 0 THROUGH 3 TO OUTPUT AND
          OUT     (3),A        ;PINS 4 THROUGH 7 TO INPUT
;SET HAMMER PULSE, PW READY AND PW REL HIGH
;SET START RIBBON MOTION LOW
          LD      A,7
          OUT     (2),A
;PRINT CYCLE PROGRAM
;IN BETWEEN PRINT CYCLES TEST FFI (BIT 5 OF I/O PORT B)
;FOR A ZERO VALUE
START:    IN      A,(2)        ;INPUT I/O PORT B TO ACCUMULATOR
          BIT     5,A          ;TEST BIT 5
          JR      NZ,START     ;IF NOT ZERO, RETURN TO START
;INITIALIZE PRINT CYCLE. OUTPUT 0 TO BITS 0 AND 1 OF I/O PORT B
;OUTPUT 1 TO BITS 2 AND 3 OF I/O PORT B
          LD      A,0CH        ;LOAD MASK INTO ACCUMULATOR
          OUT     (2),A        ;OUTPUT TO I/O PORT B
;OUTPUT 0 TO BIT 3 OF I/O PORT B, COMPLETING START RIBBON
;MOTION PULSE
          RES     3,A          ;RESET BIT 3 OF MASK IN ACCUMULATOR
          OUT     (2),A        ;OUTPUT TO I/O PORT B
;TEST FOR END OF PRINTWHEEL POSITIONING
;BIT 5 OF I/O PORT B (FFI) WILL BE 1
PWPOS:    IN      A,(2)        ;INPUT I/O PORT B TO ACCUMULATOR
          BIT     5,A          ;TEST BIT 5
          JR      Z,PWPOS      ;IF 0 RETURN TO CHECK AGAIN
;EXECUTE PRINTWHEEL SETTLING 2 MILLISECOND DELAY
          LD      A,0FAH       ;LOAD INITIAL TIME DELAY CONSTANT
PWSET:    DEC     A            ;DECREMENT ACCUMULATOR
          JR      NZ,PWSET     ;RE-DECREMENT IF NOT ZERO
;TEST PRINTHAMMER FIRING CONDITIONS
PHFIR:    IN      A,(2)        ;INPUT I/O PORT B TO ACCUMULATOR
          BIT     7,A          ;TEST BIT 7 (HAMMER ENABLE)
          JP      Z,PWRDY      ;IF IT IS 0, BYPASS PRINTHAMMER FIRING
          BIT     4,A          ;TEST HAMMER INTERLOCK
          JR      Z,PHFIR      ;WAIT FOR NONZERO VALUE BEFORE FIRING
;FIRE PRINTHAMMER
          RES     2,A          ;SET HAMMER PULSE LOW:
          OUT     (2),A        ;OUTPUT 0 TO BIT 2 OF I/O PORT B
          IN      A,(0)        ;INPUT ASCII CHARACTER TO ACCUMULATOR
          RES     7,A          ;RESET HIGH ORDER BIT
;COMPARE ASCII CODE WITH LOWEST LEGAL VALUE
          CP      20H
          JP      M,PWRDY      ;IF CODE IS 1FH OR LESS, BYPASS HAMMER
                               ;FIRING
```

Figure 4-6. A Simple Print Cycle Program

```
        ;COMPARE ASCII CODE WITH HIGHEST LEGAL VALUE
                CP      7BH
                JP      P,PWRDY         ;IF CODE IS 7BH OR GREATER, BYPASS HAMMER
                                        ;FIRING
        ;ASCII CODE IS VALID
                SUB     20H             ;SUBTRACT 20H
                LD      HL,INDX         ;LOAD INDEX TABLE BASE ADDRESS TO HL
                ADD     L               ;ADD ACCUMULATOR CONTENTS TO HL
                LD      L,A
                LD      A,(HL)          ;LOAD INDEX INTO ACCUMULATOR
                ADD     A               ;MULTIPLY BY 2
                LD      HL,DELY         ;LOAD DELAY TABLE BASE ADDRESS INTO HL
                ADD     L               ;ADD ACCUMULATOR CONTENTS TO HL
                LD      L,A
                LD      E,(HL)          ;LOAD DELAY CONSTANT INTO D,E
                INC     HL
                LD      D,(HL)
        PRDLY:  DEC     DE              ;EXECUTE PRINTING DELAY
                LD      A,D
                OR      E
                JR      NZ,PRDLY
                IN      A,(2)           ;AT END OF DELAY OUTPUT 1 TO BIT 2 OF I/O
                SET     2,A             ;PORT B. THIS SETS HAMMER PULSE HIGH
                OUT     (2),A
        ;EXECUTE A 3 MILLISECOND PRINTWHEEL RELEASE TIME DELAY
                LD      DE,231          ;LOAD INITIAL TIME DELAY CONSTANT
        PWREL:  DEC     DE              ;EXECUTE LONG TIME DELAY
                LD      A,D
                OR      E
                JR      NZ,PWREL
        ;OUTPUT 1 TO BIT 0 OF I/O PORT B. THIS SETS PW REL HIGH
                IN      A,(2)           ;INPUT I/O PORT B TO ACCUMULATOR
                SET     0,A             ;SET BIT 0 TO 1
                OUT     (2),A           ;OUTPUT RESULT
        ;EXECUTE A 2 MILLISECOND PRINTWHEEL READY DELAY
        PWRDY:  LD      A,0FAH          ;LOAD TIME DELAY CONSTANT
        RDYDLY: DEC     A               ;DECREMENT ACCUMULATOR
                JR      NZ,RDYDLY       ;RE-DECREMENT IF NOT ZERO
        ;TEST FOR EOR DET (BIT 6 OF I/O PORT B) EQUAL TO 1 AS A PREREQUISITE
        ;FOR ENDING THE PRINT CYCLE
        EORCHK  IN      A,(2)           ;INPUT I/O PORT B TO ACCUMULATOR
                BIT     6,A             ;TEST BIT 6
                JR      Z,EORCHK        ;RETURN AND RETEST IF 0
        ;AT END OF PRINT CYCLE SET BIT 1 OF I/O PORT B TO 1
        ;THIS SETS CH READY HIGH
                SET     1,A             ;SET BIT 1 OF PORT B (IN ACCUMULATOR)
                OUT     (2),A           ;OUTPUT RESULT
                JP      START           ;JUMP TO NEW PRINT CYCLE TEST
        ;INDEX TABLE FOLLOWS HERE
                ORG     INDX
                Data representing 90 index entries follows here
        ;DELAY TABLE FOLLOWS HERE
                ORG     DELY
                Data representing 6 delays follows here
```

Figure 4-6. A Simple Print Cycle Program (Continued)

# Chapter 5
# A PROGRAMMER'S PERSPECTIVE

The program we developed in Chapter 4 is considerably shorter and easier to follow than the digital simulation of Chapter 3. While we came a long way in Chapter 4, we still have a way to go. The program in Figure 4-6 treats the logic to be implemented as a single transfer function, but it is not a well-written program.

To the digital logic designer, one of the most confusing things about programming is the trivial ease with which you can do the same thing in ten different ways. Does this imply that some implementations are more efficient than others? Indeed yes. To a great extent writing efficient programs is a talent, just as creating efficient digital logic is a talent; but there are certain rules which, if followed, will at least help you avoid obvious mistakes. In this chapter we are going to take the program created in Chapter 4 and look at it a little more carefully.

## SIMPLE PROGRAMMING EFFICIENCY

The first thing you should do, after writing a source program, is to go back over it, looking for elementary ways in which you can cut out instructions.

### EFFICIENT TABLE LOOKUPS

On average, you will find that it is possible to reduce a program to two-thirds of its original length, simply by writing more efficient instruction sequences. In Figure 4-6, the most obvious example of sloppy programming involves the Index Table. The program loads a value between 1 and 6 from an Index Table byte, then multiplies this value by two before adding it to the base address of the Delay Table. **Why not directly store twice the index in the Index Table?** That cuts out one instruction, as follows:

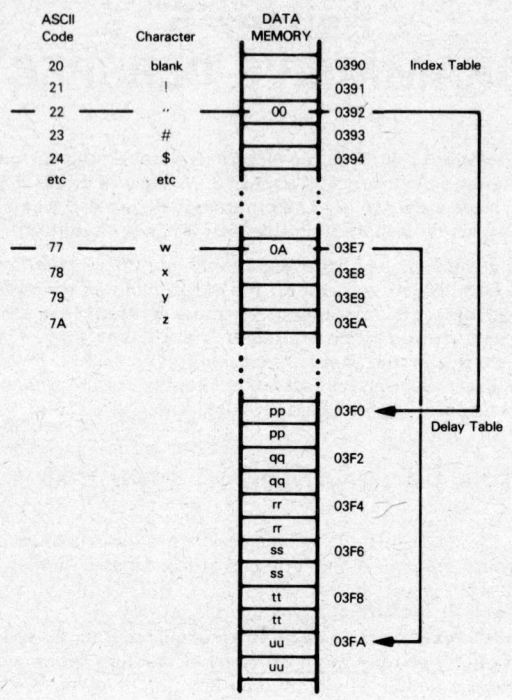

```
;ASCII CODE IS VALID
        SUB     20H         ;SUBTRACT 20H
        LD      HL,INDX     ;LOAD INDEX TABLE BASE ADDRESS TO HL
        ADD     L           ;ADD ACCUMULATOR CONTENTS TO HL
        LD      L,A
        LD      A,(HL)      ;LOAD INDEX X2 INTO ACCUMULATOR
        LD      HL,DELY     ;LOAD DELAY TABLE BASE ADDRESS INTO HL
        ADD     L           ;ADD ACCUMULATOR CONTENTS TO HL
        LD      L,A
```

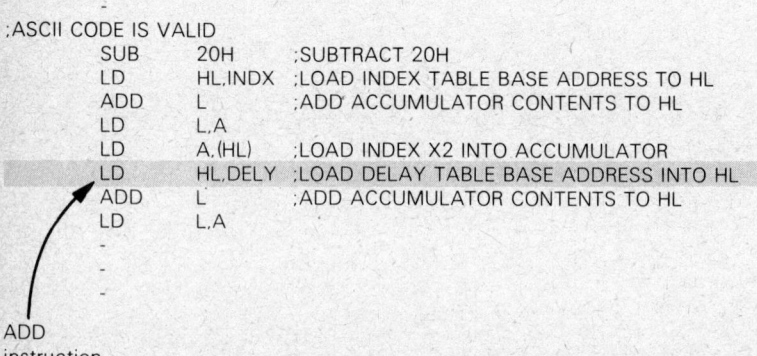

ADD instruction dropped

In the instruction sequence above, notice that one instruction has been removed following the shaded LD instruction.

There are still a number of additional ways in which we can make the Delay Table lookup more efficient. **Why subtract $20_{16}$ from the ASCII code,** for example? If we are going to add the ASCII code to a base address, there is nothing to stop us Equating this base address, represented by the symbol INDEX, to a value $20_{16}$ less than the first real Index Table byte. Our instruction sequence now collapses further, as follows:

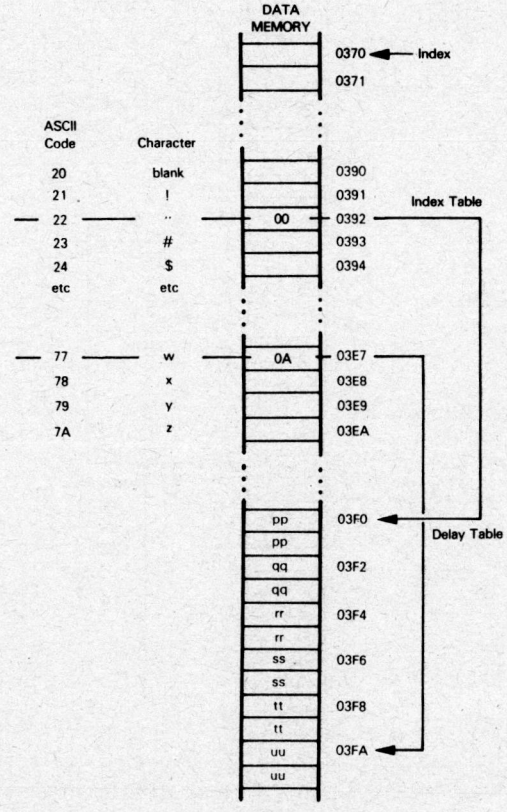

```
INDEX      EQU      0370H           ;EQUATE INDEX TABLE BASE ADDRESS - 20H
             -
             -
;ASCII CODE IS VALID
           LD       HL,INDX         ;LOAD INDEX TABLE BASE ADDRESS - 20H
           ADD      L               ;ADD ACCUMULATOR CONTENTS TO HL
           LD       L,A
           LD       A,(HL)          ;LOAD INDEX X2 INTO ACCUMULATOR
           LD       HL,DELY         ;LOAD DELAY TABLE BASE ADDRESS INTO HL
SUB        ADD      L               ;ADD ACCUMULATOR CONTENTS TO HL
instruction LD      L,A
dropped
```

Okay, so INDEX is now being equated to $0370_{16}$ — which means that we no longer need to subtract $20_{16}$ from the ASCII code. We have eliminated the SUB instruction which was above the shaded LD instruction. **Now, instead of storing twice the character density index in the Index Table, why not store the second half of the Delay Table address?** Our program will now contract further, as follows:

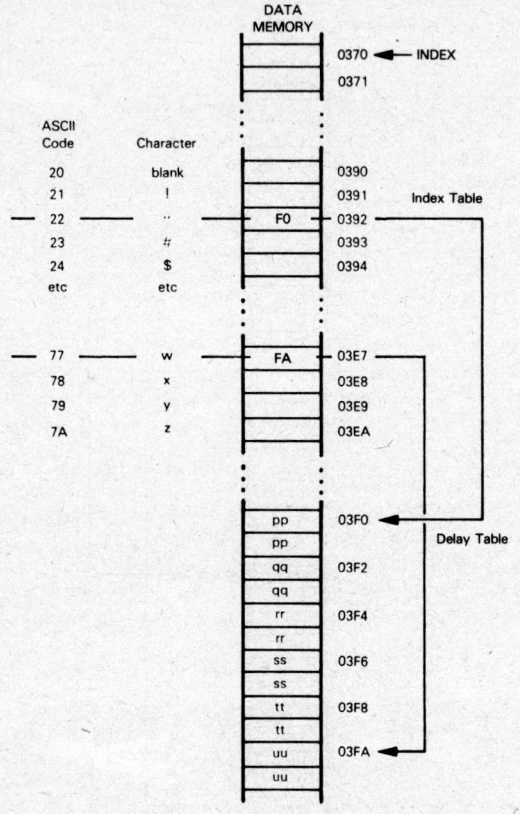

| INDEX | EQU | 0370H | ;EQUATE INDEX TO TABLE BASE |
| | | | ;ADDRESS - 20H |

```
            -
            -
            -
;ASCII CODE IS VALID
            LD      HL,INDX      ;LOAD INDEX TABLE BASE ADDRESS - 20H
            ADD     L            ;ADD ACCUMULATOR CONTENTS TO HL
            LD      L,A
            LD      L,(HL)       ;LOAD LOW ORDER BYTE OF DELAY TABLE
                                 ;ADDRESS
            LD      H,3          ;LOAD HIGH ORDER BYTE OF DELAY TABLE
                                 ;ADDRESS
```

Two more instructions have disappeared.

We have now taken out four instructions from the sequence which loads the printhammer firing initial delay constant — and we are still not done.

**Why not move the whole Index Table, so that instead of occupying memory locations $0390_{16}$ through $03EA_{16}$, it occupies memory locations $0320_{16}$ through $037A_{16}$?** The ASCII code, stripped of the parity bit, now becomes the low-order byte of the Index Table address; our instruction sequence contracts further as follows:

**TABLES POSITIONED TO SIMPLIFY ACCESS INSTRUCTION SEQUENCE**

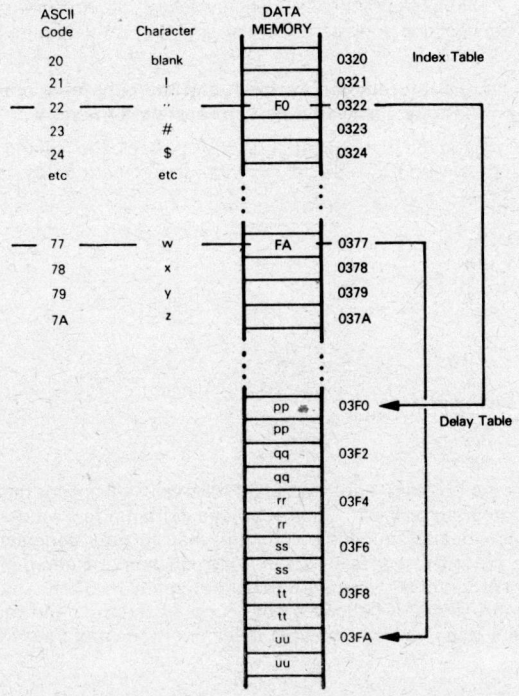

```
;ASCII CODE IS VALID
        LD      H,3             ;LOAD INDEX TABLE ADDRESS HIGH ORDER BYTE
        LD      L,A             ;MOVE LOW ORDER BYTE OF ADDRESS TO L
        LD      L,(HL)          ;LOAD LOW ORDER BYTE OF DELAY TABLE
                                ;ADDRESS
```

Suppose a "w" character is to be printed. Before the first of the above three instructions is executed, the Accumulator contains $77_{16}$, as a result of the previous:

```
        IN      A,(0)
        RES     7,A
```

instructions' execution. Following execution of the:

```
        LD      H,3
```

instruction, the H register will contain $03_{16}$; this is the upper half of the implied memory address. Next the instruction:

        LD        L,A

moves $77_{16}$ from the Accumulator to the L register. H and L now contain $0377_{16}$; this is the effective implied address. The next instruction:

        LD        L,(HL)

moves, to the L register, the contents of the memory byte addressed by HL.

HL contains $0377_{16}$. Memory byte $0377_{16}$ contains $FA_{16}$, therefore $FA_{16}$ is moved to the L register. The new implied address is $03FA_{16}$; and that is the required Delay Table address.

**Nine instructions have been reduced to three, and the only price paid is that we have had to move the Index Table to a new area of data memory.**

To ensure that you understand how the program will now look, the old and new instruction sequences are shown side-by-side below, without comment fields:

| Old Program | | New Program | |
|---|---|---|---|
| ;ASCII CODE IS VALID | | | |
| SUB | 20H | LD | H,3 |
| LD | HL,INDX | LD | L,A |
| ADD | L | LD | L,(HL) |
| LD | L,A | | |
| LD | A,(HL) | | |
| ADD | A | | |
| LD | HL,DELY | | |
| ADD | L | | |
| LD | L,A | | |

Unfortunately there are no golden rules which, if followed, will ensure that you always write the shortest program possible. Once you have written a few programs, you will understand how individual instructions work, and that, in turn, generates efficiency. The purpose of the preceding pages has been to demonstrate the enormous difference between a compact program and a straightforward program. If your product is going to be produced in high volume, it behooves you to spend the time and money cutting down program size — then you may be able to eliminate some of your ROM chips.

# SUBROUTINES

If you look again at the program in Figure 4-6, you will notice that at two points within this program we execute identical instruction sequences to create a two-millisecond delay. Now, it only takes three instructions to execute a two-millisecond delay, so the fact that these three instructions have been repeated is no big tragedy. If you think about it, however, the potential exists for some very uneconomical memory utilization in longer programs.

We have kept our program simple in Chapter 4 because it must remain small enough to handle in a book; but project, if you will, a more complex routine where a 30-instruction sequence needs to be repeated, rather than a three-instruction sequence. We must now find some way of including the instruction sequence just once, then branching to this single sequence from a number of different locations within a program, as needed. That is what a subroutine will do for you.

Let us take the three instructions which execute a two-millisecond delay and convert them into a subroutine. This is what happens to relevant portions of the program:

```
            ORG     0
            LD      SP,08FFH   ;INITIALIZE STACK POINTER TO END OF DATA AREA
            -
            -
            -
;EXECUTE PRINTWHEEL SETTLING 2 MILLISECOND DELAY
            CALL    D2MS
            -
            -
            -
;EXECUTE A 2 MILLISECOND PRINTWHEEL READY DELAY
PWRDY       CALL    D2MS
            -
            -
            -
;AT END OF PRINT CYCLE SET BIT 1 OF I/O PORT B TO 1
;THIS SETS CH RDY HIGH
            SET     1,A
            OUT     (2),A
            JP      START
;SUBROUTINE TO EXECUTE A 2 MILLISECOND DELAY
D2MS        LD      A,0FAH     ;LOAD ACCUMULATOR WITH 0
LOPD        DEC     A          ;DECREMENT A
            JR      NZ,LOPD    ;IF A DOES NOT DECREMENT TO 0, RE-DECREMENT
            RET                ;RETURN FROM SUBROUTINE
```

In order to understand how a subroutine works, we will assign some arbitrary memory addresses for our source program's object code; we will show, step-by-step, what happens when a subroutine is called and what happens upon returning from the subroutine. **First of all, here is the assumed memory map:**

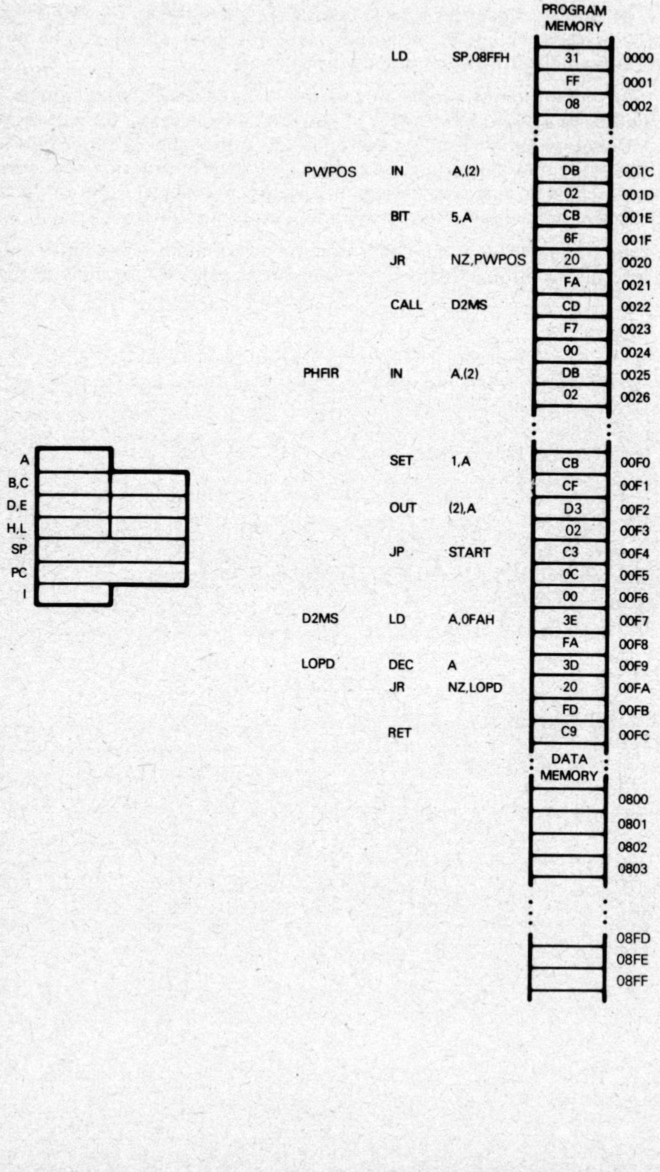

| | | | PROGRAM MEMORY | |
|---|---|---|---|---|
| | LD | SP,08FFH | 31 | 0000 |
| | | | FF | 0001 |
| | | | 08 | 0002 |
| | | | ⋮ | |
| PWPOS | IN | A,(2) | DB | 001C |
| | | | 02 | 001D |
| | BIT | 5,A | CB | 001E |
| | | | 6F | 001F |
| | JR | NZ,PWPOS | 20 | 0020 |
| | | | FA | 0021 |
| | CALL | D2MS | CD | 0022 |
| | | | F7 | 0023 |
| | | | 00 | 0024 |
| PHRR | IN | A,(2) | DB | 0025 |
| | | | 02 | 0026 |
| | | | ⋮ | |
| | SET | 1,A | CB | 00F0 |
| | | | CF | 00F1 |
| | OUT | (2),A | D3 | 00F2 |
| | | | 02 | 00F3 |
| | JP | START | C3 | 00F4 |
| | | | 0C | 00F5 |
| | | | 00 | 00F6 |
| D2MS | LD | A,0FAH | 3E | 00F7 |
| | | | FA | 00F8 |
| LOPD | DEC | A | 3D | 00F9 |
| | JR | NZ,LOPD | 20 | 00FA |
| | | | FD | 00FB |
| | RET | | C9 | 00FC |
| | | | DATA MEMORY | |
| | | | | 0800 |
| | | | | 0801 |
| | | | | 0802 |
| | | | | 0803 |
| | | | ⋮ | |
| | | | | 08FD |
| | | | | 08FE |
| | | | | 08FF |

Registers:
- A
- B,C
- D,E
- H,L
- SP
- PC
- I

# SUBROUTINE CALL

Suppose we are about to execute the first CALL D2MS instruction. At this point registers will contain the following data:

| Label | Instr | Operand | Prog Mem | Addr |
|---|---|---|---|---|
| | LD | SP,08FFH | 31 | 0000 |
| | | | FF | 0001 |
| | | | 08 | 0002 |
| PWPOS | IN | A,(2) | DB | 001C |
| | | | 02 | 001D |
| | BIT | 5,A | CB | 001E |
| | | | 6F | 001F |
| | JR | NZ,PWPOS | 20 | 0020 |
| | | | FA | 0021 |
| | CALL | D2MS | CD | 0022 |
| | | | F7 | 0023 |
| | | | 00 | 0024 |
| PHFIR | IN | A,(2) | DB | 0025 |
| | | | 02 | 0026 |
| | SET | 1,A | CB | 00F0 |
| | | | CF | 00F1 |
| | OUT | (2),A | D3 | 00F2 |
| | | | 02 | 00F3 |
| | JP | START | C3 | 00F4 |
| | | | 0C | 00F5 |
| | | | 00 | 00F6 |
| D2MS | LD | A,0FAH | 3E | 00F7 |
| | | | FA | 00F8 |
| LOPD | DEC | A | 3D | 00F9 |
| | JR | NZ,LOPD | 20 | 00FA |
| | | | FD | 00FB |
| | RET | | C9 | 00FC |

| Reg | Value |
|---|---|
| A | 00 |
| B,C | |
| D,E | |
| H,L | |
| SP | 08FF |
| PC | 0022 |
| I | 28 |

DATA MEMORY: 0800, 0801, 0802, 0803, ... 08FD, 08FE, 08FF

The Program Counter (PC) addresses the first byte of the Call instruction's object code; this address is $0022_{16}$. The Instruction register holds the object code for the most recently executed instruction; this is a JR instruction located at byte $0020_{16}$. The Stack Pointer, you will notice, was initialized at the beginning of the program; it contains $08FF_{16}$. According to Figure 4-2, this is the address of the first byte of read/write memory. Since the stack has not been used, the Stack Pointer will still contain $08FF_{16}$.

The Accumulator contains 00 because this was the condition which caused execution to break out of the holding loop starting at PWPOS.

## Now when the Call instruction is executed, steps occur as follows:

The Call instruction object code is loaded into the Instruction register and the Program Counter is incremented:

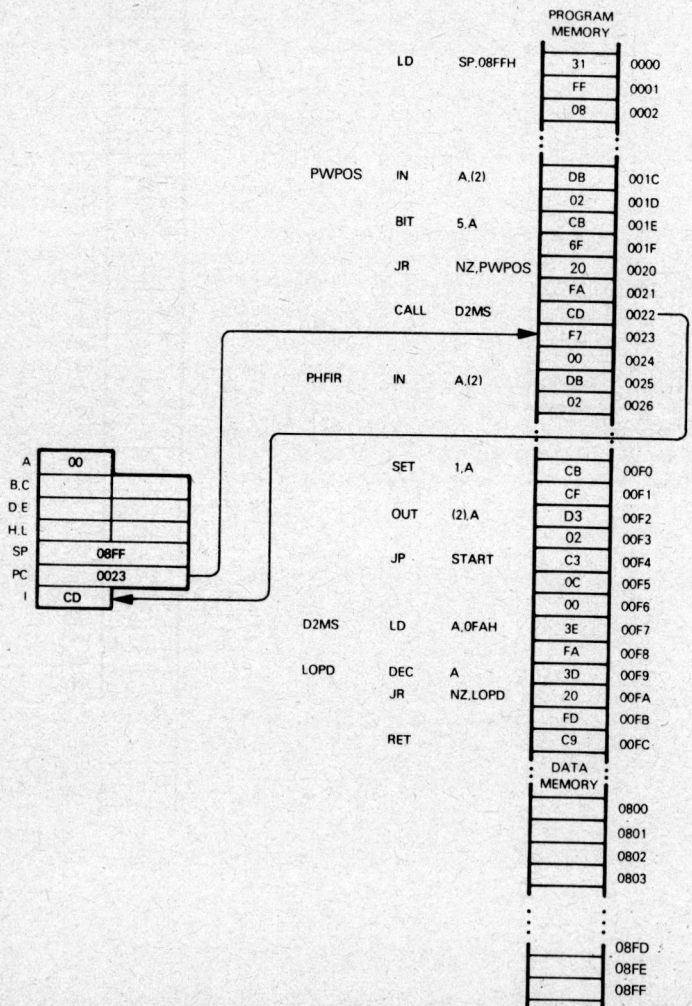

The Program Counter is incremented by 2 to bypass the CALL address. This incremented value is saved in the first two stack bytes. The CALL address is then loaded into the Program Counter. The Stack Pointer is decremented by 2 so that it addresses the first free stack byte:

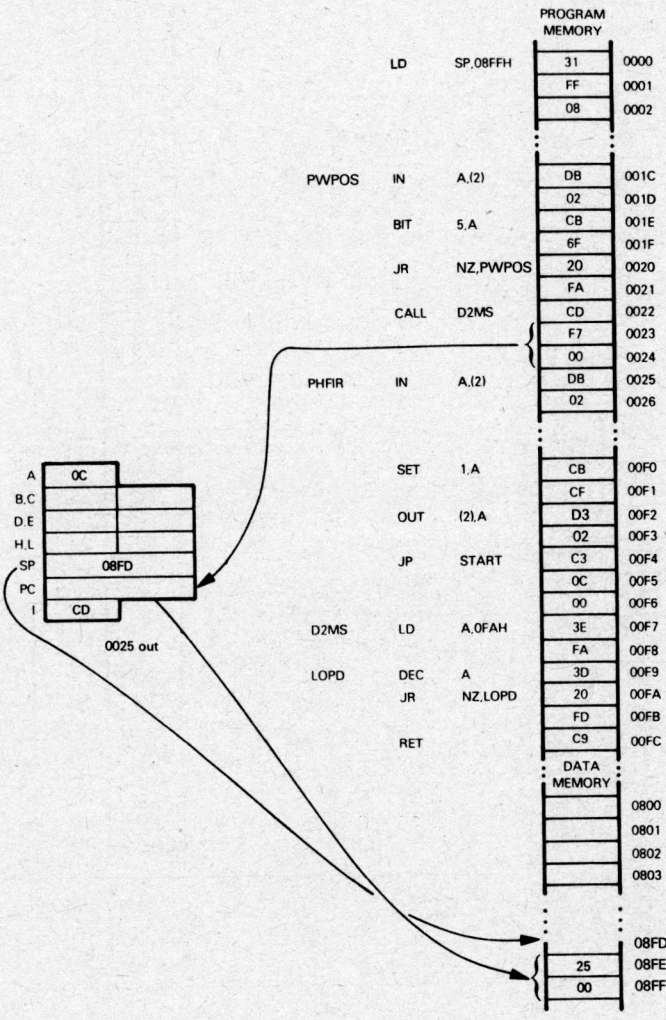

The next instruction executed has its object code stored in memory byte $00F7_{16}$; this is the memory byte now addressed by the Program Counter:

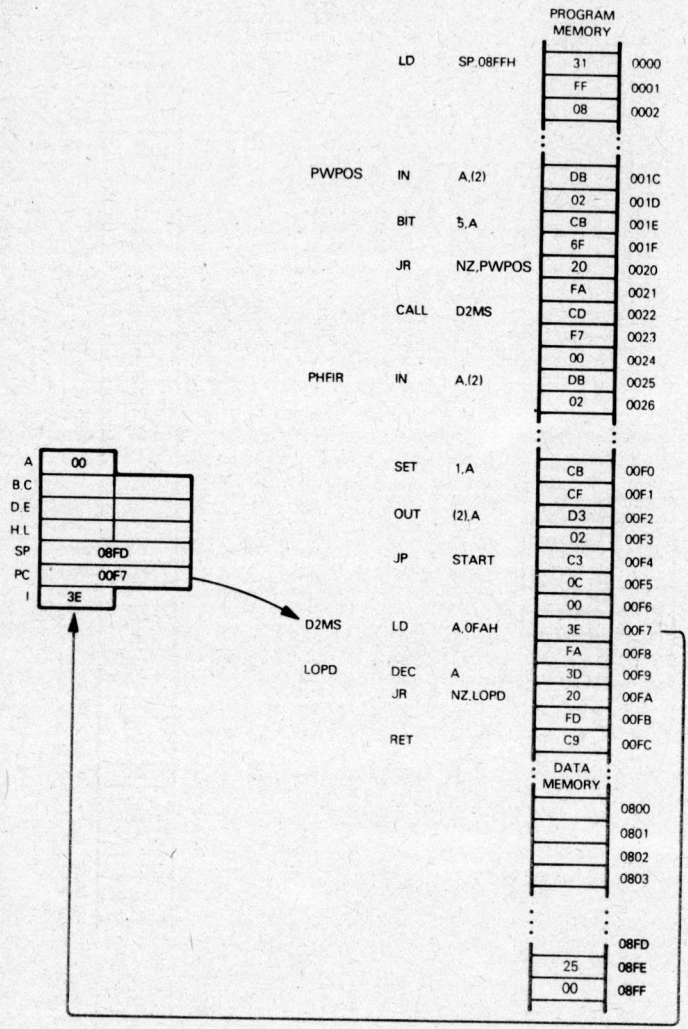

Instructions within the two-millisecond delay loop are now executed repetitively until the Accumulator contents decrement from 01 to 00.

## SUBROUTINE RETURN

**When the Accumulator finally decrements from 01 to 00, execution passes to the Return (RET) instruction. This instruction increments the contents of the Stack Pointer by 2, then moves the contents of the two top stack bytes into the Program Counter. Thus, program execution returns to the instruction that follows the Call:**

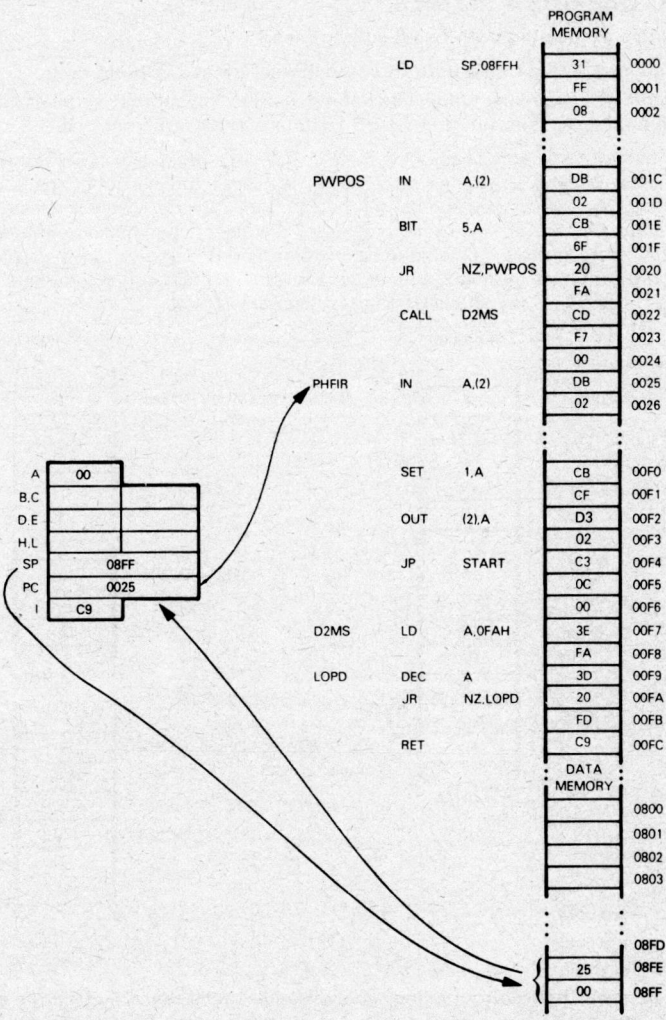

In summary, this is what happened:

When the Call instruction was executed, the address of the next instruction was saved in the stack. The Call instruction provided the address of the next instruction to be executed.

The next instruction to be executed was the first instruction of the subroutine.

The last instruction of the subroutine merely caused the address saved at the top of the stack to be returned to the Program Counter, and this, in turn, caused execution to branch back to the instruction following the Call.

## WHEN TO USE SUBROUTINES

**There is a price associated with using subroutines:**

1) **Each Call instruction represents three additional bytes of object code.**
2) **The instruction sequence which has been moved to the subroutine must have an appended Return instruction which costs one byte of object code.**

**Let us first look at our specific case.** The three instructions which constitute the two-millisecond delay occupy five bytes of object code. These three instructions occur twice, therefore, combined, they occupy ten bytes of object code. When moved to a subroutine, adding the Return instruction increases the object code bytes from five to six. In addition, there are two Call instructions, and each requires three bytes of object code —which means that the two Call instructions, plus the subroutine, generate 12 bytes of object code. This may be illustrated as follows:

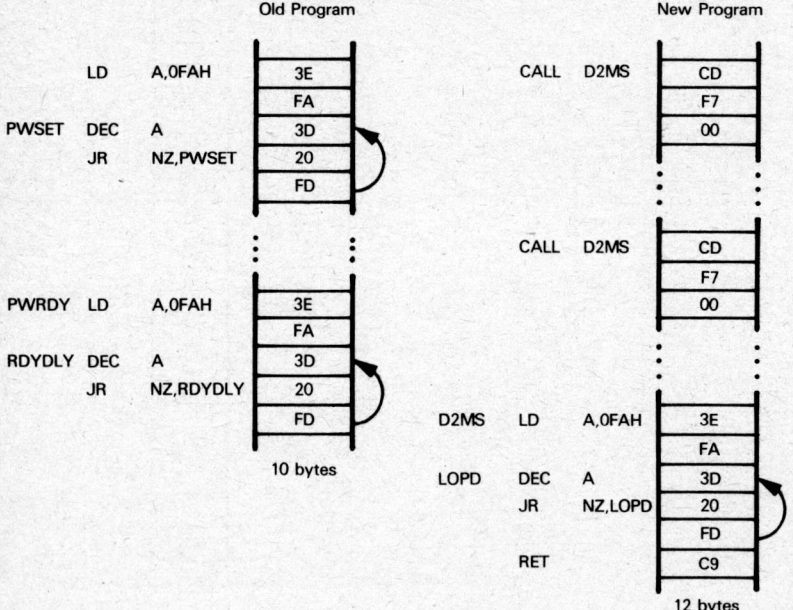

**In our specific case, therefore, moving the two-millisecond delay instruction sequence into a subroutine has cost us two bytes of object code. It has cost us three additional bytes of object code — those required to initialize the Stack Pointer; and our microcomputer system is now going to require RAM memory.**

A stack can only exist if read/write memory is present.

Now, these comments do not imply that subroutines are a dubious programming feature to be used sparingly; on the contrary, it is hard to conceive of any program which, when well-written, will not include some subroutines. But bear in mind that **there is a minimum subroutine size below which subroutines in general become uneconomical.**

**Suppose there are n bytes of object code** in an instruction sequence which you are planning to convert into a subroutine.

Suppose the **n bytes of object code occur m times;** that means when the n bytes of object code become a subroutine, it will be called by m CALL instructions.

**Without subroutines, m x n bytes will be consumed repeating n bytes m times.**

**With subroutines, the number of bytes consumed is:**

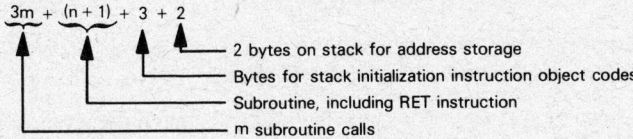

- 2 bytes on stack for address storage
- Bytes for stack initialization instruction object codes
- Subroutine, including RET instruction
- m subroutine calls

**For the subroutine to be worthwhile, 3m + n + 6 must be less than m x n.**

**Table 5-1 shows the minimum economic subroutine length as a function of the number of subroutine calls.**

Table 5-1. The Shortest Economic Subroutine Length as a Function of the Number of Times the Subroutine Is Called

| Number of Subroutine Calls (m) | Minimum Economic Subroutine Length (n) |
|---|---|
| 2 | 12 Bytes |
| 3 | 8 Bytes |
| 4 | 6 Bytes |
| 5 | 6 Bytes |
| 10 | 4 Bytes |
| 20 | 4 Bytes |

## CONDITIONAL SUBROUTINE RETURNS

Even though none of the repeated instruction sequences within the program in Figure 4-6 are long enough to justify being turned into a subroutine, we will nonetheless explore the potential of subroutines further.

**Just as there are conditional Jump instructions, which we use frequently within a time delay loop, so there are conditional subroutine Call instructions and conditional Return from Subroutine instructions.**

Conditional subroutine Call and Return instructions are particularly useful in longer subroutines within which there are variable execution paths.

**Consider the printhammer firing instruction sequence in Figure 4-6.** Given the program as illustrated, this instruction sequence occurs just once, which means that converting it into a subroutine would make no sense. **It is possible to imagine a more extensive program which performs a wide variety of printer interface operations, such that printhammer firing logic might be triggered for a number of different**

reasons. Since the printhammer firing logic consists of a fairly long set of instructions, putting these instructions in a subroutine would be absolutely mandatory. Consider the following subroutine implementation:

```
;PRINTHAMMER FIRING SUBROUTINE
PHFIR:  IN      A,(2)       ;INPUT I/O PORT B TO ACCUMULATOR
        BIT     7,A         ;TEST BIT 7 (HAMMER ENABLE)
        RET     Z           ;IF IT IS 0, RETURN
        BIT     4,A         ;TEST HAMMER INTERLOCK
        RET     Z           ;IF IT IS 0, RETURN
;FIRE PRINTHAMMER
        RES     2,A         ;SET HAMMER PULSE LOW
        OUT     (2),A       ;OUTPUT 0 TO BIT 2 OF I/O PORT B
        IN      A,(0)       ;INPUT ASCII CHARACTER TO ACCUMULATOR
        RES     7,A         ;RESET HIGH ORDER BIT
;COMPARE ASCII CODE WITH LOWEST LEGAL VALUE
        CP      20H
        RET     M           ;IF CODE IS 1FH OR LESS, BYPASS HAMMER FIRING
;COMPARE ASCII CODE WITH HIGHEST LEGAL VALUE
        CP      7BH
        RET     P           ;IF CODE IS 7BH OR GREATER BYPASS HAMMER
                            ;FIRING
;ASCII CODE IS VALID
        LD      H,03H       ;LOAD INDEX TABLE ADDRESS HIGH ORDER BYTE
        LD      L,A         ;MOVE LOW ORDER BYTE OF ADDRESS TO L
        LD      L,(HL)      ;LOAD LOW ORDER BYTE OF DELAY TABLE
                            ;ADDRESS
        CALL    LDLY
        IN      A,(2)       ;AT END OF DELAY OUTPUT 1 TO BIT 2 OF I/O
        SET     2,A         ;PORT B. THIS SETS HAMMER PULSE HIGH
        OUT     (2),A
;EXECUTE A 3 MILLISECOND PRINTWHEEL RELEASE TIME DELAY
        LD      HL,MS3
        CALL    LDLY
;OUTPUT 1 TO BIT 0 OF I/O PORT B. THIS SETS PW REL HIGH
        IN      A,(2)       ;INPUT I/O PORT B TO ACCUMULATOR
        SET     0,A         ;SET BIT 0 TO 1
        OUT     (2),A       ;OUTPUT RESULT
        RET                 ;RETURN FROM SUBROUTINE
;LONG DELAY SUBROUTINE. ASSUME H AND L
;ADDRESS THE FIRST OF TWO DATA BYTES WHICH
;HOLD THE INITIAL DELAY CONSTANT
LDLY    LD      E,(HL)      ;LOAD DELAY CONSTANT INTO D,E
        INC     HL
        LD      D,(HL)
LDLP:   DEC     DE
        LD      A,D         ;EXECUTE PRINTING DELAY
        OR      E
        JR      NZ,LDLP
        RET                 ;RETURN AT END OF LONG DELAY
MS3     DEFW    231         ;PRINTWHEEL RELEASE TIME DELAY CONSTANT
```

The subroutine illustrated above only fires the printhammer if all necessary conditions have been met; a quick exit is executed if any firing condition has not been met. The conditional Return instructions are shaded.

**CONDITIONAL RETURN**

Note that **we have added a subroutine within the subroutine. The long delay instruction sequence has been moved to a subroutine, the first instruction of which is labeled LDLY. This is referred to as a "nested subroutine".**

> **NESTED SUBROUTINES**

One novel feature of subroutine LDLY is that it requires the initial delay constant to be stored in two bytes of memory, the first of which is addressed by the H and L registers when LDLY is called.

> **SUBROUTINE PARAMETER**

**Instructions within subroutine LDLY will actually load the initial delay constant into the D and E registers. The initial delay constant becomes a parameter,** which allows one subroutine to implement a complete spectrum of time delays. Subroutine parameters are a very important feature of subroutine use.

The second time subroutine LDLY is called, instead of loading the required initial constant (231) into the D and E registers, we load an address represented by the symbol MS3 into the H and L registers. The symbol MS3 will become the address of two data bytes, somewhere in memory; within these two data bytes the value 231 must be stored.

## MULTIPLE SUBROUTINE RETURNS

**Subroutine PHFIR is not as useful as it could be. There are four conditional returns from this subroutine, each of which is triggered by a different invalid condition. There is also a subroutine return following valid printhammer firing.**

**How is the calling program to know whether the printhammer was or was not fired after PHFIR was called?** Testing statuses is not very safe, since we cannot be certain what happens to status conditions during execution of the printhammer firing instructions themselves.

**Subroutines which contain a large number of conditional error exits, in addition to a standard return, will often contain logic which returns to a number of different instructions in the calling program. Take the case of subroutine PHFIR. The instruction sequence which calls this subroutine may appear as follows:**

```
           -
           -
           -
RT0     CALL    PHFIR   ;CALL PRINTHAMMER FIRING SUBROUTINE
        JR      RT1     ;RETURN HERE FOR PRINTWHEEL REPOSITIONING
        JR      RT0     ;RETURN HERE FOR HAMMER INTERLOCK LOW
        JR      RT2     ;RETURN HERE FOR ASCII CODE LESS THAN 20H
        JR      RT3     ;RETURN HERE FOR ASCII CODE GREATER THAN 7AH
;INSTRUCTIONS WHICH FOLLOW ARE EXECUTED AFTER VALID
;PRINTHAMMER FIRING
           -
           -
           -
;INSTRUCTIONS WHICH FOLLOW ARE EXECUTED FOR PRINTWHEEL
;REPOSITIONING
RT1
           -
           -
           -
;INSTRUCTIONS WHICH FOLLOW ARE EXECUTED FOR ASCII CODE
;LESS THAN 20H
RT2        -
           -
           -
```

```
;INSTRUCTIONS WHICH FOLLOW ARE EXECUTED FOR ASCII CODE
;GREATER THAN 7AH
RT3     -
        -
        -
```

Now, for this scheme to work, subroutine PHFIR must increment the return address, which is stored in the top two bytes of the stack every time a conditional Return is executed. Subroutine PHFIR is therefore modified as follows:

```
;PRINTHAMMER FIRING SUBROUTINE
PHFIR:  IN      A,(2)       ;INPUT I/O PORT B TO ACCUMULATOR
        BIT     7,A         ;TEST BIT 7 (HAMMER ENABLE)
        RET     Z           ;IF IT IS 0, RETURN
        CALL    INCR        ;INCREMENT RETURN ADDRESS
        BIT     4,A         ;TEST HAMMER INTERLOCK
        RET     Z           ;IF IT IS 0, RETURN
        CALL    INCR        ;INCREMENT RETURN ADDRESS
;FIRE PRINTHAMMER
        RES     2,A         ;SET HAMMER PULSE LOW:
        OUT     (2),A       ;OUTPUT 0 TO BIT 2 OF I/O PORT B
        IN      A,(0)       ;INPUT ASCII CHARACTER TO ACCUMULATOR
        RES     7,A         ;RESET HIGH ORDER BIT
;COMPARE ASCII CODE WITH LOWEST LEGAL VALUE
        CP      20H
        RET     M           ;IF CODE IS 1FH OR LESS BYPASS HAMMER FIRING
        CALL    INCR        ;INCREMENT RETURN ADDRESS
;COMPARE ASCII CODE WITH HIGHEST LEGAL VALUE
        CP      7BH
        RET     P           ;IF CODE IS 7BH OR GREATER BYPASS HAMMER
                            ;FIRING
        CALL    INCR        ;INCREMENT RETURN ADDRESS
;ASCII CODE IS VALID
        LD      H,03H       ;LOAD INDEX TABLE ADDRESS, HIGH ORDER BYTE
        LD      L,A         ;MOVE LOW ORDER BYTE OF ADDRESS TO L
        LD      L,(HL)      ;LOAD LOW ORDER BYTE OF DELAY TABLE
                            ;ADDRESS
        CALL    LDLY
        IN      A,(2)       ;AT END OF DELAY OUTPUT 1 TO BIT 2 OF I/O
        SET     2,A         ;PORT B. THIS SETS HAMMER PULSE HIGH
        OUT     (2),A
;EXECUTE A 3 MILLISECOND PRINTWHEEL RELEASE TIME DELAY
        LD      HL,MS3
        CALL    LDLY
;OUTPUT 1 TO BIT 0 OF I/O PORT B. THIS SETS PW REL HIGH
        IN      A,(2)       ;INPUT I/O PORT B TO ACCUMULATOR
        SET     0,A         ;SET BIT 0 TO 1
        OUT     (2),A       ;OUTPUT RESULT
        RET                 ;RETURN FROM SUBROUTINE
;LONG DELAY SUBROUTINE. ASSUME H AND L
;ADDRESS THE FIRST OF TWO DATA BYTES WHICH HOLD THE
;INITIAL DELAY CONSTANT
LDLY    LD      E,(HL)      ;LOAD DELAY CONSTANT INTO D,E
        INC     HL
        LD      D,(HL)
```

```
LDLP:    DEC    DE           ;EXECUTE PRINTING DELAY
         LD     A,D
         OR     E
         JR     NZ,LDLP
         RET                 ;RETURN AT END OF LONG DELAY
MS3      DEFW   231          ;PRINTWHEEL RELEASE TIME DELAY CONSTANT
;SUBROUTINE TO INCREMENT THE RETURN ADDRESS
;OF THE CALLING SUBROUTINE
INCR     INC    SP           ;INCREMENT STACK POINTER TWICE
         INC    SP           ;TO ACCESS PHFIR RETURN ADDRESS
         EX     (SP),HL      ;EXCHANGE HL WITH PHFIR RETURN ADDRESS
         INC    HL           ;ADD 2 TO RETURN ADDRESS
         INC    HL
         EX     (SP),HL      ;RESTORE RETURN ADDRESS
         DEC    SP           ;DECREMENT STACK POINTER TWICE
         DEC    SP
         RET                 ;RETURN
```

**Subroutine INCR is interesting; it shows how the stack may be manipulated. Let us take a look at what happens.**

**STACK MANIPULATION**

As soon as subroutine INCR is entered, the Stack Pointer contents are increased by two. This has the effect of addressing the PHFIR return address rather than the INCR return address:

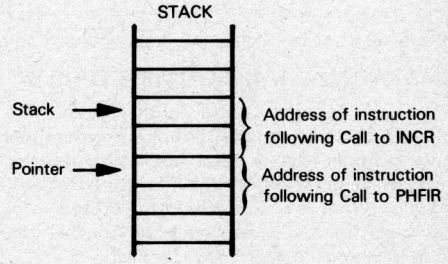

The EX (SP),HL instruction simply saves the contents of the H and L registers at what is now the top of the stack, while moving what was at the top of the stack to the H and L registers:

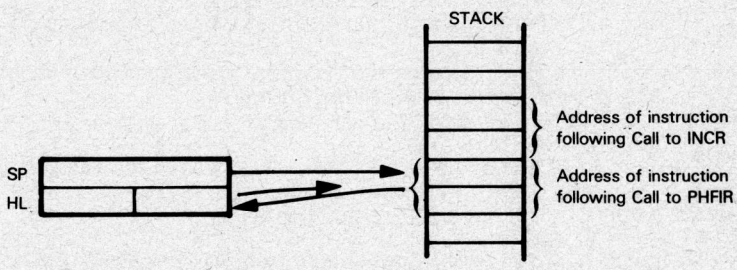

The next two instructions add 2 to the contents of the H and L registers, which now hold the PHFIR return address. We add 2 to the return address because, if you look at the calling sequence, a series of Jump (JR) instructions follow. Each JR instruction occupies two bytes, which means that each time we bypass a Conditional Return we must increment the return address by 2:

| | | |
|---|---|---|
| CALL | PHFIR | CD |
| | | xx |
| | | xx |
| JR | RT1 | 18 |
| | | yy |
| JR | RT0 | 18 |
| | | zz |
| JR | RT2 | 18 |
| • | | xx |
| • | | pp |
| • | | : |

The next EX (SP),HL simply restores the incremented PHFIR return address to the top of the stack.

Finally we must restore the Stack Pointer to its original contents, so that the INCR Return instruction will fetch the correct return address.

## CONDITIONAL SUBROUTINE CALLS

We are now going to create another subroutine which fires the printhammer but makes no tests to ensure that the printhammer should be fired. This subroutine simply assumes that a valid ASCII character is in the Accumulator and that the printhammer must be fired. All logic to determine whether printhammer firing is valid is external to the printhammer firing subroutine; therefore, this subroutine is called conditionally — so long as all printhammer firing conditions have been met. This is how our program now looks:

```
;TEST PRINTHAMMER FIRING CONDITIONS
PHFIR:  IN      A,(2)       ;INPUT I/O PORT B TO ACCUMULATOR
        BIT     7,A         ;TEST BIT 7 (HAMMER ENABLE)
        JP      Z,PWRDY     ;IF IT IS 0, BYPASS PRINTHAMMER FIRING
        BIT     4,A         ;TEST HAMMER INTERLOCK
        JR      Z,PHFIR     ;WAIT FOR NONZERO VALUE BEFORE FIRING
;INPUT CHARACTER TO BE PRINTED
        IN      A,(0)       ;INPUT ASCII CHARACTER TO ACCUMULATOR
        RES     7,A         ;RESET HIGH ORDER BIT
;COMPARE ASCII CODE WITH LOWEST LEGAL VALUE
        CP      20H
        JP      M,PWRDY     ;IF CODE IS 1FH OR LESS BYPASS HAMMER
                            ;FIRING
;COMPARE ASCII CODE WITH HIGHEST LEGAL VALUE
        CP      7BH
        CALL    M,FIRE      ;IF CODE VALID, CALL FIRING SUBROUTINE
;EXECUTE A 2 MILLISECOND PRINTWHEEL READY DELAY
PWRDY   LD      A,0FAH      ;LOAD TIME DELAY CONSTANT
```

**Notice that the Conditional Return instruction reflects OR programming logic, whereas the Conditional Call instruction reflects AND logic.** Thus, subroutine PHFIR includes a number of Conditional Return instructions, each of which will execute providing any one invalid condition is encountered. Subroutine FIRE, on the other hand, is called conditionally only when the last of the necessary valid conditions has been tested.

Subroutine FIRE is not shown in detail, since writing it out would add little to the understanding of the Conditional Call instruction. With reference to Figure 4-6, subroutine FIRE would consist of instructions to:

>Set the hammer pulse signal low
>Execute the hammer firing pulse delay
>Set the printhammer firing pulse high
>Execute the 3 millisecond printwheel release time delay
>Output PW REL high

## MACROS

When talking about subroutines, we glossed over one consideration — you, the programmer. Subroutines have an additional value, in that if they can reduce the number of source program instructions then they will also reduce the amount of time you spend writing the source program, since program writing time will be directly proportional to program length.

Let us take another look at the two-millisecond time delay subroutine. Although in subroutine form the program required more object code bytes, it did not require more instructions:

```
        Old Program                              New Program
        LD    A,0FAH                             CALL   D2MS
PWSET   DEC   A                                  -
        JR    NZ,PWSET                           CALL   D2MS
        -                                        -
        -                                        -
PWRDY   LD    A,0FAH            D2MS    LD      A,0FAH
RDYDLY  DEC   A                 LOPD    DEC     A
        JR    NZ,RDYDLY                 JR      NZ,LOPD
                                        RET

        6 instructions                   6 instructions
          (10 bytes)                    (12 bytes, excluding
                                         stack and initialization
                                              instructions)
```

**Subroutines can decrease the length of your source program, while increasing the length of your object program and the program's execution time.**

**Macros decrease the length of your source program, but have absolutely no effect on your object program.**

### WHAT IS A MACRO?

**A macro is a form of programming "short hand"; it allows you to define an instruction sequence with a single mnemonic.**   | MACRO DEFINITION

Consider the two-millisecond time delay instruction sequence: we can define it as a macro, labeled D2MS, as follows:

```
        D2MS    MACRO
                LD      A,0FAH
        LOPD    DEC     A
                JR      NZ,LOPD
                ENDM
```

The two shaded instructions above are, in reality, assembler directives: they bracket a sequence of instructions which henceforth can be identified as a group, using the label of the MACRO assembler directive.

**MACRO ASSEMBLER DIRECTIVES**

This is how we would use the two-millisecond time delay in our print cycle program.

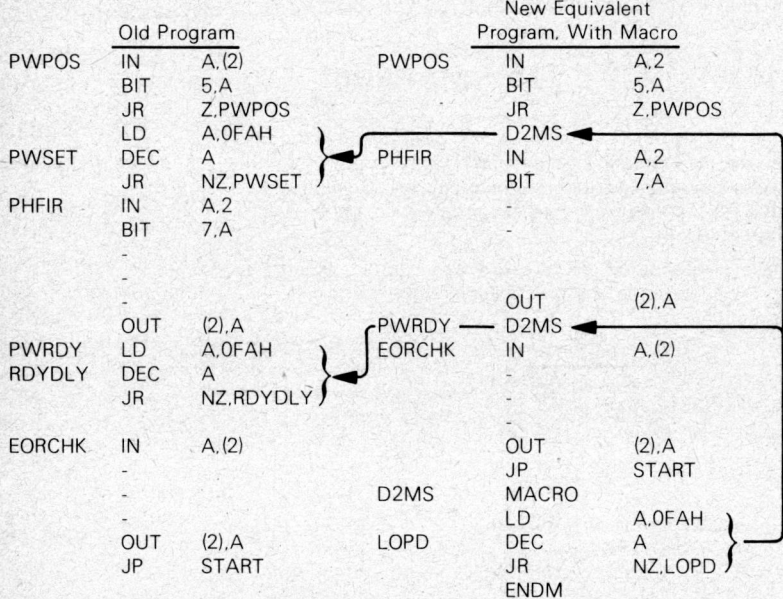

When the Assembler encounters the symbol D2MS in the mnemonic field, what it does is replace this symbol with the instructions bracketed by directives MACRO and ENDM. The Assembler knows which macro to use in the event that your program has more than one macro, since the symbol in the mnemonic field must be identical to the label of a MACRO directive.

Notice that the Assembler can also do a certain amount of housekeeping associated with the use of macros. The "Old Program" illustrated above has labels PWSET and RDYDLY for the two DEC instructions. The "New Program" has a single label, LOPD, within the macro. The Assembler is smart enough to know that a label appearing within a macro definition must become a series of separate labels when the macro subsequently is inserted a number of times into the source program.

To summarize, you simply take a sequence of repeated instructions, bracket them with MACRO and ENDM directives, then give the macro directive a unique label. Now use the MACRO's label as though it were an instruction mnemonic. **The macro definition must appear once and only once, somewhere in the source program.** It is a good idea to collect all of your macros and insert them at the beginning or at the end of the entire source program.

> **MACRO DEFINITION LOCATION IN A SOURCE PROGRAM**

## MACROS WITH PARAMETERS

**Instructions within a macro can have variable operands;** for example, we can create a variable time delay macro as follows:

```
DVMS   MACRO   TIME
       LD      A,TIME
LOPD   DEC     A
       JR      NZ,LOPD
       ENDM
```

Symbols appearing in the MACRO directive's operand field are assumed by the Assembler to be "dummy" symbols; the macro reference in the body of the source program must include an equivalent operand field. The Assembler will equate the macro reference's operand field to the MACRO directive's operand field, and make substitutions accordingly.

**This is how the substitution works:**

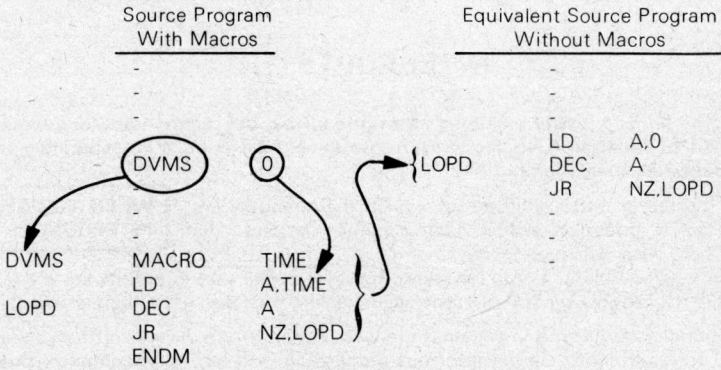

Here is another example; the macro reference:

```
       DVMS    80H
```

is equivalent to:

```
       LD      A,80H
LOPD   DEC     A
       JR      NZ,LOPD
```

Depending on whose Assembler you are using, you can play interesting games with the macro parameter list; in theory (but not always in practice), there are no restrictions on the length or nature of the macro parameter list. Suppose you want to vary the register used in the time delay instruction sequence; some assemblers will let you do so as follows:

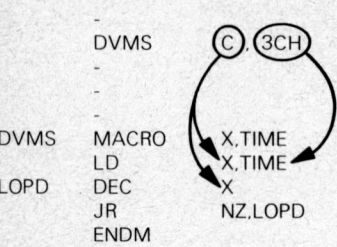

```
              DVMS    C,3CH

DVMS   MACRO   X,TIME
       LD      X,TIME
LOPD   DEC     X
       JR      NZ,LOPD
       ENDM
```

The Assembler will substitute:

```
              DVMS    C,3CH
```

with:

```
       LD      C,3CH
LOPD   DEC     C
       JR      NZ,LOPD
```

**You will have to read the Assembler manual that accompanies your development system in order to know the exact macro features available to you.**

# INTERRUPTS

**It would be hard to justify including interrupts within the microcomputer system developed in Chapter 4. In fact, interrupts should be used quite sparingly in microcomputer applications.**

We will not enter into a long discussion on the strengths and weaknesses of interrupts within microcomputer systems; that subject has been adequately covered in <u>An Introduction to Microcomputers: Volume I</u>. To summarize, however, recall that interrupts are a valid tool within microcomputer systems only when dealing with fast, asynchronous events.

**WHEN TO USE INTERRUPTS**

**Now, having issued a warning against the indiscriminate use of interrupts, we will proceed to incorporate simple interrupt processing into our microcomputer program in the interests of demonstrating how it is done.**

## INTERRUPT HARDWARE CONSIDERATIONS

**For an interrupt to be processed within a Z80 microcomputer system, an interrupt request signal must be input low to the CPU at a time when interrupts have been enabled.**

**Interrupts are enabled and disabled by executing EI and DI instructions, respectively.** Any interrupt request will simply be ignored by the CPU while interrupts have been disabled.

**INTERRUPT ENABLE**

Note that there is an exception to our last statement: the Z80 has a non-maskable interrupt input that is always enabled. This interrupt request line is typically used for special situations such as power-fail conditions, and is not relevant to our discussion here.

If an interrupt request is received while interrupts have been enabled, then upon completing execution of the current instruction, the CPU will output an interrupt acknowledge signal (IORQ during M1 time).

**INTERRUPT ACKNOWLEDGE**

The response of the external logic to this interrupt acknowledge is dependent on the mode in which the Z80 CPU is being operated. There are three possible modes: 0, 1 and 2.

If the CPU is operating in Mode 0, external logic is expected to input an 8-bit interrupt vector which is going to be interpreted as the instruction code to be executed next. Usually one of the eight possible Restart instruction codes will be fetched. These instructions are equivalent to single byte subroutine calls; they cause the contents of the Program Counter to be pushed onto the stack, after which program execution continues at a low memory address which may be computed as follows:

**Z80 CPU INTERRUPT MODE 0**

```
RST n instruction code:  1 1 1 x x x 1 1
                                 0 0 0     n=0
                                 0 0 1     n=1
                                 0 1 0     n=2
                                 0 1 1     n=3
                                 1 0 0     n=4
                                 1 0 1     n=5
                                 1 1 0     n=6
                                 1 1 1     n=7

New Program
Counter Contents:  0 0 0 0 0 0 0 0 0 0 x x x 0 0 0
```

Z80 interrupt response logic in **Mode 1** automatically assumes that the first instruction executed following the interrupt response will be a Restart, branching to memory location $0056_{16}$. If the Z80 is in Mode 1, no interrupt vector is needed.

**Z80 CPU INTERRUPT MODE 1**

When you operate the Z80 in **Mode 2, you must create a table of 16-bit interrupt address vectors,** which can reside anywhere in addressable memory. These 16-bit addresses identify the first executable instruction of interrupt service routines. When an interrupt is acknowledged by the CPU in Mode 2, **the acknowledged external logic must place an interrupt response vector on the Data Bus. The Z80 CPU will combine the I register contents with the interrupt acknowledge vector to form a 16-bit address, which accesses the interrupt address vector table.** Since 16-bit addresses must lie at even memory address boundaries, only seven of the eight bits pro-

**Z80 CPU INTERRUPT MODE 2**

vided by the acknowledged external logic will be used to create the table address; the low order bit will be set to 0. Thus, the table of 16-bit interrupt address vectors will be accessed as follows:

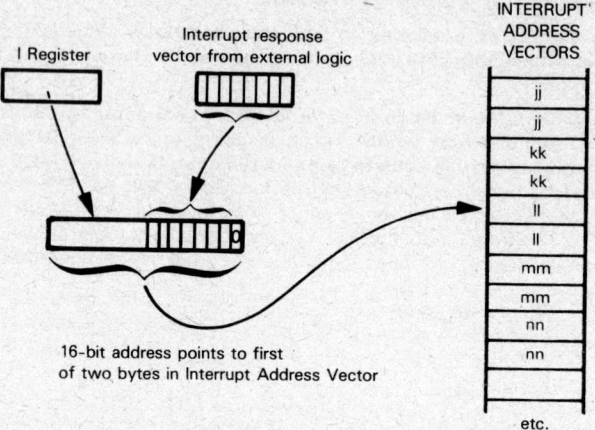

The Z80 CPU will execute a Call to the memory location obtained from the interrupt address vector table.

**Let us clarify this logic with a simple example.** Suppose that you have 64 possible external interrupts; each interrupt has its own interrupt service routine, therefore 64 starting addresses will be stored in 128 bytes of memory. Let us arbitrarily assume that these 128 bytes are stored in a table with memory addresses $0F00_{16}$ through $0F7F_{16}$. Now, in order to use Mode 2, you must initially load the value of $0F_{16}$ into the Z80 I register. Subsequently, an external interrupt request is acknowledged and the acknowledged external logic returns the vector $2E_{16}$ on the Data Bus; this is what will happen:

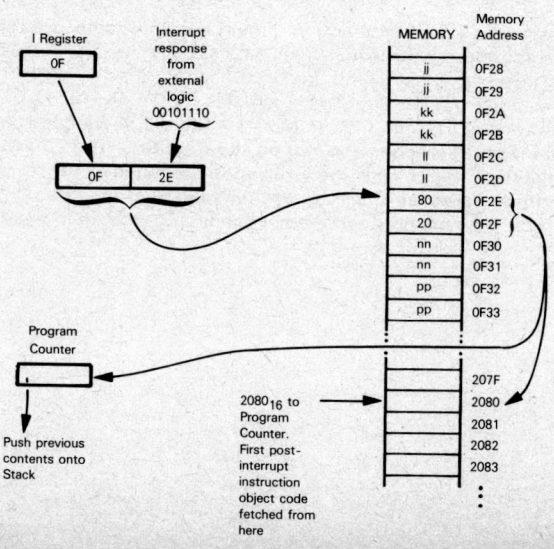

5-26

From our descriptions of the Z80 CPU interrupt modes, it is obvious that **Mode 1 is the most straightforward:** it merely requires that the first instruction of our interrupt program begin at memory location $0056_{16}$. No external logic is required to generate a vector in response to the CPU's interrupt acknowledgment. **However, some external logic must still be provided to sense the conditions required to generate an interrupt, to actually generate the interrupt request signal, and to reset the interrupt request signal once the interrupt has been acknowledged. All of these functions can be performed by the PIO that is already included in our system shown in Figure 4-2.** The only hardware change needed is to connect the $\overline{INT}$ signal from the PIO to the CPU, as shown in Figure 5-1.

**Now, having pointed out the simplicity of the CPU's Mode 1 operation — we will proceed to disregard that and operate the CPU in interrupt Mode 2: we do this because the PIO has been specifically designed to operate with the CPU using the Mode 2 interrupt response.** As we shall see, this mode of operation turns out to be more straightforward than it would first appear — this is so because of the logic provided by the PIO.

**Let us now examine how the PIO responds to the Mode 2 interrupt acknowledge from the CPU.**

Each port (A and B) of the PIO has an independent interrupt vector that can be loaded with the desired vector value. The vector is loaded by writing a control word to the control register of the port in the following format:

| **Z80-PIO** |
| **INTERRUPT** |
| **ACKNOWLEDGE** |
| **RESPONSE** |

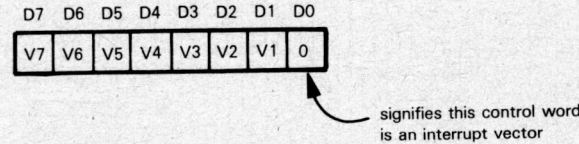

signifies this control word is an interrupt vector

D0 is used as a flag bit which, when low, causes V7 through V1 to be loaded into the Vector register. At interrupt acknowledge time, the vector of the interrupting port will be input to the CPU exactly in the format shown above. For example, if you refer back to our discussion of the Z80-CPU Mode 2 interrupt operation, we had external logic provide an interrupt vector of $2E_{16}$. The binary format for this vector which would be loaded into the PIO register is:

In summary, this is what happens when external logic (the PIO) requests an interrupt:

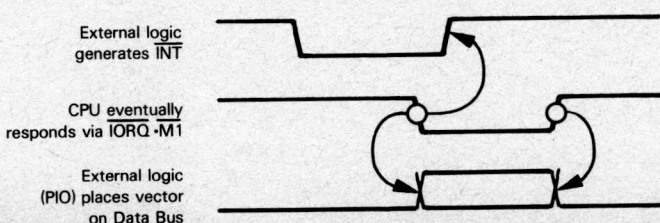

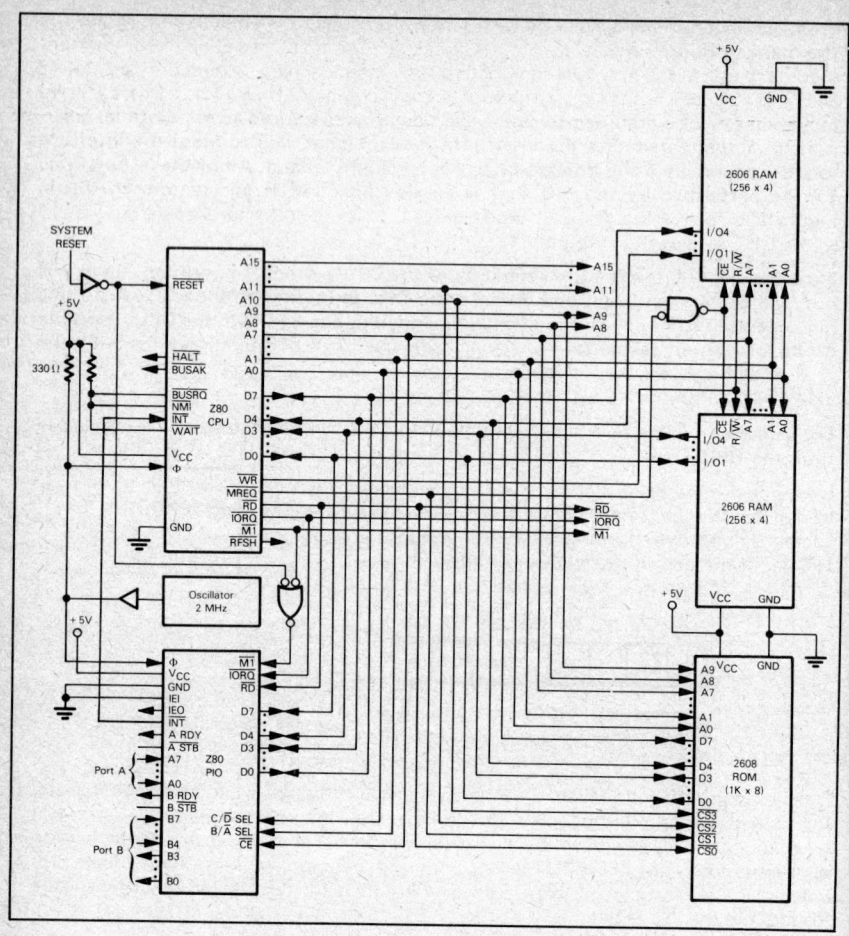

Figure 5-1. Z80 Microcomputer Configuration Using a PIO to Generate an Interrupt

You, as a logic designer or programmer, do not need to concern yourself with Data Bus timing. The $\overline{IORQ} \cdot \overline{M1}$ combination is an interrupt acknowledgment signal and will also correctly strobe the interrupt vector into the CPU. As a programmer, of course, you must concern yourself with the steps required to place the CPU and PIO in the proper interrupt and operating modes and to load the PIO with the desired vector for the interrupt service routine. As the system designer, you must also concern yourself with the logic required to initiate the interrupt request. We will now examine this point — after this has been defined, we will summarize all of the programming considerations resulting from this use of interrupts.

In order to determine what will initiate the interrupt, we must decide first of all, **how we are going to use the interrupt.**

We could assume that the microcomputer system is being used to do more than implement print cycle logic. **Suppose there is a great deal of routine housekeeping logic required by the printer interface, with the result that the entire print cycle can be looked upon as an intermittent asynchronous event. Now, instead of having our program execute an "in between print cycles" instruction loop, we will assume that some other program is being continuously executed in between print cycles. Execution of the print cycle program is triggered by the VELOCITY DECODE signal. This is the instruction execution pattern which results:**

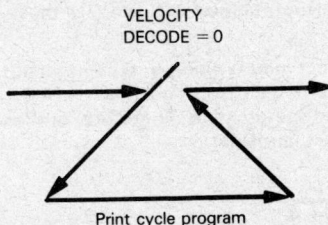

Referring back to pin assignments in Chapter 4, you will see that the VELOCITY DECODE signal is an input to bit 5 of Port B in the Z80-PIO. **Because of the design of the PIO, we can use the VELOCITY DECODE signal directly, without any additional logic beyond the PIO, to initiate an interrupt request and thus trigger the print cycle program.**

The PIO has an interrupt control word for each port (A and B) that determines the conditions under which an interrupt request will be sent to the CPU. In our system, we would specify the desired interrupt conditions by writing a word to the control register of PIO Port B. The interrupt control word has the following format:

**INITIATING INTERRUPTS VIA THE PIO**

**PIO INTERRUPT CONTROL WORD**

| D7 | D6 | D5 | D4 | D3 | D2 | D1 | D0 |
|---|---|---|---|---|---|---|---|
| Enable Interrupt | AND/ OR | High/ Low | Mask follows | 0 | 1 | 1 | 1 |

signifies interrupt control word

Bit D7 is used to enable the port to generate an interrupt: if bit 7 = 1, interrupts can be generated. Bit D6 defines the logical operation to be performed in determining whether or not an interrupt request should be generated. If D6 = 1, an AND function is specified; all selected bits of the port must go high (or low, depending on bit D5) before an interrupt request will be generated. If D6 = 0, then an OR function is specified and an interrupt will be generated if any specified bit goes to the active (high or low) state.

Bit D5 defines the active polarity of the port Data Bus line to be monitored. If bit D5 = 1, the port data lines are monitored for a high state; if bit D5 = 0, the data lines are monitored for a low state.

If bit D4 = 0, then all bits will be monitored according to the rules defined by bits D6 and D5 of the interrupt control word.

If D4 = 1, then the next control word sent to the PIO must define a mask as follows:

| D7 | D6 | D5 | D4 | D3 | D2 | D1 | D0 |
|----|----|----|----|----|----|----|----|
| MB7 | MB6 | MB5 | MB4 | MB3 | MB2 | MB1 | MB0 |

Only those port lines whose mask bit is zero will be monitored for generating an interrupt.

**Now, having described all of the possible situations and combinations where the PIO could generate an interrupt request, let us relate these capabilities to our particular example.**

**Recall that we are only concerned with bit 5, which is input to Port B of the PIO as VELOCITY DECODE.** Now, when the signal goes low, we want to generate an interrupt request to trigger the print cycle program. **Therefore, our interrupt control word to the PIO (Port B) would look like this:**

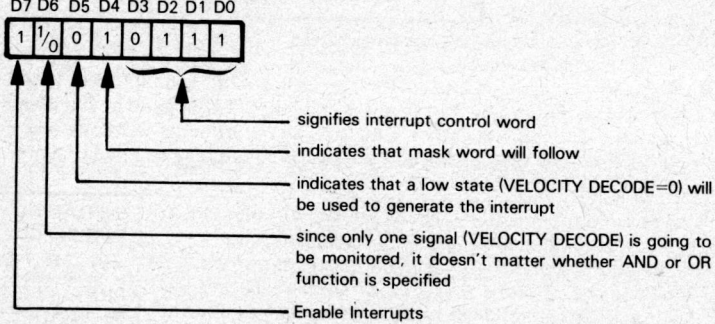

And the mask word that follows specifies that only bit D5 (VELOCITY DECODE) be monitored. The format of the mask word would be:

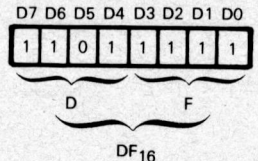

**Now, the only step remaining is to set up our interrupt vector.** If you refer back to our discussion of the CPU Mode 2 interrupt operations, **you will see this simply requires that the CPU I register and PIO Interrupt Vector register each be loaded with values that will be combined to produce a 16-bit address. We must also load the location specified by that address and the adjacent memory location with the address of the first instruction of the print program.** Once again let us illustrate the Mode 2 interrupt operation using arbitrary addresses.

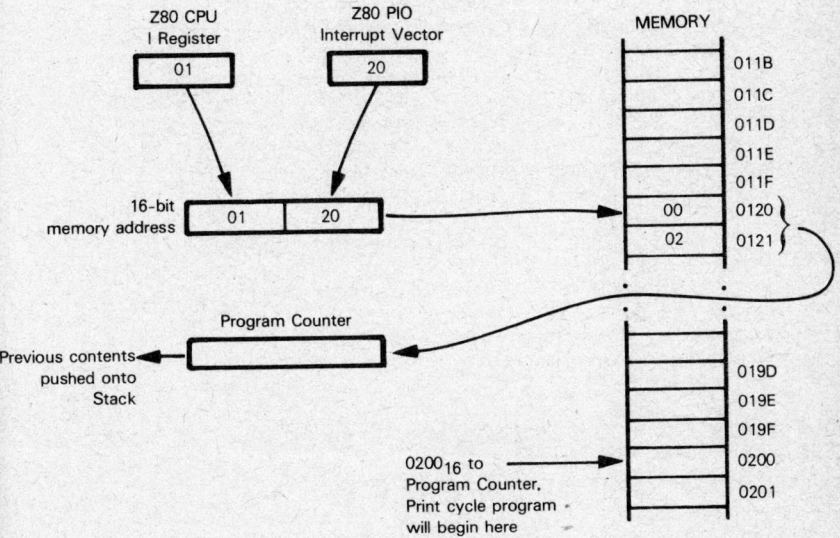

**Note that the actual beginning location (or origin) specified for the print cycle program is unimportant.** We do not know what other programs are being executed within the microcomputer system, or where these other programs may reside in program memory; therefore we cannot assign memory space to the print cycle program at this time. When you actually implement the entire microcomputer system you must carefully map out exactly where in memory every program resides, but for the purposes of our current illustration this is a completely unimportant consideration.

**INTERRUPT PROGRAM ORIGIN**

**Let us now summarize the changes we must make to our program if we are to use an interrupt to initiate our print cycle program.** As we shall see, the changes are rather minimal and mostly consist of adding instructions to the initialization portion of our program.

```
            ORG     0
;FIRST OUTPUT CONTROL CODES TO I/O PORT A CONTROL REGISTER
            LD      A,0FFH      ;SET MODE 3
            OUT     (1),A
            OUT     (1),A       ;ALL LINES INPUT
;NEXT OUTPUT CONTROL CODES TO I/O PORT B CONTROL REGISTER
            OUT     (3),A       ;SET MODE 3
            LD      A,0F0H      ;SET PINS 0 THROUGH 3 TO OUTPUT AND
            OUT     (3),A       ;PINS 4 THROUGH 7 TO INPUT
            LD      A,097H      ;LOAD INTERRUPT CONTROL WORD
            OUT     (3),A
            LD      A,0DFH      ;SET INTERRUPT MASK WORD
            OUT     (3),A
            LD      A,020H      ;LOAD INTERRUPT VECTOR (20)
            OUT     (3),A       ;INTO PORT B VECTOR REGISTER
;THEN SET UP Z80-CPU FOR INTERRUPT MODE 2
            IM2                 ;SET INTERRUPT MODE 2
            LD      A,010H      ;LOAD THE CPU I REGISTER
            LD      I,A         ;WITH INTERRUPT VECTOR (01)
            LD      HL,0002H    ;LOAD INTERRUPT VECTOR LOCATION (0120) WITH
            LD      (0120H),HL  ;STARTING ADDRESS (0200) OF PRINT CYCLE
                                ;PROGRAM
;SET HAMMER PULSE, PW READY AND PW REL HIGH
;SET START RIBBON MOTION LOW
            LD      A,7
            OUT     (2),A
;ALL INITIAL CONDITIONS HAVE NOW BEEN ESTABLISHED.
;INTERRUPTS CAN NOW BE ENABLED.
            EI
             -
             -
             -
            ORG     0200H
;ORIGIN PRINT CYCLE PROGRAM INTERRUPT SERVICE ROUTINE
;AT 0200H, SINCE THIS IS THE EXECUTION ADDRESS STORED
;AT INTERRUPT VECTOR LOCATION 0120.
;PRINT CYCLE PROGRAM
;INITIALIZE PRINT CYCLE. OUTPUT 0 TO BITS 0 AND 1 OF I/O PORT B
;OUTPUT 1 TO BITS 2 AND 3 OF I/O PORT B
START       LD      A,0CH       ;LOAD MASK INTO ACCUMULATOR
            OUT     (2),A       ;OUTPUT TO I/O PORT B
             -
             -
             -
;AT END OF PRINT CYCLE SET BIT 1 OF I/O PORT B TO 1
;THIS SETS CH READY HIGH
            SET     1,A         ;SET BIT 1 OF PORT B (IN ACCUMULATOR)
            OUT     (2),A       ;OUTPUT RESULT
            RET
```

**The instructions that we have added to the program illustrated in Figure 4-6 are shaded, and consist primarily of steps necessary to set up the CPU and PIO to operate in the desired interrupt mode.** Once all of the required initial conditions have been established, the EI instruction is executed, enabling the CPU to respond to interrupt requests.

The print cycle program now begins at memory location $0200_{16}$ and will be initiated as a result of an interrupt triggered by VELOCITY DECODE = 0. Notice that the "in between print cycles" instructions from the beginning of Figure 4-6 have been removed; START now identifies the first instruction of the print cycle itself. The final JP START instruction is replaced by a simple RETURN instruction, since the entire print cycle program was, in effect, called as a subroutine.

**The method we have just described for processing an interrupt is fairly simple: there is only one problem with it — the program will not work. We have shown a background program being interrupted in order to execute the print cycle routine;**

**SAVING REGISTERS AND STATUS**

**but when does the background program get interrupted?** Remember, the program which is interrupted is sharing the same CPU and the same registers with the print cycle program. We have to assume that the interrupted program has useful information stored in the registers, and perhaps the status flags have meaning which must be preserved. Given the interrupt service program illustrated thus far, when we return from the print cycle program to the interrupted program, we are giving the interrupted program whatever arbitrary register contents the print cycle program finishes up with. That will never do. **We must therefore bracket the print cycle execution program with instructions that save the contents of registers and status — before modifying a single register or status; at the end of the program, original registers and status contents must be restored.** Typically, the contents of registers and status are saved by pushing them onto the stack, and restored at the end of the program by popping them off the stack. The sequence of instructions would be as follows:

```
        ORG     0200H
;ORIGIN PRINT CYCLE PROGRAM INTERRUPT SERVICE ROUTINE
;AT 0200H, SINCE THIS IS THE EXECUTION ADDRESS STORED
;AT INTERRUPT VECTOR LOCATION 0120.
START   PUSH    AF              ;SAVE ACCUMULATOR AND FLAGS
        PUSH    BC              ;SAVE B AND C REGISTERS
        PUSH    DE              ;SAVE D AND E REGISTERS
        PUSH    HL              ;SAVE H AND L REGISTERS
;INITIALIZE PRINT CYCLE. OUTPUT 0 TO BITS 0 AND 1 OF I/O PORT B
;OUTPUT 1 TO BITS 2 AND 3 OF I/O PORT B
        LD      A,0CH           ;LOAD MASK INTO ACCUMULATOR
        OUT     (2),A           ;OUTPUT TO I/O PORT B
         -
         -
         -
;AT END OF PRINT CYCLE SET BIT 1 OF I/O PORT B TO 1
;THIS SETS CH READY HIGH
        SET     1,A             ;SET BIT 1 OF PORT B (IN ACCUMULATOR)
        OUT     (2),A           ;OUTPUT RESULT
        POP     HL              ;RESTORE H AND L REGISTERS
        POP     DE              ;RESTORE D AND E REGISTERS
        POP     BC              ;RESTORE B AND C REGISTERS
        POP     AF              ;RESTORE ACCUMULATOR AND FLAGS
        RET
```

The entire save/restore sequence adds a total of eight instructions to our print program. So long as you remember to pop registers and status contents in the reverse order from which you pushed them, you will have no problems.

As we stated at the beginning of this discussion, the push/pop sequence is the typical method used to save/restore status and register contents. The Z80-CPU, however, provides one quite atypical architectural feature that can be used to simplify this save/restore process. You will recall that the Z80-CPU provides two matched sets of general purpose registers as shown below.

**USING Z80 CPU AUXILIARY REGISTERS**

| | | | | | |
|---|---|---|---|---|---|
| | F | Program Status Words | | | F' |
| | A | Primary Accumulators | | | A' |
| B | C | Secondary Accumulators/Data Counters | | B' | C' |
| D | E | Secondary Accumulators/Data Counters | | D' | E' |
| H | L | Secondary Accumulators/Data Counters | | H' | L' |
| SP | | Stack Pointer | | | |
| PC | | Program Counter | | | |
| IX | | Index Register X | | | |
| IY | | Index Register Y | | | |
| | IV | Interrupt Vector | | | |
| | R | Memory Refresh Counter | | | |

**Now, the Z80 instruction set includes two instructions that allow the contents of these duplicate sets of registers to be exchanged.** The instruction EX AF,AF' exchanges the contents of the registers A and F with the contents of A' and F'. The instruction EXX exchanges the contents of register pairs BC, DE, and HL with the contents of register pairs B'C', D'E', and H'L' respectively. **Therefore, our sequence of four PUSH instructions to save registers and four POP instructions to restore registers can be replaced by using the EX AF,AF' and EXX instructions as follows:**

```
        Old Program                    New Program
START     PUSH    AF         START     EX AF,AF'
          PUSH    BC                   EXX
          PUSH    DE                   -
          PUSH    HL                   -
            -                          EX AF,AF'
            -                          EXX
            -                          RET
            -
          POP     HL
          POP     DE
          POP     BC
          POP     AF
          RET
```

**Using the Exchange instructions instead of the PUSH/POP instructions has saved us a total of four instructions, and also results in a much faster response to an interrupt** since execution of the two Exchange instructions requires only one fifth of the time that is needed to execute the four PUSH instructions. **Another advantage of the Exchange instructions is that no read/write memory has been used by this sequence, while the PUSH/POP sequence uses eight bytes of stack memory.**

Of course, the Exchange instructions can only be used for one level of interrupts; if multiple, nested interrupts must be serviced, then the stack must be used to save register contents. Now let us see what other demands multiple interrupts would make upon our system.

# MULTIPLE INTERRUPTS

**What if your microcomputer system is connected to more than one external logic device that is capable of requesting interrupts? For example, a single Z80 microcomputer system might be driving a number of printers.** Without going into the economics of microcomputer multiple interrupt configurations, let us examine the ways in which multiple interrupts can be handled.

**The one thing that changes when we go from single interrupts to multiple interrupts is the fact that** the interrupt service routine is no longer unique. **There must be a different interrupt service routine for every external device capable of requesting an interrupt.** In turn that means that, following an interrupt acknowledge, we must have some means of knowing which interrupt service routine is to execute. Also, **if more than one device simultaneously requests interrupt service, which are we going to acknowledge — and in what order?** These are problems of interrupt vectoring and priority arbitration, subjects which have been covered in some detail in <u>An Introduction to Microcomputers: Volume I — Basic Concepts</u>. We will not repeat discussion of these basic concepts in this book; rather, we will look at practical ways in which multiple interrupts can be serviced within a Z80 microcomputer system. We will see that the design of the Z80-CPU and Z80-PIO makes servicing of multiple interrupts quite straightforward.

**There are innumerable ways in which multiple interrupts could be implemented in a Z80-type microcomputer system,** and it is certainly beyond the scope of this book to explore them all. Therefore **we will limit our discussion to the most obvious and straightforward method — the method that is supported by the design of the Z80-CPU and Z80-PIO** (as well as other Z80 parts that we have not needed to describe in this book).

As we have just stated, **the two main problems that must be solved in systems utilizing multiple interrupts are: 1) interrupt vectoring and 2) priority arbitration.**

Interrupt vectoring has already been described earlier in this chapter when we discussed **the Z80-CPU Mode 2 interrupt operation.** This mode of operation **allows vectoring for a nearly unlimited number of interrupting devices.** The only requirement placed on the interrupting device is that it respond to the CPU's interrupt acknowledgment by placing a 7-bit vector on the system Data Bus. This is performed automatically by the Z80-PIO, but could also be performed quite easily by logic of your own design.

**Interrupt priority arbitration is also provided by the Z80-PIO, and a discussion of how this device performs the arbitration will also serve as an example of the general theory involved. The Z80-PIO uses a typical daisy chain scheme to set interrupt priorities. Interrupt Enable In (IEI) and Interrupt Enable Out (IEO) are standard daisy chain interrupt priority signals.** When more than one PIO is present in a system, the highest priority PIO (i.e., the one electrically closest to the CPU) will have IEI tied to +5V and will connect its IEO to the IEI for the next highest priority PIO in the daisy chain:

**INTERRUPT PRIORITY ARBITRATION**

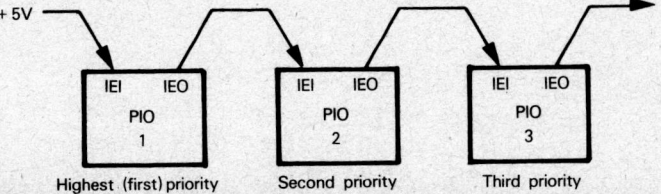

**Daisy chaining has been described in good detail in Volume I. If you are unsure of daisy chain priority networks after reading this paragraph, refer back to Volume I for clarification.** When more than one device is requesting an interrupt, an acknowledge ripples down the daisy chain until trapped by the interrupt requesting device electrically closest to the CPU. As soon as the interrupt acknowledge process has ceased, an interrupt service routine is executed for the acknowledged interrupt; acknowledged external logic will now remove its interrupt request. In most microcomputer systems, unless the CPU disables further interrupts, a lower priority device can immediately interrupt the service routine of a higher priority device. With the Z80 system, that is not the case. A device which has its interrupt request acknowledged continues to suppress interrupt requests from all lower priority devices in a daisy chain, until the second object code byte for an RETI or RETN instruction is detected on the Data Bus. The acknowledged device responds to an RETI or RETN instruction's object code by re-enabling interrupts for devices with lower priority in the daisy chain.

Providing a Z80 microcomputer system has been designed to make correct use of the RETI or RETN instruction, interrupt priority arbitration logic will allow an interrupt service routine to be interrupted only by a higher priority interrupt request.

Here is an illustration of the Z80 interrupt priority arbitration scheme:

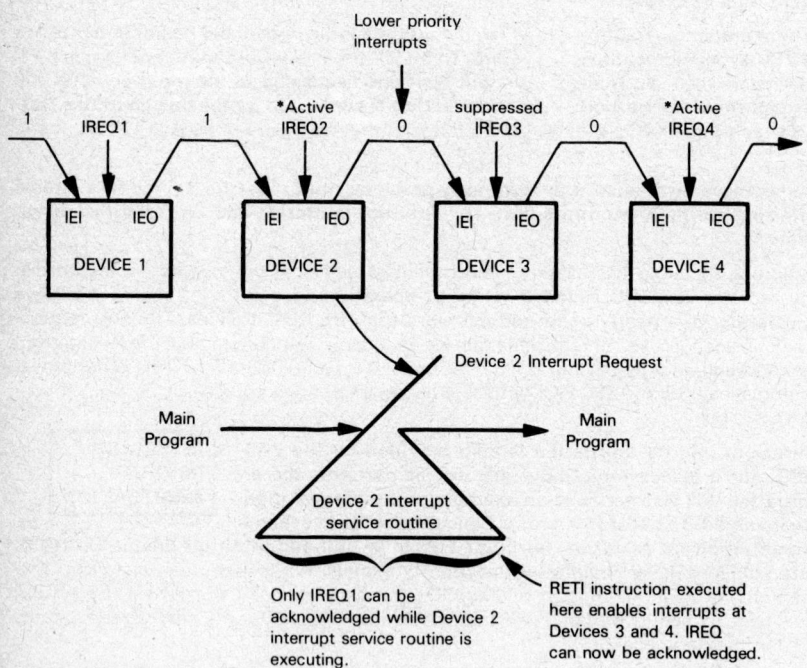

## JUSTIFYING INTERRUPTS

Minicomputer programmers and large computer programmers make indiscriminate use of interrupts simply to share the cost of the Central Processing Unit among a number of different applications.

**INTERRUPT ECONOMICS**

You, as a microcomputer user, are going to have to justify sharing a cost which may range between $5 and $20. Against this cost you must charge the cost of external logic needed to create interrupt request signals — as well as the extra cost of programming. **The economic tradeoff makes it far from obvious that interrupts are viable within microcomputer systems.** You must examine your application with care before assuming out of hand that interrupts represent the way to go. A second CPU, or an entire second microcomputer system, will frequently be cheaper than using interrupts to share a single microcomputer system between a number of different applications.

**Assuming that interrupts look economical for your application, timing considerations are also important.**

**INTERRUPT TIMING CONSIDERATIONS**

Certainly, interrupts look very attractive when your application is handling asynchronous events. In our case, **suppose the average print cycle lasts approximately 10 milliseconds; also, suppose it is impossible to say whether the time interval between print cycles will be 1 millisecond or 100 milliseconds. Under these circumstances, in order to execute some other program in the time in between print cycles, we must use interrupts to initiate the print cycle** — since we have no idea when the next print cycle is to begin.

**In reality,** the time which elapses between print cycles will be very accurately known. **A printer will have some advertised character printing rate.** If this rate is 45 characters per second, then 22.2 milliseconds will be required per printed character. If 10 of the 22 milliseconds are needed to execute the actual print cycle routine, then 12 milliseconds will remain in between print cycles. **We no longer need interrupts.** So long as the program which executes in between print cycles is broken into segments, each of which executes in 12 milliseconds or less, then each segment can terminate with an instruction loop which tests the status of the velocity decode input in order to initiate the next print cycle:

```
START:  IN    A,(2)      ;INPUT I/O PORT B TO ACCUMULATOR
        BIT   5,A        ;TEST BIT 5
        JR    NZ,START   ;IF NOT ZERO RETURN TO START
```

# Chapter 6
# THE Z80 INSTRUCTION SET

**Instructions falsely frighten microcomputer users who are new to programming. Taken as an isolated event, operations associated with the execution of a single instruction are easy enough to follow — and that is the purpose of this chapter.**

Why are the instructions of a microcomputer referred to as an instruction "set"? The answer is that the instructions selected by the designers of any microcomputer are selected with great care; it must be easy to execute complex operations as a sequence of simple events — each of which is represented by one instruction from a well-designed instruction "set".

**Remaining consistent with <u>An Introduction to Microcomputers, Volume II</u>, Table 6-1 summarizes the Z80 microcomputer instruction set, with similar instructions grouped together.**

**Individual instructions are described next in alphabetical order of instruction mnemonic.**

In addition to simply stating what each instruction does, the purpose of the instruction within normal programming logic is identified.

# ABBREVIATIONS
**These are the abbreviations used in this chapter:**

| | |
|---|---|
| A,F,B,C,D,E,H,L | The 8-bit registers. A is the Accumulator and F is the Program Status Word. |
| AF',BC',DE',HL' | The alternative register pairs |
| addr | A 16-bit memory address |
| x(b) | Bit b of 8-bit register or memory location x |
| cond | Condition for program branching. Conditions are:<br>NZ - Non-Zero (Z=0)<br>Z  - Zero (Z=1)<br>NC - Non-carry (C=0)<br>C  - Carry (C=1)<br>PO - Parity Odd (P=0)<br>PE - Parity Even (P=1)<br>P  - Sign Positive (S=0)<br>M  - Sign Negative (S=1) |
| data | An 8-bit binary data unit |
| data16 | A 16-bit binary data unit |
| disp | An 8-bit signed binary address displacement |
| xx(HI) | The high-order 8 bits of a 16-bit quantity xx |
| IV | Interrupt vector register (8 bits) |
| IX,IY | The Index registers (16 bits each) |
| xy | Either one of the Index registers (IX or IY) |
| LSB | Least Significant Bit (Bit 0) |
| label | A 16-bit instruction memory address |
| xx(LO) | The low-order 8 bits of a 16-bit quantity xx |
| MSB | Most Significant Bit (Bit 7) |
| PC | Program Counter |
| port | An 8-bit I/O port address |
| pr | Any of the following register pairs:<br>    BC<br>    DE<br>    HL<br>    AF |
| R | The Refresh register (8 bits) |
| reg | Any of the following registers:<br>    A<br>    B<br>    C<br>    D<br>    E<br>    H<br>    L |
| rp | Any of the following register pairs:<br>    BC<br>    DE<br>    HL<br>    SP |

| | |
|---|---|
| SP | Stack Pointer (16 bits) |
| Statuses | The Z80 has the following status flags:<br>C - Carry status<br>Z - Zero status<br>S - Sign status<br>P/O - Parity/Overflow status<br>$A_C$ - Auxiliary Carry status<br>N - Subtract status<br><br>The following symbols are used in the status columns:<br>X - flag is affected by operation<br>(blank) - flag is not affected by operation<br>1 - flag is set by operation<br>0 - flag is reset by operation<br>? - flag is unknown after operation<br>P - flag shows parity status<br>O - flag shows overflow status<br>I - flag shows interrupt enabled/disabled status |
| [ ] | Contents of location enclosed within brackets. If a register designation is enclosed within the brackets, then the designated register's contents are specified. If an I/O port number is enclosed within the brackets, then the I/O port contents are specified. If a memory address is enclosed within the brackets, then the contents of the addressed memory location are specified. |
| [[ ]] | Implied memory addressing; the contents of the memory location designated by the contents of a register. |
| $\Lambda$ | Logical AND |
| V | Logical OR |
| ⊻ | Logical Exclusive-OR |
| ← | Data is transferred in the direction of the arrow |
| ⟷ | Data is exchanged between the two locations designated on either side of the arrow. |

## STATUS

The six status flags are stored in a Flag register (F) as follows:

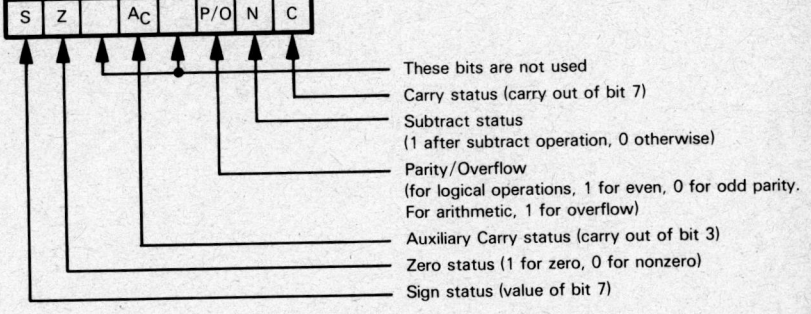

F and A are sometimes treated as a register pair.

**The effect of instruction execution on status is illustrated as follows:**

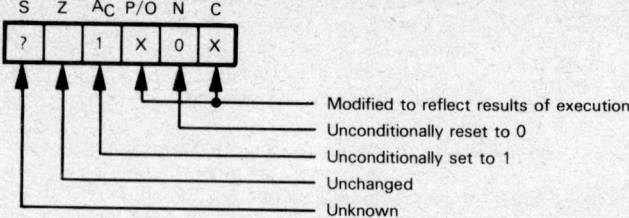

Within instruction execution illustrations, an X identifies a status that is set or reset. A 0 identifies a status that is always cleared. A 1 identifies a status that is always set. A blank means the status does not change. A question mark (?) means the status is not known.

**STATUS CHANGES WITH INSTRUCTION EXECUTION**

## INSTRUCTION MNEMONICS

The fixed part of an assembly language instruction is shown in UPPER CASE.

The variable part (immediate data, I/O device number, register name, label or address) is shown in lower case.

## INSTRUCTION OBJECT CODES

Instruction object codes are represented as two hexadecimal digits for instructions without variations.

Instruction object codes are represented as eight binary digits for instructions with variations; the binary digit representation of variations is then identifiable.

## INSTRUCTION EXECUTION TIMES AND CODES

Table 6-2 lists instructions in alphabetical order, showing object codes and execution times expressed as machine cycles.

Where two instruction cycles are shown the first is for "condition not met", whereas the second is for "condition met".

*Address Bus: A0-A7: [C]
A8-A15: [B]

Table 6-1. A Summary of the Z80 Instruction Set

| TYPE | MNEMONIC | OPERAND(S) | BYTES | STATUS ||||| OPERATION PERFORMED |
|------|----------|------------|-------|---|---|---|---|---|---------------------|
|      |          |            |       | C | Z | S | P/O | AC | N |
| O/I  | IN       | A,port     | 2     |   | X | X | P | X | 0 | [A]←[port]<br>Input to Accumulator from directly addressed I/O port.<br>Address Bus: A0-A7: port<br>A8-A15: [A] |
|      | IN       | reg,(C)    | 2     |   | ? | ? | ? | ? | ? | [reg]←[[C]]<br>Input to register from I/O port addressed by the contents of C.*<br>If second byte is 70₁₆ only the flags will be affected. |
|      | INIR     |            | 2     |   | 1 | ? | ? | ? | 1 | Repeat until [B]=0:<br>[[HL]]←[[C]]<br>[B]←[B]-1<br>[HL]←[HL]+1<br>Transfer a block of data from I/O port addressed by contents of C to memory location addressed by contents of HL, going from low addresses to high. Contents of B serve as a count of bytes remaining to be transferred.* |
|      | INDR     |            | 2     |   | 1 | ? | ? | ? | 1 | Repeat until [B]=0:<br>[[HL]]←[[C]]<br>[B]←[B]-1<br>[HL]←[HL]-1<br>Transfer a block of data from I/O port addressed by contents of C to memory location addressed by contents of HL, going from high addresses to low. Contents of B serve as a count of bytes remaining to be transferred.* |
|      | INI      |            | 2     |   | X | ? | ? | ? | 1 | [[HL]]←[[C]]<br>[B]←[B]-1<br>[HL]←[HL]+1<br>Transfer a byte of data from I/O port addressed by contents of C to memory location addressed by contents of HL. Decrement byte count and increment destination address.* |

*Address Bus: A0-A7: [C]
         A8-A15: [B]

Table 6-1. A Summary of the Z80 Instruction Set (Continued)

| TYPE | MNEMONIC | OPERAND(S) | BYTES | STATUS C | Z | S | P/O | A_C | N | OPERATION PERFORMED |
|---|---|---|---|---|---|---|---|---|---|---|
| I/O (Continued) | IND | | 2 | | X | ? | ? | ? | 1 | [[HL]]←[[C]]<br>[B]←[B]-1<br>[HL]←[HL]-1<br>Transfer a byte of data from I/O port addressed by contents of C to memory location addressed by contents of HL. Decrement both byte count and destination address.* |
| | OUT | port,A | 2 | | | | | | | Output from Accumulator to directly addressed I/O port.<br>Address Bus: A0-A7: port<br>         A8-A15: [A]<br>[port]←[A] |
| | OUT | (C),reg | 2 | | | | | | | [[C]]←[reg]<br>Output from register to I/O port addressed by the contents of C.* |
| | OTIR | | 2 | | 1 | ? | ? | ? | 1 | Repeat until [B]=0:<br>[[C]]←[[HL]]<br>[B]←[B]-1<br>[HL]←[HL]+1<br>Transfer a block of data from memory location addressed by contents of HL to I/O port addressed by contents of C, going from low memory to high. Contents of B serve as a count of bytes remaining to be transferred.* |
| | OTDR | | 2 | | 1 | ? | ? | ? | 1 | Repeat until [B]=0:<br>[[C]]←[[HL]]<br>[B]←[B]-1<br>[HL]←[HL]-1<br>Transfer a block of data from memory location addressed by contents of HL to I/O port addressed by contents of C, going from high memory to low. Contents of B serve as a count of bytes remaining to be transferred.* |

*Address Bus: A0-A7: [C]
A8-A15: [B]

Table 6-1. A Summary of the Z80 Instruction Set (Continued)

| TYPE | MNEMONIC | OPERAND(S) | BYTES | STATUS C | STATUS Z | STATUS S | STATUS P/O | STATUS Ac | STATUS N | OPERATION PERFORMED |
|---|---|---|---|---|---|---|---|---|---|---|
| I/O (Continued) | OUTI | | 2 | | X | ? | ? | ? | 1 | [[C]]←[[HL]]<br>[B]←[B] - 1<br>[HL]←[HL] + 1<br>Transfer a byte of data from memory location addressed by contents of HL to I/O port addressed by contents of C. Decrement byte count and increment source address.* |
| | OUTD | | 2 | | X | ? | ? | ? | 1 | [[C]]←[[HL]]<br>[B]←[B] - 1<br>[HL]←[HL] - 1<br>Transfer a byte of data from memory location addressed by contents of HL to I/O port addressed by contents of C. Decrement both byte count and source address.* |
| PRIMARY MEMORY REFERENCE | LD | A,(addr) | 3 | | | | | | | [A]←[addr]<br>Load Accumulator from directly addressed memory location. |
| | LD | HL,(addr) | 3 | | | | | | | [H]←[addr + 1], [L]←[addr]<br>Load HL from directly addressed memory. |
| | LD | rp,(addr)<br>xy,(addr) | 4 | | | | | | | [rpHI]]←[addr + 1], [rpLO)]←[addr] or<br>[xyHI)]←[addr + 1], [xy(LO)]←[addr]<br>Load register pair or Index register from directly addressed memory. |
| | LD | (addr),A | 3 | | | | | | | [addr]←[A]<br>Store Accumulator contents in directly addressed memory location. |
| | LD | (addr),HL | 3 | | | | | | | [addr + 1]←[H], [addr]←[L]<br>Store contents of HL to directly addressed memory location. |
| | LD | (addr),rp<br>(addr),xy | 4 | | | | | | | [addr + 1]←[rpHI)], [addr]←[rp(LO)] or<br>[addr + 1]←[xy(HI)], [addr]←[xy(LO)]<br>Store contents of register pair or Index register to directly addressed memory. |
| | LD | A,(BC)<br>A,(DE) | 1 | | | | | | | [A]←[[BC]] or [A]←[[DE]]<br>Load Accumulator from memory location addressed by the contents of the specified register pair. |

Table 6-1. A Summary of the Z80 Instruction Set (Continued)

| TYPE | MNEMONIC | OPERAND(S) | BYTES | STATUS | | | | | | OPERATION PERFORMED |
|---|---|---|---|---|---|---|---|---|---|---|
| | | | | C | Z | S | P/O | A_C | N | |
| **PRIMARY MEMORY REFERENCE (Continued)** | LD | reg,(HL) | 1 | | | | | | | [reg]←[[HL]]<br>Load register from memory location addressed by contents of HL. |
| | LD | (BC),A<br>(DE),A | 1 | | | | | | | [[BC]]←[A] or [[DE]]←[A]<br>Store Accumulator to memory location addressed by the contents of the specified register pair. |
| | LD | (HL),reg | 1 | | | | | | | [[HL]]←[reg]<br>Store register contents to memory location addressed by the contents of HL. |
| | LD | reg,(xy + disp) | 3 | | | | | | | [reg]←[[xy]+disp]<br>Load register from memory location using base relative addressing. |
| | LD | (xy + disp),reg | 3 | | | | | | | [[xy]+disp]←[reg]<br>Store register to memory location addressed relative to contents of index register. |
| **BLOCK TRANSFER AND SEARCH** | LDIR | | 2 | | | | 0 | 0 | 0 | Repeat until [BC]=0:<br>[[DE]]←[[HL]]<br>[DE]←[DE]+1<br>[HL]←[HL]+1<br>[BC]←[BC]-1<br>Transfer a block of data from the memory location addressed by the contents of HL to the memory location addressed by the contents of DE, going from low addresses to high. Contents of BC serve as a count of bytes to be transferred. |
| | LDDR | | 2 | | | | 0 | 0 | 0 | Repeat until [BC]=0:<br>[[DE]]←[[HL]]<br>[DE]←[DE]-1<br>[HL]←[HL]-1<br>[BC]←[BC]-1<br>Transfer a block of data from the memory location addressed by the contents of HL to the memory location addressed by the contents of DE, going from high addresses to low. Contents of BC serve as a count of bytes to be transferred. |

Table 6-1. A Summary of the Z80 Instruction Set (Continued)

| TYPE | MNEMONIC | OPERAND(S) | BYTES | STATUS ||||| OPERATION PERFORMED |
|------|----------|------------|-------|---|---|---|---|---|---|
| | | | | C | Z | S | P/O | A_C | N | |
| BLOCK TRANSFER AND SEARCH (Continued) | LDI | | 2 | | | | X | 0 | 0 | [[DE]]←[[HL]]<br>[DE]←[DE]+1<br>[HL]←[HL]+1<br>[BC]←[BC]-1<br>Transfer one byte of data from the memory location addressed by the contents of HL to the memory location addressed by the contents of DE. Increment source and destination addresses and decrement byte count. |
| | LDD | | 2 | | | | X | 0 | 0 | [[DE]]←[[HL]]<br>[DE]←[DE]-1<br>[HL]←[HL]-1<br>[BC]←[BC]-1<br>Transfer one byte of data from the memory location addressed by the contents of HL to the memory location addressed by the contents of DE. Decrement source and destination addresses and byte count. |
| | CPIR | | 2 | | X | X | X | X | 1 | Repeat until [A]=[[HL]] or [BC]=0:<br>[A] - [[HL]] (only flags are affected)<br>[HL]←[HL]+1<br>[BC]←[BC]-1<br>Compare contents of Accumulator with those of memory block addressed by contents of HL, going from low addresses to high. Stop when a match is found or when the byte count becomes zero. |
| | CPDR | | 2 | | X | X | X | X | 1 | Repeat until [A]=[[HL]] or [BC]=0:<br>[A] - [[HL]] (only flags are affected)<br>[HL]←[HL]-1<br>[BC]←[BC]-1<br>Compare contents of Accumulator with those of memory block addressed by contents of HL, going from high addresses to low. Stop when a match is found or when the byte count becomes zero. |

Table 6-1. A Summary of the Z80 Instruction Set (Continued)

| TYPE | MNEMONIC | OPERAND(S) | BYTES | C | Z | S | P/O | A_C | N | OPERATION PERFORMED |
|---|---|---|---|---|---|---|---|---|---|---|
| BLOCK TRANSFER AND SEARCH (Continued) | CPI | | 2 | | x | x | x | x | 1 | [A] - [[HL]] (only flags are affected)<br>[HL]←[HL]+1<br>[BC]←[BC]-1<br>Compare contents of Accumulator with those of memory location addressed by contents of HL. Increment address and decrement byte count. |
| | CPD | | 2 | | x | x | x | x | 1 | [A] - [[HL]] (only flags are affected)<br>[HL]←[HL]-1<br>[BC]←[BC]-1<br>Compare contents of Accumulator with those of memory location addressed by contents of HL. Decrement address and byte count. |
| SECONDARY MEMORY REFERENCE | ADD | (HL) | 1 | x | x | x | O | x | 0 | [A]←[A]+[[HL]] or [A]←[A]+[[xy]+disp] |
| | | (xy+disp) | 3 | x | x | x | O | x | 0 | Add to Accumulator using implied addressing or base relative addressing. |
| | ADC | (HL) | 1 | x | x | x | O | x | 0 | [A]←[A]+[[HL]]+C or [A]←[A]+[[xy]+disp]+C |
| | | (xy+disp) | 3 | x | x | x | O | x | 0 | Add with Carry using implied addressing or base relative addressing. |
| | SUB | (HL) | 1 | x | x | x | O | x | 1 | [A]←[A]-[[HL]] or [A]←[A]-[[xy]+disp] |
| | | (xy+disp) | 3 | x | x | x | O | x | 1 | Subtract from Accumulator using implied addressing or base relative addressing. |
| | SBC | (HL) | 1 | x | x | x | O | x | 1 | [A]←[A]-[[HL]]-C or [A]←[A]-[[xy]+disp]-C |
| | | (xy+disp) | 3 | x | x | x | O | x | 1 | Subtract with Carry using implied addressing or base relative addressing. |
| | AND | (HL) | 1 | 0 | x | x | P | 1 | 0 | [A]←[A]∧[[HL]] or [A]←[A]∧[[xy]+disp] |
| | | (xy+disp) | 3 | 0 | x | x | P | 1 | 0 | AND with Accumulator using implied addressing or base relative addressing. |
| | OR | (HL) | 1 | 0 | x | x | P | 1 | 0 | [A]←[A]∨[[HL]] or [A]←[A]∨[[xy]+disp] |
| | | (xy+disp) | 3 | 0 | x | x | P | 1 | 0 | OR with Accumulator using implied addressing or base relative addressing. |
| | XOR | (HL) | 1 | 0 | x | x | P | 1 | 0 | [A]←[A]⊻[[HL]] or [A]←[A]⊻[[xy]+disp] |
| | | (xy+disp) | 3 | 0 | x | x | P | 1 | 0 | Exclusive-OR with Accumulator using implied addressing or base relative addressing. |
| | CP | (HL) | 1 | x | x | x | O | x | 1 | [A] - [[HL]] or [A] - [[xy]+disp] |
| | | (xy+disp) | 3 | x | x | x | O | x | 1 | Compare with Accumulator using implied addressing or base relative addressing. Only the flags are affected. |
| | INC | (HL) | 1 | | x | x | O | x | 0 | [[HL]]←[[HL]]+1 or [[xy]+disp]←[[xy]+disp]+1 |
| | | (xy+disp) | 3 | | x | x | O | x | 0 | Increment using implied addressing or base relative addressing. |
| | DEC | (HL) | 1 | | x | x | O | x | 1 | [[HL]]←[[HL]]-1 or [[xy]+disp]←[[xy]+disp]-1 |
| | | (xy+disp) | 3 | | x | x | O | x | 1 | Decrement using implied addressing or base relative addressing. |

Table 6-1. A Summary of the Z80 Instruction Set (Continued)

| TYPE | MNEMONIC | OPERAND(S) | BYTES | STATUS C | Z | S | P/O | $A_C$ | N | OPERATION PERFORMED |
|---|---|---|---|---|---|---|---|---|---|---|
| MEMORY SHIFT AND ROTATE | RLC | (HL)<br>(xy + disp) | 2<br>4 | X | X | X | P | 0 | 0 | Rotate contents of memory location (implied or base relative addressing) left with branch Carry. [[HL]] or [[xy]+disp] |
| | RL | (HL)<br>(xy + disp) | 2<br>4 | X | X | X | P | 0 | 0 | Rotate contents of memory location left through Carry. [[HL]] or [[xy]+disp] |
| | RRC | (HL)<br>(xy + disp) | 2<br>4 | X | X | X | P | 0 | 0 | Rotate contents of memory location right with branch Carry. [[HL]] or [[xy]+disp] |
| | RR | (HL)<br>(xy + disp) | 2<br>4 | X | X | X | P | 0 | 0 | Rotate contents of memory location right through Carry. [[HL]] or [[xy]+disp] |
| | SLA | (HL)<br>(xy + disp) | 2<br>4 | X | X | X | P | 0 | 0 | Shift contents of memory location left and clear LSB (Arithmetic Shift). [[HL]] or [[xy]+disp] |
| | SRA | (HL)<br>(xy + disp) | 2<br>4 | X | X | X | P | 0 | 0 | Shift contents of memory location right and preserve MSB (Arithmetic Shift). [[HL]] or [[xy]+disp] |
| | SRL | (HL)<br>(xy + disp) | 2<br>4 | X | X | X | P | 0 | 0 | Shift contents of memory location right and clear MSB (Logical Shift). [[HL]] or [[xy]+disp] |

Table 6-1. A Summary of the Z80 Instruction Set (Continued)

| TYPE | MNEMONIC | OPERAND(S) | BYTES | STATUS | | | | | | OPERATION PERFORMED |
|------|----------|------------|-------|---|---|---|---|---|---|---------------------|
| | | | | C | Z | S | P/O | AC | N | |
| IMMEDIATE | LD | reg,data | 2 | | | | | | | [reg]←data<br>Load immediate into register. |
| | LD | rp,data16 | 3 | | | | | | | [rp]←data16 or [xy]←data16<br>Load 16 bits of immediate data into register pair or Index register. |
| | LD | xy,data16 | 4 | | | | | | | |
| | LD | (HL),data | 2 | | | | | | | [[HL]]←data or [[xy]+disp]←data<br>Load immediate into memory location using implied or base relative addressing. |
| | LD | (xy+disp),data | 4 | | | | | | | |
| JUMP | JP | label | 3 | | | | | | | [PC]←label<br>Jump to instruction at address represented by label. |
| | JR | disp | 2 | | | | | | | [PC]←[PC]+2+disp<br>Jump relative to present contents of Program Counter. |
| | JP | (HL) | 1 | | | | | | | [PC]←[HL] or [PC]←[xy]<br>Jump to address contained in HL or Index register. |
| | JP | (xy) | 2 | | | | | | | |
| SUBROUTINE CALL AND RETURN | CALL | label | 3 | | | | | | | [[SP]-1]←[PCHI]<br>[[SP]-2]←[PCLO]<br>[SP]←[SP]-2<br>[PC]←label<br>Jump to subroutine starting at address represented by label. |
| | CALL | cond,label | 3 | | | | | | | Jump to subroutine if condition is satisfied; otherwise, continue in sequence. |
| | RET | | 1 | | | | | | | [PCLO]←[[SP]]<br>[PCHI]←[[SP]+1]<br>[SP]←[SP]+2<br>Return from subroutine. |
| | RET | cond | 1 | | | | | | | Return from subroutine if condition is satisfied; otherwise, continue in sequence. |

Table 6-1. A Summary of the Z80 Instruction Set (Continued)

| TYPE | MNEMONIC | OPERAND(S) | BYTES | STATUS C | STATUS Z | STATUS S | STATUS P/O | STATUS Ac | STATUS N | OPERATION PERFORMED |
|---|---|---|---|---|---|---|---|---|---|---|
| IMMEDIATE OPERATE | ADD | data | 2 | x | x | x | O | x | 0 | [A]←[A] + data. Add immediate to Accumulator. |
| | ADC | data | 2 | x | x | x | O | x | 0 | [A]←[A] + data + C. Add immediate with Carry. |
| | SUB | data | 2 | x | x | x | O | x | 1 | [A]←[A] - data. Subtract immediate from Accumulator. |
| | SBC | data | 2 | x | x | x | O | x | 1 | [A]←[A] - data - C. Subtract immediate with Carry. |
| | AND | data | 2 | 0 | x | x | P | 1 | 0 | [A]←[A] ∧ data. AND immediate with Accumulator. |
| | OR | data | 2 | 0 | x | x | P | 1 | 0 | [A]←[A] ∨ data. OR immediate with Accumulator. |
| | XOR | data | 2 | 0 | x | x | P | 1 | 0 | [A]←[A]⊻data. Exclusive-OR immediate with Accumulator. |
| | CP | data | 2 | x | x | x | O | x | 1 | [A] - data. Compare immediate data with Accumulator contents; only the flags are affected. |
| JUMP ON CONDITION | JP | cond,label | 3 | | | | | | | If cond, then [PC]←label. Jump to instruction at address represented by label if the condition is true. |
| | JR | C,disp | 2 | | | | | | | If C=1, then [PC]←[PC] + 2 + disp. Jump relative to contents of Program Counter if Carry flag is set. |
| | JR | NC,disp | 2 | | | | | | | If C=0, then [PC]←[PC] + 2 + disp. Jump relative to contents of Program Counter if Carry flag is reset. |
| | JR | Z,disp | 2 | | | | | | | If Z=1, then [PC]←[PC] + 2 + disp. Jump relative to contents of Program Counter if Zero flag is set. |
| | JR | NZ,disp | 2 | | | | | | | If Z=0, then [PC]←[PC] + 2 + disp. Jump relative to contents of Program Counter if Zero flag is reset. |
| | DJNZ | disp | 2 | | | | | | | [B]←[B] - 1. If [B]≠0, then [PC]←[PC] + 2 + disp. Decrement contents of B and Jump relative to contents of Program Counter if result is not 0. |

Table 6-1. A Summary of the Z80 Instruction Set (Continued)

| TYPE | MNEMONIC | OPERAND(S) | BYTES | STATUS ||||| OPERATION PERFORMED |
|---|---|---|---|---|---|---|---|---|---|
| | | | | C | Z | S | P/O | AC | N |
| REGISTER-REGISTER MOVE | LD | dst,src | 1 | | | | | | | [dst]←[src]<br>Move contents of source register to destination register. Register designations src and dst may each be A, B, C, D, E, H or L. |
| | LD | A,IV | 2 | | x | x | 1 | 0 | 0 | [A]←[IV]<br>Move contents of Interrupt Vector register to Accumulator. |
| | LD | A,R | 2 | | x | x | 1 | 0 | 0 | [A]←[R]<br>Move contents of Refresh register to Accumulator. |
| | LD | IV,A | 2 | | | | | | | [IV]←[A]<br>Load Interrupt Vector register from Accumulator. |
| | LD | R,A | 2 | | | | | | | [R]←[A]<br>Load Refresh register from Accumulator. |
| | LD | SP,HL | 1 | | | | | | | [SP]←[HL]<br>Move contents of HL to Stack Pointer. |
| | LD | SP,xy | 2 | | | | | | | [SP]←[xy]<br>Move contents of Index register to Stack Pointer. |
| | EX | DE,HL | 1 | | | | | | | [DE]←→[HL]<br>Exchange contents of DE and HL. |
| | EX | AF,AF' | 1 | | | | | | | [AF]←→[AF']<br>Exchange program status and alternate program status. |
| | EXX | | 1 | | | | | | | $\begin{pmatrix}[BC]\\[DE]\\[HL]\end{pmatrix} \longleftrightarrow \begin{pmatrix}[BC']\\[DE']\\[HL']\end{pmatrix}$<br>Exchange register pairs and alternate register pairs. |

Table 6-1. A Summary of the Z80 Instruction Set (Continued)

| TYPE | MNEMONIC | OPERAND(S) | BYTES | STATUS ||||||  OPERATION PERFORMED |
|---|---|---|---|---|---|---|---|---|---|---|
| | | | | C | Z | S | P/O | A_C | N | |
| REGISTER-REGISTER OPERATE | ADD | reg | 1 | X | X | X | O | X | 0 | [A]←[A] + [reg]<br>Add contents of register to Accumulator. |
| | ADC | reg | 1 | X | X | X | O | X | 0 | [A]←[A] + [reg] + C<br>Add contents of register and Carry to Accumulator. |
| | SUB | reg | 1 | X | X | X | O | X | 1 | [A]←[A] - [reg]<br>Subtract contents of register from Accumulator. |
| | SBC | reg | 1 | X | X | X | O | X | 1 | [A]←[A] - [reg] - C<br>Subtract contents of register and Carry from Accumulator. |
| | AND | reg | 1 | 0 | X | X | P | 1 | 0 | [A]←[A] ∧ [reg]<br>AND contents of register with contents of Accumulator. |
| | OR | reg | 1 | 0 | X | X | P | 1 | 0 | [A]←[A] ∨ [reg]<br>OR contents of register with contents of Accumulator. |
| | XOR | reg | 1 | 0 | X | X | P | 1 | 0 | [A]←[A] ⊻ [reg]<br>Exclusive-OR contents of register with contents of Accumulator. |
| | CP | reg | 1 | X | X | X | O | X | 1 | [A] - [reg]<br>Compare contents of register with contents of Accumulator. Only the flags are affected. |
| | ADD | HL,rp | 1 | X | | | | ? | 0 | [HL]←[HL] + [rp]<br>16-bit add register pair contents to contents of HL. |
| | ADC | HL,rp | 2 | X | X | X | O | ? | 0 | [HL]←[HL] + [rp] + C<br>16-bit add with Carry register pair contents to contents of HL. |
| | SBC | HL,rp | 2 | X | X | X | O | ? | 1 | [HL]←[HL] - [rp] - C<br>16-bit subtract with Carry register pair contents from contents of HL. |
| | ADD | IX,pp | 2 | X | | | | ? | 0 | [IX]←[IX] + [pp]<br>16-bit add register pair contents to contents of Index register IX (pp=BC, DE, IX, SP). |
| | ADD | IY,rr | 2 | X | | | | ? | 0 | [IY]←[IY] + [rr]<br>16-bit add register pair contents to contents of Index register IY (rr=BC, DE, IY, SP). |

Table 6-1. A Summary of the Z80 Instruction Set (Continued)

| TYPE | MNEMONIC | OPERAND(S) | BYTES | STATUS ||||| OPERATION PERFORMED |
|---|---|---|---|---|---|---|---|---|---|
| | | | | C | Z | S | P/O | Ac | N | |
| REGISTER OPERATE | DAA | | 1 | X | X | X | P | X | 1 | Decimal adjust Accumulator, assuming that Accumulator contents are the sum or difference of BCD operands. |
| | CPL | | 1 | | | | | 1 | 1 | Complement Accumulator (ones complement). [A]←[Ā] |
| | NEG | | 2 | X | X | X | O | X | 1 | Negate Accumulator (twos complement). [A]←[Ā]+1 |
| | INC | reg | 1 | | X | X | O | X | 0 | Increment register contents. [reg]←[reg]+1 |
| | INC | rp<br>xy | 1<br>1 | | | | | | | Increment register contents. [rp]←[rp]+1 or [xy]←[xy]+1 Increment contents of register pair or Index register. |
| | DEC | reg | 1 | | X | X | O | X | 1 | Decrement register contents. [reg]←[reg]-1 |
| | DEC | rp<br>xy | 1<br>2 | | | | | | | Decrement register contents. [rp]←[rp]-1 or [xy]←[xy]-1 Decrement contents of register pair or Index register. |
| REGISTER SHIFT AND ROTATE | RLCA | | 1 | X | | | | 0 | 0 | [A]<br>Rotate Accumulator left with branch Carry. |
| | RLA | | 1 | X | | | | 0 | 0 | [A]<br>Rotate Accumulator left through Carry. |
| | RRCA | | 1 | X | | | | 0 | 0 | [A]<br>Rotate Accumulator right with branch Carry. |

Table 6-1. A Summary of the Z80 Instruction Set (Continued)

| TYPE | MNEMONIC | OPERAND(S) | BYTES | STATUS ||||||  OPERATION PERFORMED |
|---|---|---|---|---|---|---|---|---|---|---|
| | | | | C | Z | S | P/O | $A_C$ | N | |
| REGISTER SHIFT AND ROTATE (Continued) | RRA | | 1 | X | | | | 0 | 0 | Rotate Accumulator right through Carry. [A] |
| | RLC | reg | 2 | X | X | X | P | 0 | 0 | Rotate contents of register left with branch Carry. [reg] |
| | RL | reg | 2 | X | X | X | P | 0 | 0 | Rotate contents of register left through Carry. [reg] |
| | RRC | reg | 2 | X | X | X | P | 0 | 0 | Rotate contents of register right with branch Carry. [reg] |
| | RR | reg | 2 | X | X | X | P | 0 | 0 | Rotate contents of register right through Carry. [reg] |
| | SLA | reg | 2 | X | X | X | P | 0 | 0 | Shift contents of register left and clear LSB (Arithmetic Shift). [reg] |
| | SRA | reg | 2 | X | X | X | P | 0 | 0 | Shift contents of register right and preserve MSB (Arithmetic Shift). [reg] |

Table 6-1. A Summary of the Z80 Instruction Set (Continued)

| TYPE | MNEMONIC | OPERAND(S) | BYTES | STATUS C | STATUS Z | STATUS S | STATUS P/O | STATUS $A_C$ | STATUS N | OPERATION PERFORMED |
|---|---|---|---|---|---|---|---|---|---|---|
| REGISTER SHIFT AND ROTATE (Continued) | SRL | | 2 | X | X | X | P | 0 | 0 | Shift contents of register right and clear MSB (Logical Shift). |
| | RLD | | 2 | | X | X | P | 0 | 0 | Rotate one BCD digit left between the Accumulator and memory location (implied addressing). Contents of the upper half of the Accumulator are not affected. |
| | RRD | reg | 2 | | X | X | P | 0 | 0 | Rotate one BCD digit right between the Accumulator and memory location (implied addressing). Contents of the upper half of the Accumulator are not affected. |
| BIT MANIPULATION | BIT | b,reg | 2 | | X | ? | ? | 1 | 0 | $Z \leftarrow \overline{reg(b)}$ Zero flag contains complement of the selected register bit. |
| | BIT | b,(HL) b,(xy + disp) | 2 4 | | X | ? | ? | 1 | 0 | $Z \leftarrow \overline{[[HL]](b)}$ or $Z \leftarrow \overline{[[xy] + disp](b)}$ Zero flag contains complement of selected bit of the memory location (implied addressing or base relative addressing). |
| | SET | b,reg | 2 | | | | | | | reg(b)←1 Set indicated register bit. |
| | SET | b,(HL) b,(xy + disp) | 2 4 | | | | | | | [[HL]](b)←1 or [[xy] + disp](b)←1 Set indicated bit of memory location (implied addressing or base relative addressing). |
| | RES | b,reg | 2 | | | | | | | reg(b)←0 Reset indicated register bit. |
| | RES | b,(HL) b,(xy + disp) | 2 4 | | | | | | | [[HL]](b)←0 or [[xy] + disp](b)←0 Reset indicated bit in memory location (implied addressing or base relative addressing). |

Table 6-1. A Summary of the Z80 Instruction Set (Continued)

| TYPE | MNEMONIC | OPERAND(S) | BYTES | STATUS C | Z | S | P/O | AC | N | OPERATION PERFORMED |
|---|---|---|---|---|---|---|---|---|---|---|
| STACK | PUSH | pr | 1 | | | | | | | [[SP]-1]←[prHI] |
| | | xy | 2 | | | | | | | [[SP]-2]←[prLO] |
| | | | | | | | | | | [SP]←[SP]-2 |
| | | | | | | | | | | Put contents of register pair or Index register on top of Stack and decrement Stack Pointer. |
| | POP | pr | 1 | | | | | | | [prLO)]←[[SP]] |
| | | xy | 2 | | | | | | | [prHI]←[[SP]+1] |
| | | | | | | | | | | [SP]←[SP]+2 |
| | | | | | | | | | | Put contents of top of Stack in register pair or Index register and increment Stack Pointer. |
| | EX | (SP),HL | 1 | | | | | | | [H]←[[SP]+1] |
| | | (SP),xy | 2 | | | | | | | [L]←[[SP]] |
| | | | | | | | | | | Exchange contents of HL or Index register and top of Stack. |
| INTERRUPT | DI | | 1 | | | | | | | Disable interrupts. |
| | EI | | 1 | | | | | | | Enable interrupts. |
| | RST | n | 1 | | | | | | | [[SP]-1]←[PCHI] |
| | | | | | | | | | | [[SP]-2]←[PCLO] |
| | | | | | | | | | | [SP]←[SP]-2 |
| | | | | | | | | | | [PC]←(8-n)16 |
| | | | | | | | | | | Restart at designated location. |
| | RETI | | 2 | | | | | | | Return from interrupt. |
| | RETN | | 2 | | | | | | | Return from nonmaskable interrupt. |
| | IM | 0 | 2 | | | | | | | Set interrupt mode 0, 1, or 2. |
| | | 1 | | | | | | | | |
| | | 2 | | | | | | | | |
| STATUS | SCF | | 1 | 1 | | | | 0 | 0 | C←1 |
| | | | | | | | | | | Set Carry flag. |
| | CCF | | 1 | X | | | | ? | 0 | C←$\bar{C}$ |
| | | | | | | | | | | Complement Carry flag. |
| | NOP | | 1 | | | | | | | No operation — volatile memories are refreshed. |
| | HALT | | 1 | | | | | | | CPU halts, executes NOPs to refresh volatile memories. |

## Table 6-2. A Summary of Instruction Object Codes and Execution Cycles

| INSTRUCTION | | OBJECT CODE | BYTES | CLOCK PERIODS |
|---|---|---|---|---|
| ADC | data | CE yy | 2 | 7 |
| ADC | (HL) | 8E | 1 | 7 |
| ADC | HL,rp | ED 01xx1010 | 2 | 15 |
| ADC | (IX + disp) | DD 8E yy | 3 | 19 |
| ADC | (IY + disp) | FD 8E yy | 3 | 19 |
| ADC | reg | 10001xxx | 1 | 4 |
| ADD | data | C6 yy | 2 | 7 |
| ADD | (HL) | 86 | 1 | 7 |
| ADD | HL,rp | 00xx1001 | 1 | 11 |
| ADD | (IX + disp) | DD 86 yy | 3 | 19 |
| ADD | IX,pp | DD 00xx1001 | 2 | 15 |
| ADD | (IY + disp) | FD 86 yy | 3 | 19 |
| ADD | IY,rr | FD 00xx1001 | 2 | 15 |
| ADD | reg | 10000xxx | 1 | 4 |
| AND | data | E6 yy | 2 | 7 |
| AND | (HL) | A6 | 1 | 7 |
| AND | (IX + disp) | DD A6 yy | 3 | 19 |
| AND | (IY + disp) | FD A6 yy | 3 | 19 |
| AND | reg | 10100xxx | 1 | 4 |
| BIT | b,(HL) | CB<br>01bbb110 | 2 | 12 |
| BIT | b,(IX + disp) | DD CB yy<br>01bbb110 | 4 | 20 |
| BIT | b,(IY + disp) | FD CB yy<br>01bbb110 | 4 | 20 |
| BIT | b,reg | CB<br>01bbbxxx | 2 | 8 |
| CALL | label | CD ppqq | 3 | 17 |
| CALL | C,label | DC ppqq | 3 | 10/17 |
| CALL | M,label | FC ppqq | 3 | 10/17 |
| CALL | NC,label | D4 ppqq | 3 | 10/17 |
| CALL | NZ,label | C4 ppqq | 3 | 10/17 |
| CALL | P,label | F4 ppqq | 3 | 10/17 |
| CALL | PE,label | EC ppqq | 3 | 10/17 |
| CALL | PO,label | E4 ppqq | 3 | 10/17 |
| CALL | Z,label | CC ppqq | 3 | 10/17 |
| CCF | | 3F | 1 | 4 |
| CP | data | FE yy | 2 | 7 |
| CP | (HL) | BE | 1 | 7 |
| CP | (IX + disp) | DD BE yy | 3 | 19 |
| CP | (IY + disp) | FD BE yy | 3 | 19 |
| CP | reg | 10111xxx | 1 | 4 |
| CPD | | ED A9 | 2 | 16 |
| CPDR | | ED B9 | 2 | 21/16* |
| CPI | | ED A1 | 2 | 16 |
| CPIR | | ED B1 | 2 | 21/16* |
| CPL | | 2F | 1 | 4 |
| DAA | | 27 | 1 | 4 |
| DEC | (HL) | 35 | 1 | 11 |
| DEC | IX | DD 2B | 2 | 10 |
| DEC | (IX + disp) | DD 35 yy | 3 | 23 |
| DEC | IY | FD 2B | 2 | 10 |
| DEC | (IY + disp) | FD 35 yy | 3 | 23 |
| DEC | rp | 00xx1011 | 1 | 6 |

Table 6-2. A Summary of Instruction Object Codes and Execution Cycles (Continued)

| INSTRUCTION | | OBJECT CODE | BYTES | CLOCK PERIODS |
|---|---|---|---|---|
| DEC | reg | 00xxx101 | 1 | 4 |
| DI | | F3 | 1 | 4 |
| DJNZ | disp | 10 yy | 2 | 8/13 |
| EI | | FB | 1 | 4 |
| EX | AF,AF' | 08 | 1 | 4 |
| EX | DE,HL | EB | 1 | 4 |
| EX | (SP),HL | E3 | 1 | 19 |
| EX | (SP),IX | DD E3 | 2 | 23 |
| EX | (SP),IY | FD E3 | 2 | 23 |
| EXX | | D9 | 1 | 4 |
| HALT | | 76 | 1 | 4 |
| IM | 0 | ED 46 | 2 | 8 |
| IM | 1 | ED 56 | 2 | 8 |
| IM | 2 | ED 5E | 2 | 8 |
| IN | A,port | DB yy | 2 | 10 |
| IN | reg,(C) | ED 01ddd000 | 2 | 11 |
| INC | (HL) | 34 | 1 | 11 |
| INC | IX | DD 23 | 2 | 10 |
| INC | (IX + disp) | DD 34 yy | 3 | 23 |
| INC | IY | FD 23 | 2 | 10 |
| INC | (IY + disp) | FD 34 yy | 3 | 23 |
| INC | rp | 00xx0011 | 1 | 6 |
| INC | reg | 00xxx100 | 1 | 4 |
| IND | | ED AA | 2 | 15 |
| INDR | | ED BA | 2 | 20/15 |
| INI | | ED A2 | 2 | 15 |
| INIR | | ED B2 | 2 | 20/15 |
| JP | label | C3 ppqq | 3 | 10 |
| JP | C,label | DA ppqq | 3 | 10 |
| JP | (HL) | E9 | 1 | 4 |
| JP | (IX) | DD E9 | 2 | 8 |
| JP | (IY) | FD E9 | 2 | 8 |
| JP | M,label | FA ppqq | 3 | 10 |
| JP | NC,label | D2 ppqq | 3 | 10 |
| JP | NZ,label | C2 ppqq | 3 | 10 |
| JP | P,label | F2 ppqq | 3 | 10 |
| JP | PE,label | EA ppqq | 3 | 10 |
| JP | PO,label | E2 ppqq | 3 | 10 |
| JP | Z,label | CA ppqq | 3 | 10 |
| JR | C,disp | 38 yy | 2 | 7/12 |
| JR | disp | 18 yy | 2 | 12 |
| JR | NC,disp | 30 yy | 2 | 7/12 |
| JR | NZ,disp | 20 yy | 2 | 7/12 |
| JR | Z,disp | 28 yy | 2 | 7/12 |
| LD | A,(addr) | 3A ppqq | 3 | 13 |
| LD | A,(BC) | 0A | 1 | 7 |
| LD | A,(DE) | 1A | 1 | 7 |
| LD | A,I | ED 57 | 2 | 9 |
| LD | A,R | ED 5F | 2 | 9 |
| LD | (addr),A | 32 ppqq | 3 | 13 |
| LD | (addr),BC | ED 43 ppqq | 4 | 20 |
| LD | (addr),DE | ED 53 ppqq | 4 | 20 |

Table 6-2. A Summary of Instruction Object Codes and Execution Cycles (Continued)

| INSTRUCTION | | OBJECT CODE | BYTES | CLOCK PERIODS |
|---|---|---|---|---|
| LD | (addr),HL | 22 ppqq | 3 | 16 |
| LD | (addr),IX | DD 22 ppqq | 4 | 20 |
| LD | (addr),IY | FD 22 ppqq | 4 | 20 |
| LD | (addr),SP | ED 73 ppqq | 4 | 20 |
| LD | (BC),A | 02 | 1 | 7 |
| LD | (DE),A | 12 | 1 | 7 |
| LD | HL,(addr) | 2A ppqq | 3 | 16 |
| LD | (HL),data | 36 yy | 2 | 10 |
| LD | (HL),reg | 01110sss | 1 | 7 |
| LD | I,A | ED 47 | 2 | 9 |
| LD | IX,(addr) | DD 2A ppqq | 4 | 20 |
| LD | IX,data16 | DD 21 yyyy | 4 | 14 |
| LD | (IX + disp),data | DD 36 yy yy | 4 | 19 |
| LD | (IX + disp),reg | DD 01110sss yy | 3 | 19 |
| LD | IY,(addr) | FD 2A ppqq | 4 | 20 |
| LD | IY,data16 | FD 21 yyyy | 4 | 14 |
| LD | (IY + disp),data | FD 36 yyyy | 4 | 19 |
| LD | (IY + disp),reg | FD 01110sss yy | 3 | 19 |
| LD | R,A | ED 4F | 2 | 9 |
| LD | reg,data | 00ddd110 yy | 2 | 7 |
| LD | reg,(HL) | 01ddd110 | 1 | 7 |
| LD | reg,(IX + disp) | DD 01ddd110 yy | 3 | 19 |
| LD | reg,(IY + disp) | FD 01dddd110 yy | 3 | 19 |
| LD | reg,reg | 01dddsss | 1 | 4 |
| LD | rp,(addr) | ED 01xx1011 ppqq | 4 | 20 |
| LD | rp,data16 | 00xx0001 yyyy | 3 | 10 |
| LD | SP,HL | F9 | 1 | 6 |
| LD | SP,IX | DD F9 | 2 | 10 |
| LD | SP,IY | FD F9 | 2 | 10 |
| LDD | | ED A8 | 2 | 16 |
| LDDR | | ED B8 | 2 | 21/16* |
| LDI | | ED A0 | 2 | 16 |
| LDIR | | ED B0 | 2 | 21/16* |
| NEG | | ED 44 | 2 | 8 |
| NOP | | 00 | 1 | 4 |
| OR | data | F6 yy | 2 | 7 |
| OR | (HL) | B6 | 1 | 7 |
| OR | (IX + disp) | DD B6 yy | 3 | 19 |
| OR | (IY + disp) | FD B6 yy | 3 | 19 |
| OR | reg | 10110xxx | 1 | 4 |
| OTDR | | ED BB | 2 | 20/15* |
| OTIR | | ED B3 | 2 | 20/15* |
| OUT | (C),reg | ED 01sss001 | 2 | 12 |

Table 6-2. A Summary of Instruction Object Codes and Execution Cycles (Continued)

| INSTRUCTION | | OBJECT CODE | BYTES | CLOCK PERIODS |
|---|---|---|---|---|
| OUT | port,A | D3 yy | 2 | 11 |
| OUTD | | ED AB | 2 | 15 |
| OUTI | | ED A3 | 2 | 15 |
| POP | IX | DD E1 | 2 | 14 |
| POP | IY | FD E1 | 2 | 14 |
| POP | pr | 11xx0001 | 1 | 10 |
| PUSH | IX | DD E5 | 2 | 15 |
| PUSH | IY | FD E5 | 2 | 15 |
| PUSH | pr | 11xx0101 | 1 | 11 |
| RES | b,(HL) | CB 10bbb110 | 2 | 15 |
| RES | b,(IX + disp) | DD CB yy 10bbb110 | 4 | 23 |
| RES | b,(IY + disp) | FD CB yy 10bbb110 | 4 | 23 |
| RES | b,reg | CB 10bbbxxx | 2 | 8 |
| RET | | C9 | 1 | 10 |
| RET | C | D8 | 1 | 5/11 |
| RET | M | F8 | 1 | 5/11 |
| RET | NC | D0 | 1 | 5/11 |
| RET | NZ | C0 | 1 | 5/11 |
| RET | P | F0 | 1 | 5/11 |
| RET | PE | E8 | 1 | 5/11 |
| RET | PO | E0 | 1 | 5/11 |
| RET | Z | C8 | 1 | 5/11 |
| RETI | | ED 4D | 2 | 14 |
| RETN | | ED 45 | 2 | 14 |
| RL | (HL) | CB 16 | 2 | 15 |
| RL | (IX + disp) | DD CB yy 16 | 4 | 23 |
| RL | (IY + disp) | FD CB yy 16 | 4 | 23 |
| RL | reg | CB 00010xxx | 2 | 8 |
| RLA | | 17 | 1 | 4 |
| RLC | (HL) | CB 06 | 2 | 15 |
| RLC | (IX + disp) | DD CB yy 06 | 4 | 23 |
| RLC | (IY + disp) | FD CB yy 06 | 4 | 23 |
| RLC | reg | CB 00000xxx | 2 | 8 |
| RLCA | | 07 | 1 | 4 |
| RLD | | ED 6F | 2 | 18 |
| RR | (HL) | CB 1E | 2 | 15 |
| RR | (IX + disp) | DD CB yy 1E | 4 | 23 |
| RR | (IY + disp) | FD CB yy 1E | 4 | 23 |
| RR | reg | CB 00011xxx | 2 | 8 |
| RRA | | 1F | 1 | 4 |
| RRC | (HL) | CB 0E | 2 | 15 |
| RRC | (IX + disp) | DD CB yy 0E | 4 | 23 |
| RRC | (IY + disp) | FD CB yy 0E | 4 | 23 |

Table 6-2. A Summary of Instruction Object Codes and Execution Cycles (Continued)

| INSTRUCTION | | OBJECT CODE | BYTES | CLOCK PERIODS |
|---|---|---|---|---|
| RRC | reg | CB 00001xxx | 2 | 8 |
| RRCA | | 0F | 1 | 4 |
| RRD | | ED 67 | 2 | 18 |
| RST | n | 11xxx111 | 1 | 11 |
| SBC | data | DE yy | 2 | 7 |
| SBC | (HL) | 9E | 1 | 7 |
| SBC | HL,rp | ED 01xx0010 | 2 | 15 |
| SBC | (IX + disp) | DD 9E yy | 3 | 19 |
| SBC | (IY + disp) | FD 9E yy | 3 | 19 |
| SBC | reg | 10011xxx | 1 | 4 |
| SCF | | 37 | 1 | 4 |
| SET | b,(HL) | CB 11bbb110 | 2 | 15 |
| SET | b,(IX + disp) | DD CB yy 11bbb110 | 4 | 23 |
| SET | b,(IY + disp) | FD CB yy 11bbb110 | 4 | 23 |
| SET | b,reg | CB 11bbbxxx | 2 | 8 |
| SLA | (HL) | CB 26 | 2 | 15 |
| SLA | (IX + disp) | DD CB yy 26 | 4 | 23 |
| SLA | (IY + disp) | FD CB yy 26 | 4 | 23 |
| SLA | reg | CB 00100xxx | 2 | 8 |
| SRA | (HL) | CB 2E | 2 | 15 |
| SRA | (IX + disp) | DD CB yy 2E | 4 | 23 |
| SRA | (IY + disp) | FD CB yy 2E | 4 | 23 |
| SRA | reg | CB 00101xxx | 2 | 8 |
| SRL | (HL) | CB 3E | 2 | 15 |
| SRL | (IX + disp) | DD CB yy 3E | 4 | 23 |
| SRL | (IY + disp) | FD CB yy 3E | 4 | 23 |
| SRL | reg | CB 00111xxx | 2 | 8 |
| SUB | data | D6 yy | 2 | 7 |
| SUB | (HL) | 96 | 1 | 7 |
| SUB | (IX + disp) | DD 96 yy | 3 | 19 |
| SUB | (IY + disp) | FD 96 yy | 3 | 19 |
| SUB | reg | 10010xxx | 1 | 4 |
| XOR | data | EE yy | 2 | 7 |
| XOR | (HL) | AE | 1 | 7 |
| XOR | (IX + disp) | DD AE yy | 3 | 19 |
| XOR | (IY + disp) | FD AE yy | 3 | 19 |
| XOR | reg | 10101xxx | 1 | 4 |

| | |
|---|---|
| x | represents an optional binary digit. |
| bbb | represents optional binary digits identifying a bit location in a register or memory byte. |
| ddd | represents optional binary digits identifying a destination register. |
| sss | represents optional binary digits identifying a source register. |
| ppqq | represents a four hexadecimal digit memory address. |
| yy | represents two hexadecimal data digits. |
| yyyy | represents four hexadecimal data digits. |

When two possible execution times are shown (i.e., 5/11), it indicates that the number of clock periods depends on condition flags.

*Execution time shown is for one iteration.

## ADC A,data — ADD IMMEDIATE WITH CARRY TO ACCUMULATOR

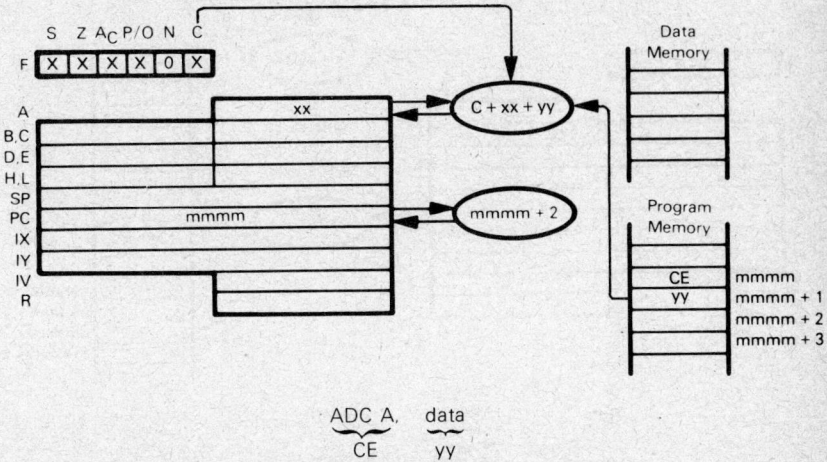

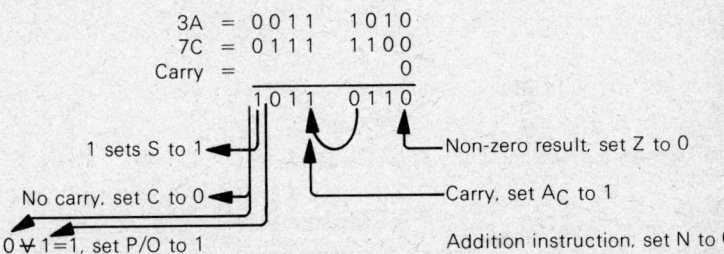

Add the contents of the next program memory byte and the Carry status to the Accumulator.

Suppose $xx=3A_{16}$, $yy=7C_{16}$, and Carry=0. After the instruction

$$\text{ADC} \quad \text{A,7CH}$$

has executed, the Accumulator will contain $B6_{16}$:

```
    3A  =  0011  1010
    7C  =  0111  1100
 Carry  =           0
         1,011  0110
```

1 sets S to 1 ◀──────────┐                            ┌──── Non-zero result, set Z to 0

No carry, set C to 0 ◀───┘                            └──── Carry, set $A_C$ to 1

$0 ⊻ 1 = 1$, set P/O to 1                       Addition instruction, set N to 0

The ADC instruction is frequently used in multibyte addition for the second and subsequent bytes.

# ADC A,reg — ADD REGISTER WITH CARRY TO ACCUMULATOR

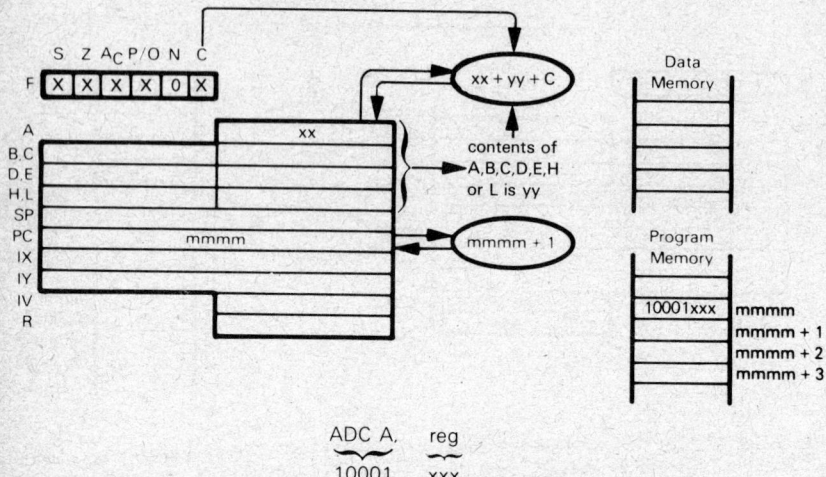

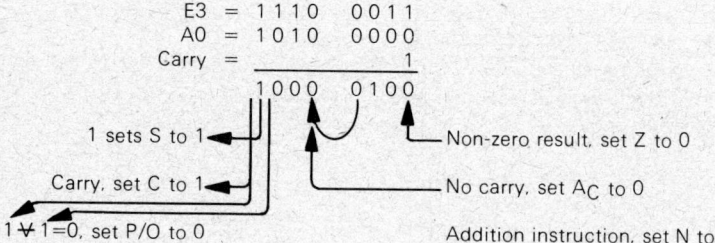

Add the contents of Register A, B, C, D, E, H or L and the Carry status to the Accumulator.

Suppose xx=E3$_{16}$, Register E contains A0$_{16}$, and Carry=1. After the instruction

$$\text{ADC A,E}$$

has executed, the Accumulator will contain 84$_{16}$:

```
   E3   =  1 1 1 0   0 0 1 1
   A0   =  1 0 1 0   0 0 0 0
 Carry  =                  1
           1 0 0 0   0 1 0 0
```

1 sets S to 1 ◄───         ───► Non-zero result, set Z to 0

Carry, set C to 1 ◄───        ───► No carry, set A$_C$ to 0

1 ⊻ 1=0, set P/O to 0                    Addition instruction, set N to 0

The ADC instruction is most frequently used in multibyte addition for the second and subsequent bytes.

# ADC A,(HL) — ADD MEMORY AND CARRY TO
# ADC A,(IX+disp)      ACCUMULATOR
# ADC A,(IY+disp)

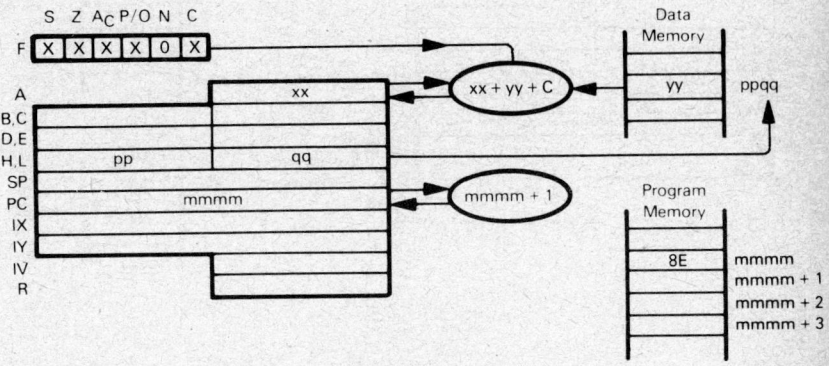

The illustration shows execution of ADC A,(HL):

$$\underbrace{\text{ADC A,(HL)}}_{\text{8E}}$$

Add the contents of memory location (specified by the contents of the HL register pair) and the Carry status to the Accumulator.

Suppose $xx = E3_{16}$, $yy = A0_{16}$, and Carry=1. After the instruction

   ADC   A,(HL)

has executed, the Accumulator will contain $84_{16}$:

```
      E3   =  1110 0011
      A0   =  1010 0000
      Carry =            1
                ─────────
               1000 0100
```

1 sets S to 1 ◄──────┐
                     │                          └── Non-zero result, set Z to 0
Carry, set C to 1 ◄──┤                          └── No carry, set $A_C$ to 0
                     │
$1 \veebar 1 = 0$, set P/O to 0                Addition instruction, set N to 0

$$\underbrace{\text{ADC A,(IX+disp)}}_{\text{DD  8E  d}}$$

Add the contents of memory location (specified by the sum of the contents of the IX register and the displacement digit d) and the Carry to the Accumulator.

$$\underbrace{\text{ADC A,(IY+disp)}}_{\text{FD  8E  d}}$$

This instruction is identical to ADC A,(IX+disp), except that it uses the IY register instead of the IX register.

The ADC instruction is most frequently used in multibyte addition for the second and subsequent bytes.

# ADC HL,rp — ADD REGISTER PAIR WITH CARRY TO H AND L

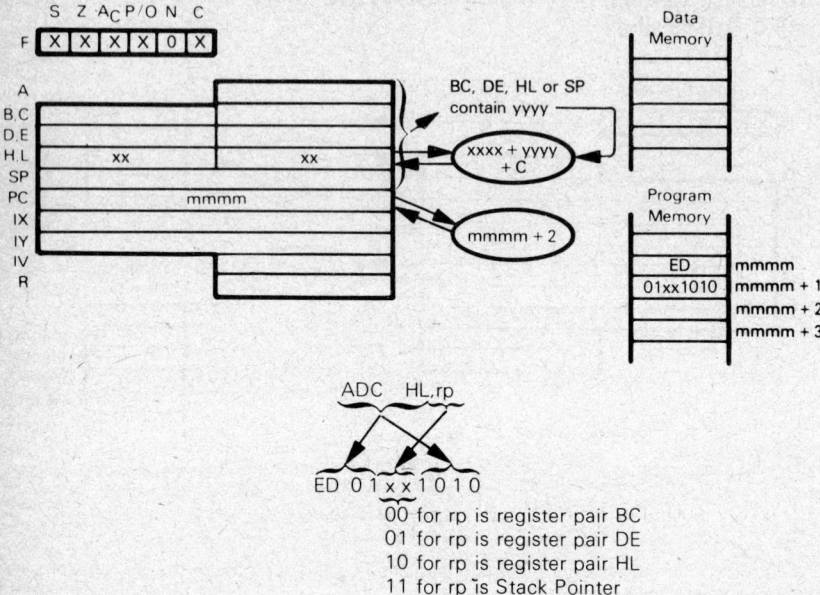

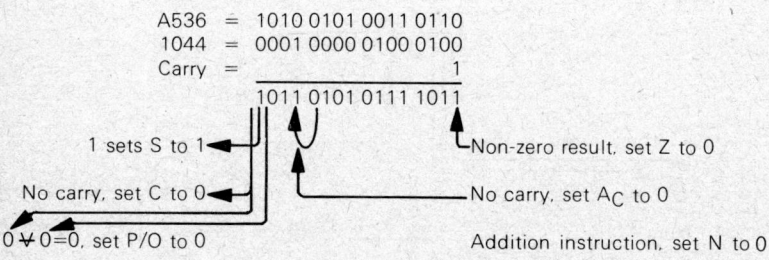

00 for rp is register pair BC
01 for rp is register pair DE
10 for rp is register pair HL
11 for rp is Stack Pointer

Add the 16-bit value from either the BC, DE, HL register pair or the Stack Pointer, and the Carry status, to the HL register pair.

Suppose HL contains $A536_{16}$, BC contains $1044_{16}$, and Carry=1. After execution of

ADC HL,BC

the HL register pair will contain:

```
A536  = 1010 0101 0011 0110
1044  = 0001 0000 0100 0100
Carry =                    1
        ───────────────────
        1011 0101 0111 1011
```

1 sets S to 1

Non-zero result, set Z to 0

No carry, set C to 0

No carry, set $A_C$ to 0

0 ⩝ 0=0, set P/O to 0

Addition instruction, set N to 0

The ADC instruction is most frequently used in multibyte addition for the second and subsequent bytes.

## ADD A,data — ADD IMMEDIATE TO ACCUMULATOR

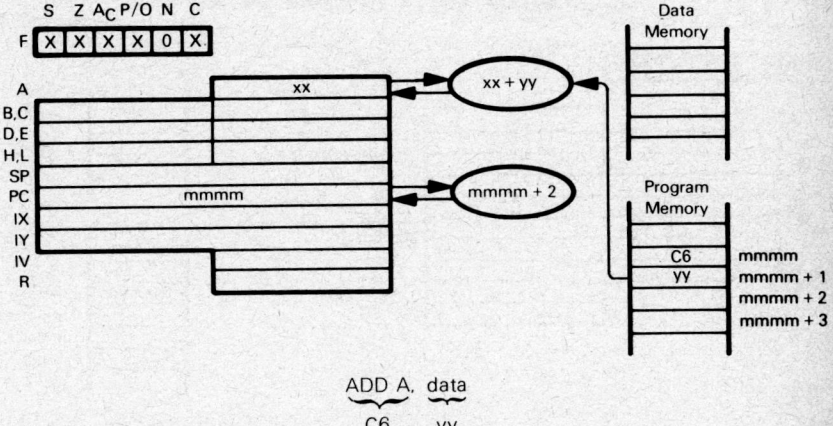

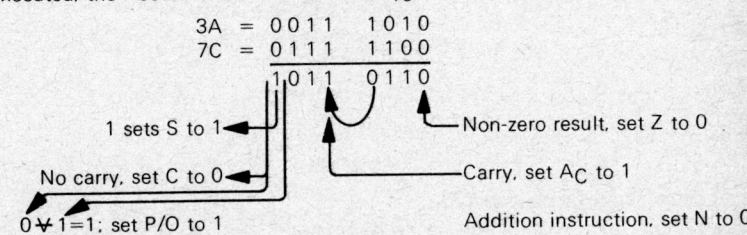

Add the contents of the next program memory byte to the Accumulator.

Suppose xx=$3A_{16}$, yy=$7C_{16}$, and Carry=0. After the instruction

$$\text{ADD} \quad \text{A,7CH}$$

has executed, the Accumulator will contain $B6_{16}$:

```
3A =  0011 1010
7C =  0111 1100
     1,011 0110
```

- 1 sets S to 1
- Non-zero result, set Z to 0
- No carry, set C to 0
- Carry, set $A_C$ to 1
- 0 ⊻ 1=1; set P/O to 1
- Addition instruction, set N to 0

This is a routine data manipulation instruction.

# ADD A,reg — ADD CONTENTS OF REGISTER TO ACCUMULATOR

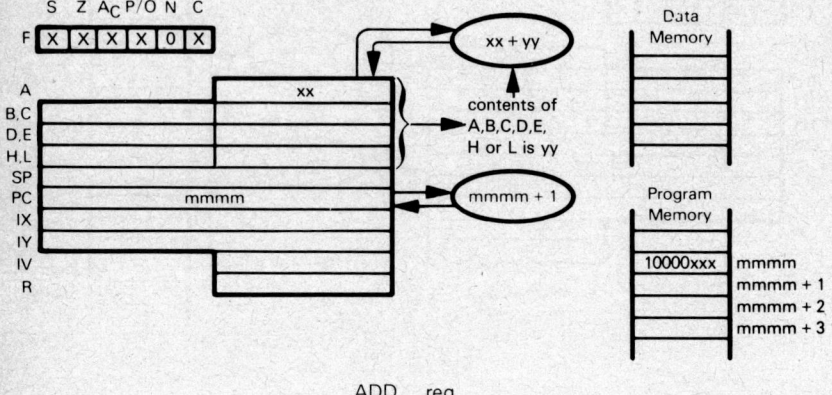

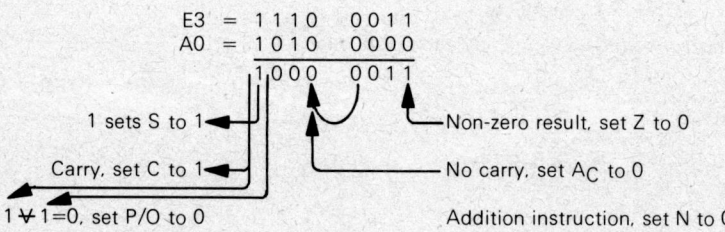

Add the contents of Register A, B, C, D, E, H or L to the Accumulator.

Suppose xx=$E3_{16}$, Register E contains $A0_{16}$. After execution of

ADD A,E

the Accumulator will contain $83_{16}$:

```
       E3 =  1 1 1 0   0 0 1 1
       A0 =  1 0 1 0   0 0 0 0
            1,0 0 0   0 0 1 1
```

- 1 sets S to 1
- Carry, set C to 1
- 1 ⊻ 1=0, set P/O to 0
- Non-zero result, set Z to 0
- No carry, set $A_C$ to 0
- Addition instruction, set N to 0

This is a routine data manipulation instruction.

# ADD A,(HL) — ADD MEMORY TO ACCUMULATOR
# ADD A,(IX+disp)
# ADD A,(IY+disp)

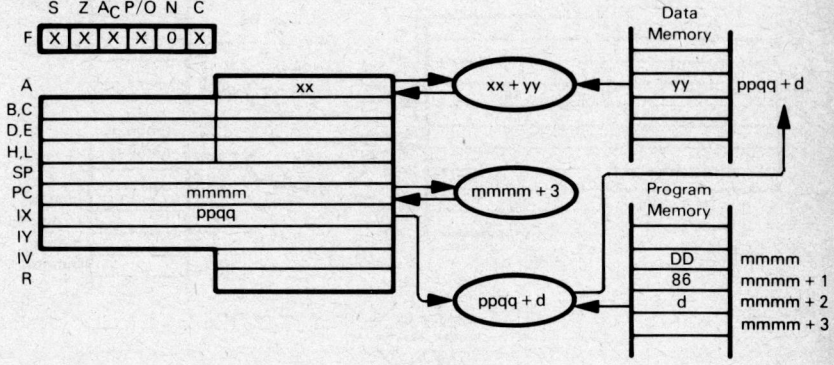

The illustration shows execution of ADD A,(IX+disp).

$$\underbrace{\text{ADD A,(IX+disp)}}_{\text{DD} \quad 86 \quad d}$$

Add the contents of memory location (specified by the sum of the contents of the IX register and the displacement digit d) to the contents of the Accumulator.

Suppose $ppqq=4000_{16}$, $xx=1A_{16}$, and memory location $400F_{16}$ contains $50_{16}$. After the instruction

$$\text{ADD A,(IX+0FH)}$$

has executed, the Accumulator will contain $6A_{16}$.

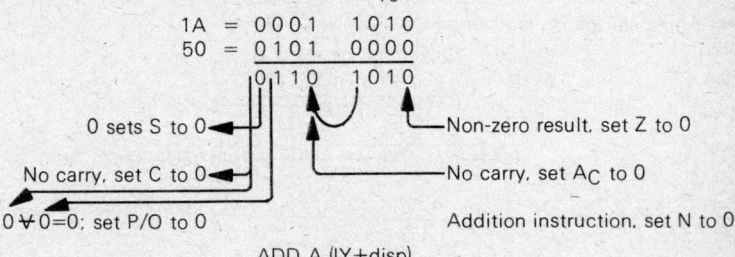

$$\underbrace{\text{ADD A,(IY+disp)}}_{\text{FD} \quad 86 \quad d}$$

This instruction is identical to ADD A,(IX+disp), except that it uses the IY register instead of the IX register.

$$\underbrace{\text{ADD A,(HL)}}_{86}$$

This version of the instruction adds the contents of memory location, specified by the contents of the HL register pair, to the Accumulator.

The ADD instruction is a routine data manipulation instruction.

# ADD HL,rp — ADD REGISTER PAIR TO H AND L

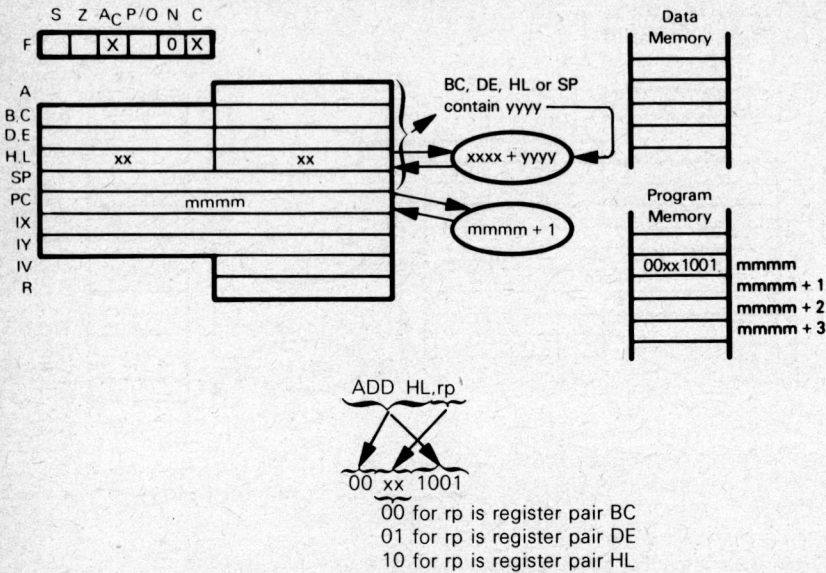

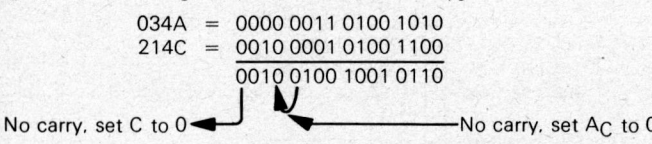

00 for rp is register pair BC
01 for rp is register pair DE
10 for rp is register pair HL
11 for rp is Stack Pointer

Add the 16-bit value from either the BC, DE, HL register pair or the Stack Pointer to the HL register pair.

Suppose HL contains $034A_{16}$ and BC contains $214C_{16}$. After the instruction

ADD HL,BC

has executed, the HL register pair will contain $2496_{16}$.

```
034A = 0000 0011 0100 1010
214C = 0010 0001 0100 1100
       0010 0100 1001 0110
```

No carry, set C to 0 ◄─────────────────── No carry, set $A_C$ to 0

Addition instruction, set N to 0

The ADD HL,HL instruction is equivalent to a 16-bit left shift.

## ADD xy,rp — ADD REGISTER PAIR TO INDEX REGISTER

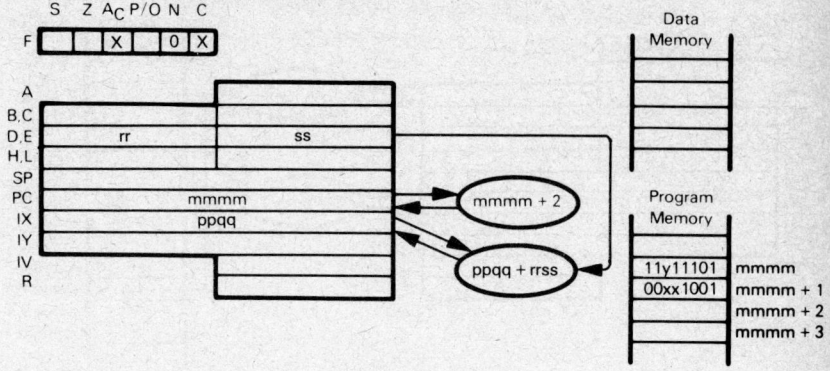

The illustration shows execution of ADD IX,DE.

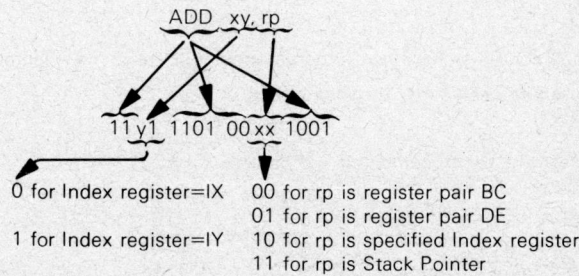

0 for Index register=IX
1 for Index register=IY

00 for rp is register pair BC
01 for rp is register pair DE
10 for rp is specified Index register
11 for rp is Stack Pointer

Add the contents of the specified register pair to the contents of the specified Index register.

Suppose IY contains $4FF0_{16}$ and BC contains $000F_{16}$. After the instruction

$$\text{ADD IY,BC}$$

has executed, Index Register IY will contain $4FFF_{16}$.

# AND data — AND IMMEDIATE WITH ACCUMULATOR

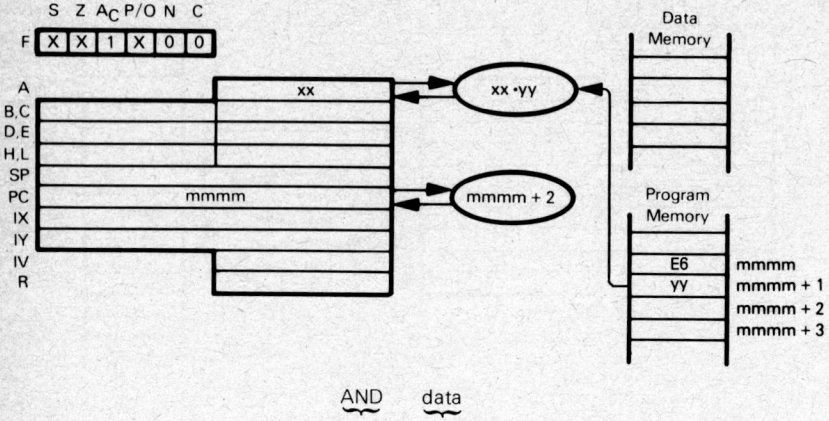

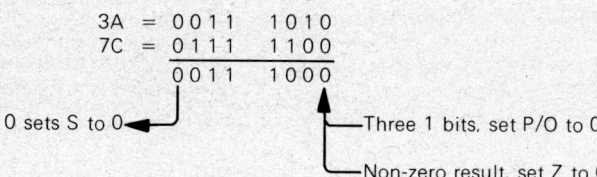

AND the contents of the next program memory byte to the Accumulator.

Suppose xx=$3A_{16}$. After the instruction

AND 7CH

has executed, the Accumulator will contain $38_{16}$.

```
3A =  0011 1010
7C =  0111 1100
      0011 1000
```

0 sets S to 0 ←

→ Three 1 bits, set P/O to 0

→ Non-zero result, set Z to 0

This is a routine logical instruction; it is often used to turn bits "off". For example, the instruction

AND 7FH

will unconditionally set the high order Accumulator bit to 0.

# AND reg — AND REGISTER WITH ACCUMULATOR

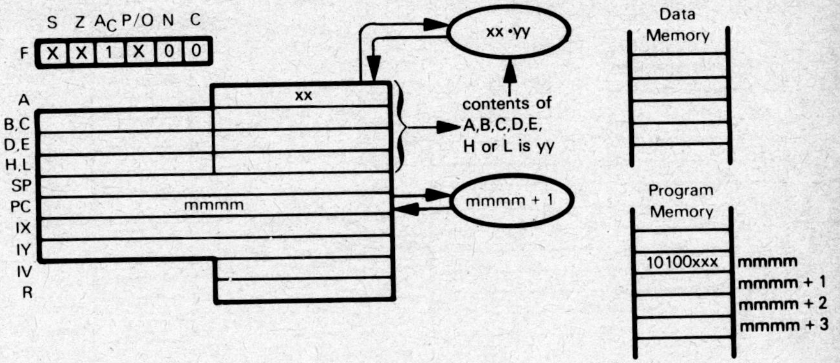

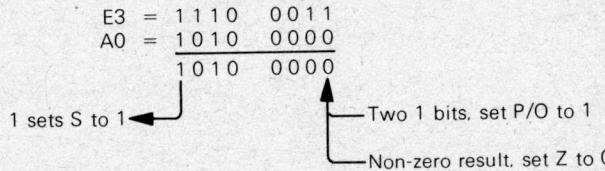

```
  AND    reg
 10100   xxx
         ───
         000  for reg=B
         001  for reg=C
         010  for reg=D
         011  for reg=E
         100  for reg=H
         101  for reg=L
         111  for reg=A
```

AND the Accumulator with the contents of Register A, B, C, D, E, H or L. Save the result in the Accumulator.

Suppose xx=$E3_{16}$, and Register E contains $A0_{16}$. After the instruction

$$\text{AND E}$$

has executed, the Accumulator will contain $A0_{16}$.

```
E3 = 1110 0011
A0 = 1010 0000
     ─────────
     1010 0000
```

1 sets S to 1 ◄─────

─── Two 1 bits, set P/O to 1

─── Non-zero result, set Z to 0

AND is a frequently used logical instruction.

# AND (HL) — AND MEMORY WITH ACCUMULATOR
# AND (IX+disp)
# AND (IY+disp)

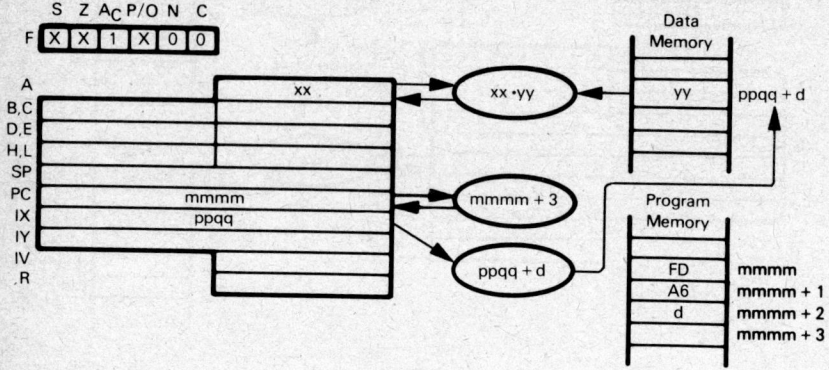

The illustration shows execution of AND (IY+disp).

$$\underbrace{\text{AND (IY+disp)}}_{\text{FD \quad A6 \quad d}}$$

AND the contents of memory location (specified by the sum of the contents of the IY register and the displacement digit d) with the Accumulator.

Suppose xx=E3$_{16}$, ppqq=4000$_{16}$, and memory location 400F$_{16}$ contains A0$_{16}$. After the instruction

$$\text{AND (IY+0FH)}$$

has executed, the Accumulator will contain A0$_{16}$.

```
E3 =  1110  0111
A0 =  1010  0000
      1010  0000
```

1 sets S to 1 ◀

Two 1 bits, set P/O to 1

Non-zero result, set Z to 0

$$\underbrace{\text{AND (IX+disp)}}_{\text{DD \quad A6 \quad d}}$$

This instruction is identical to AND (IY+disp), except that it uses the IX register instead of the IY register.

$$\underbrace{\text{AND (HL)}}_{\text{A6}}$$

AND the contents of the memory location (specified by the contents of the HL register pair) with the Accumulator.

AND is a frequently used logical instruction.

# BIT b,reg — TEST BIT b IN REGISTER reg

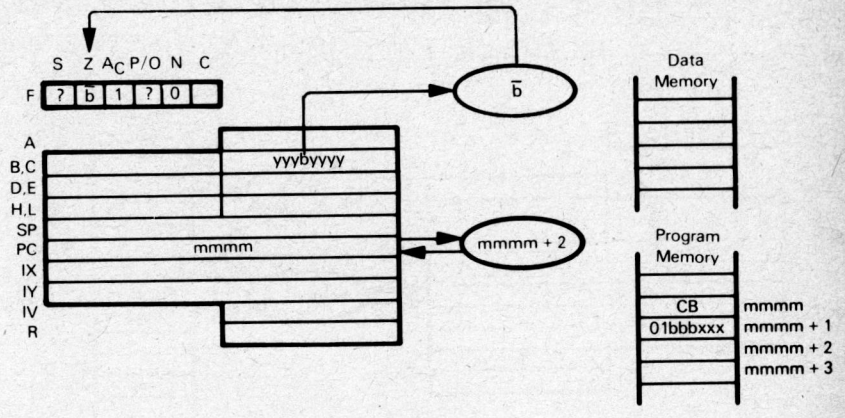

| | BIT | b, | reg | |
|---|---|---|---|---|
| | CB 01 | bbb | xxx | |
| Bit Tested | | | | Register |
| 0 | | 000 | 000 | B |
| 1 | | 001 | 001 | C |
| 2 | | 010 | 010 | D |
| 3 | | 011 | 011 | E |
| 4 | | 100 | 100 | H |
| 5 | | 101 | 101 | L |
| 6 | | 110 | 111 | A |
| 7 | | 111 | | |

Place complement of indicated register's specified bit in Z flag of F register.

Suppose Register C contains 1110 1111. The instruction BIT 4,C will then set the Z flag to 1, while bit 4 in Register C remains 0. Bit 0 is the least significant bit.

# BIT b,(HL) — TEST BIT b OF INDICATED MEMORY POSITION
# BIT b,(IX+disp)
# BIT b,(IY+disp)

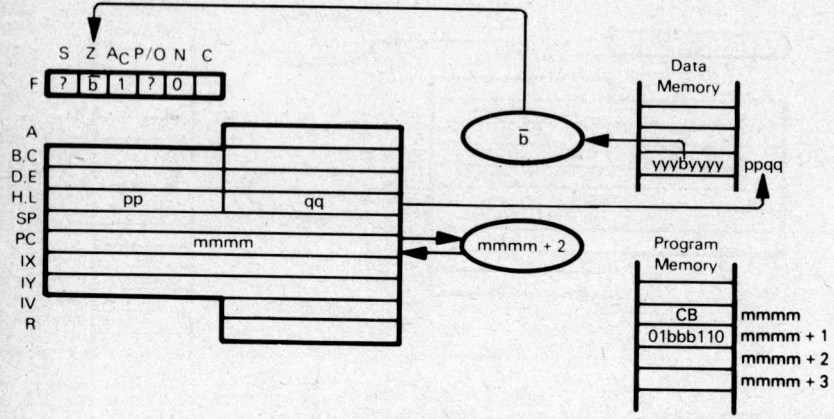

The illustration shows execution of BIT 4,(HL). Bit 0 is the least significant bit.

| BIT | b | (HL) |
|---|---|---|
| CB 01 | bbb | 110 |

| Bit Tested | bbb |
|---|---|
| 0 | 000 |
| 1 | 001 |
| 2 | 010 |
| 3 | 011 |
| 4 | 100 |
| 5 | 101 |
| 6 | 110 |
| 7 | 111 |

Test indicated bit within memory position specified by the contents of Register HL, and place bit's complement in Z flag of the F register.

Suppose HL contains 4000H and bit 3 in memory location 4000H contains 1. The instruction

BIT 3,(HL)

will then set the Z flag to 0, while bit 3 in memory location 4000H remains 1.

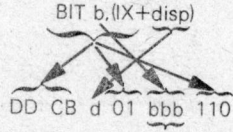

bbb is the same as in BIT b,(HL)

Examine specified bit within memory location indicated by the sum of Index Register IX and disp. Place the complement in the Z flag of the F register.

Suppose Index Register IX contains 4000H and bit 4 of memory location 4004H is 0. The instruction

$$\text{BIT 4,(IX+4H)}$$

will then set the Z flag to 1, while bit 4 of memory location 4004H remains 0.

bbb is the same as in BIT b,(HL)

This instruction is identical to BIT b,(IX+disp), except that it uses the IY register instead of the IX register.

# CALL label — CALL THE SUBROUTINE IDENTIFIED IN THE OPERAND

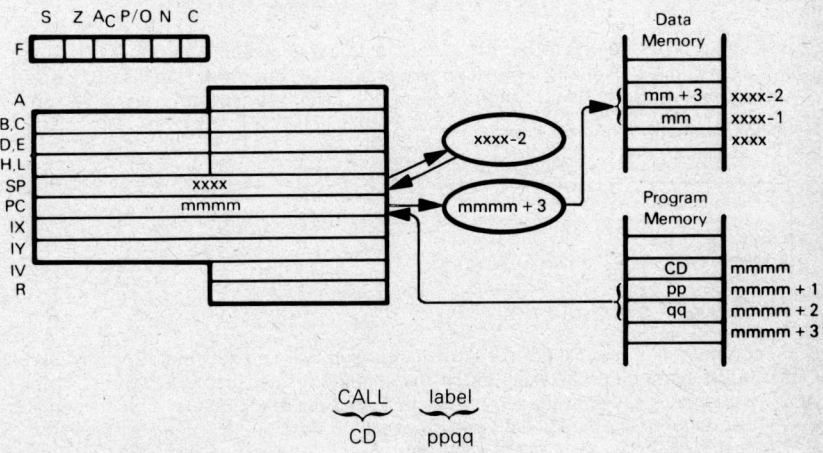

Store the address of the instruction following the CALL on the top of the stack: the top of the stack is a data memory byte addressed by the Stack Pointer. Then subtract 2 from the Stack Pointer in order to address the new top of stack. Move the 16-bit address contained in the second and third CALL instruction object program bytes to the Program Counter. The second byte of the CALL instruction is the low-order half of the address, and the third byte is the high-order byte.

Consider the instruction sequence:

```
        CALL    SUBR
        AND     7CH
        -
        -
        -
 SUBR
```

After the instruction has executed, the address of the AND instruction is saved at the top of the stack. The Stack Pointer is decremented by 2. The instruction labeled SUBR will be executed next.

# CALL condition,label — CALL THE SUBROUTINE IDENTIFIED IN THE OPERAND IF CONDITION IS SATISFIED

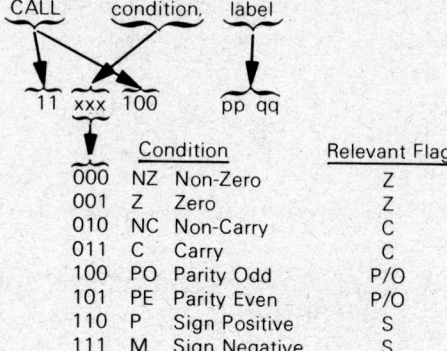

| | | Condition | Relevant Flag |
|---|---|---|---|
| 000 | NZ | Non-Zero | Z |
| 001 | Z | Zero | Z |
| 010 | NC | Non-Carry | C |
| 011 | C | Carry | C |
| 100 | PO | Parity Odd | P/O |
| 101 | PE | Parity Even | P/O |
| 110 | P | Sign Positive | S |
| 111 | M | Sign Negative | S |

This instruction is identical to the CALL instruction, except that the identified subroutine will be called only if the condition is satisfied; otherwise, the instruction sequentially following the CALL condition instruction will be executed.

Consider the instruction sequence:

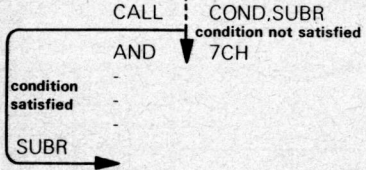

If the condition is not satisfied, the AND instruction will be executed after the CALL COND,SUBR instruction has executed. If the condition is satisfied, the address of the AND instruction is saved at the top of the stack, and the Stack Pointer is decremented by 2. The instruction labeled SUBR will be executed next.

# CCF — COMPLEMENT CARRY FLAG

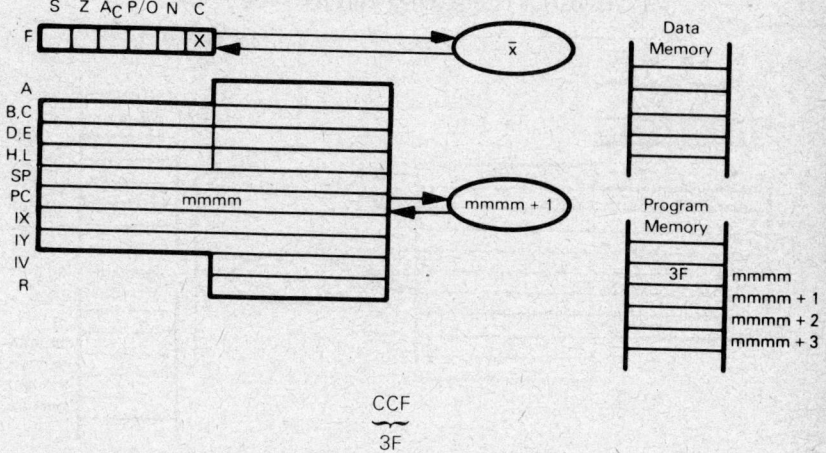

$$\frac{\text{CCF}}{\text{3F}}$$

Complement the Carry flag. No other status or register contents are affected.

# CP data — COMPARE IMMEDIATE DATA WITH ACCUMULATOR CONTENTS

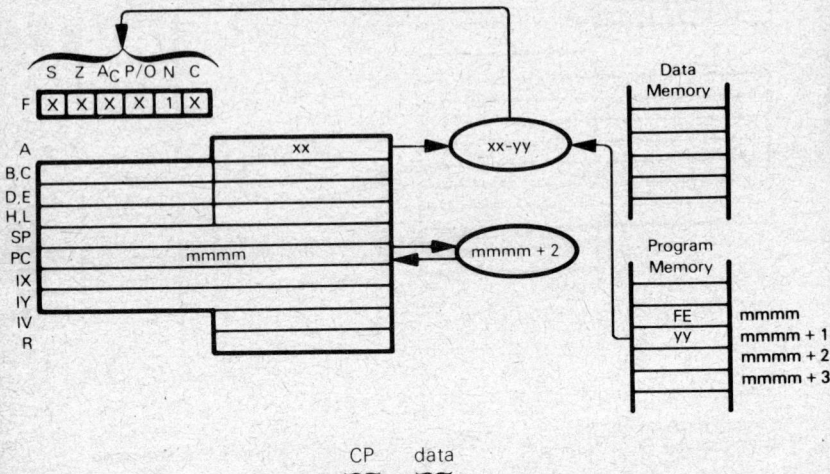

```
CP   data
⏟    ⏟
FE   yy
```

Subtract the contents of the second object code byte from the contents of the Accumulator, treating both numbers as simple binary data. Discard the result; i.e., leave the Accumulator alone, but modify the status flags to reflect the result of the subtraction.

Suppose xx=$E3_{16}$ and the second byte of the CP instruction object code contains $A0_{16}$. After the instruction

CP 0A0H

has executed, the Accumulator will still contain $E3_{16}$, but statuses will be modified as follows:

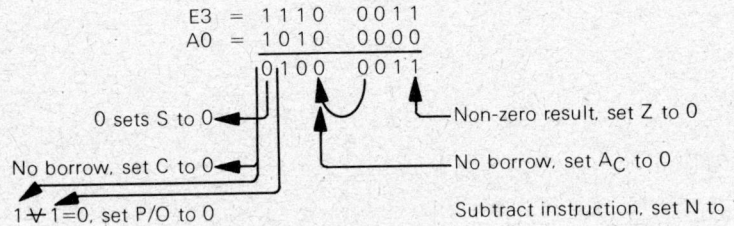

Notice that the resulting carry is complemented.

# CP reg — COMPARE REGISTER WITH ACCUMULATOR

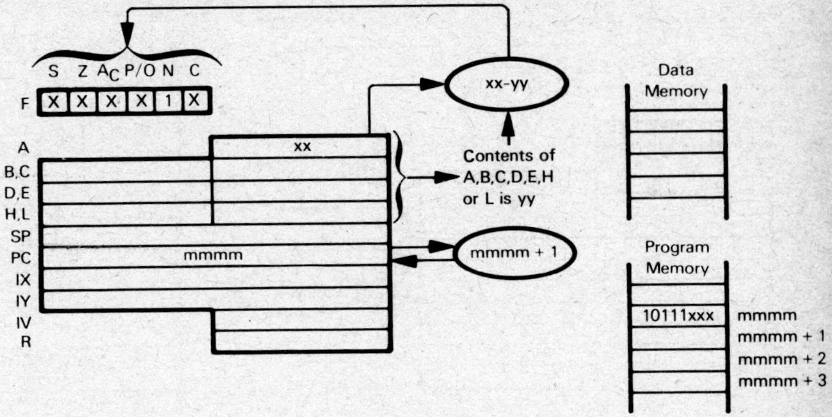

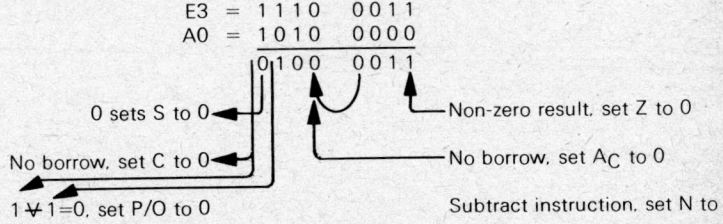

```
  CP     reg
10111    xxx
         000   for reg=B
         001   for reg=C
         010   for reg=D
         011   for reg=E
         100   for reg=H
         101   for reg=L
         111   for reg=A
```

Subtract the contents of Register A, B, C, D, E, H or L from the contents of the Accumulator, treating both numbers as simple binary data. Discard the result; i.e., leave the Accumulator alone, but modify status flags to reflect the result of the subtraction.

Suppose $xx = E3_{16}$ and Register B contains $A0_{16}$. After the instruction

CP B

has executed, the Accumulator will still contain $E3_{16}$, but statuses will be modified as follows:

```
E3 =  1110 0011
A0 =  1010 0000
    0 0100 0011
```

— 0 sets S to 0
— Non-zero result, set Z to 0
— No borrow, set C to 0
— No borrow, set $A_C$ to 0
— $1 \not\vee 1 = 0$, set P/O to 0
— Subtract instruction, set N to 1

Notice that the resulting carry is complemented.

# CP (HL) — COMPARE MEMORY WITH ACCUMULATOR
# CP (IX+disp)
# CP (IY+disp)

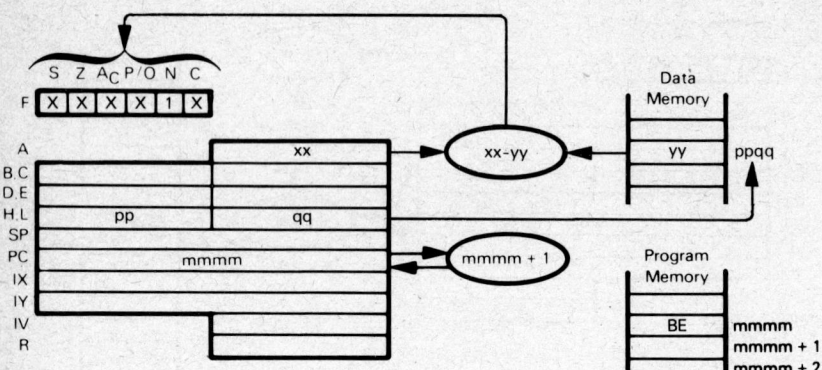

The illustration shows execution of CP (HL):

$$\underbrace{\text{CP (HL)}}_{\text{BE}}$$

Subtract the contents of memory location (specified by the contents of the HL register pair) from the contents of the Accumulator, treating both numbers as simple binary data. Discard the result; i.e., leave the Accumulator alone, but modify status flags to reflect the result of the subtraction.

Suppose xx=E3$_{16}$ and yy=A0$_{16}$. After execution of

CP (HL)

the Accumulator will still contain E3$_{16}$, but statuses will be modified as follows:

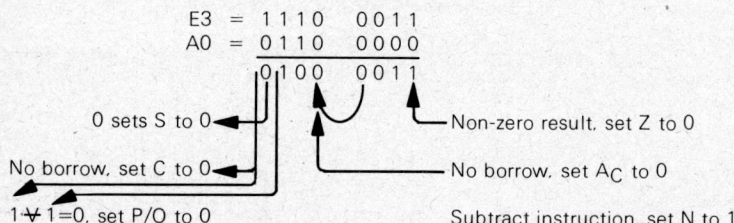

Notice that the resulting carry is complemented.

$$\underbrace{\text{CP (IX+disp)}}_{\text{DD BE}}\ \underbrace{}_{d}$$

6-44

Subtract the contents of memory location (specified by the sum of the contents of the IX register and the displacement value d) from the contents of the Accumulator, treating both numbers as simple binary data. Discard the result; i.e., leave the Accumulator alone, but modify status flags to reflect the result of the subtraction.

$$\underbrace{CP\ (IY+disp)}_{FD\ BE\ \ \ d}$$

This instruction is identical to CP (IX+disp), except that it uses the IY register instead of the IX register.

## CPD — COMPARE ACCUMULATOR WITH MEMORY. DECREMENT ADDRESS AND BYTE COUNTER

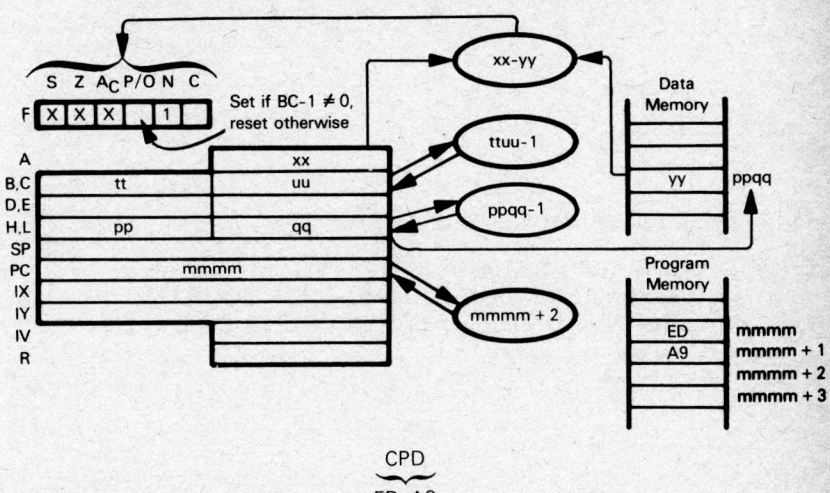

$$\underbrace{CPD}_{ED\ A9}$$

Compare the contents of the Accumulator with the contents of memory location (specified by the HL register pair). If A is equal to memory, set Z flag. Decrement the HL and BC register pairs. (BC is used as the Byte Counter.)

Suppose xx=E3$_{16}$, ppqq=4000$_{16}$, BC contains 0001$_{16}$, and yy=A0$_{16}$. After the instruction

CPD

has executed, the Accumulator will still contain E3$_{16}$, but statuses will be modified as follows:

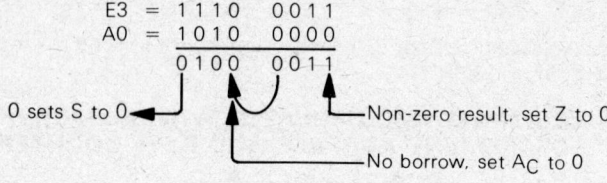

The P/O flag will be reset because BC-1=0

Subtract instruction involved, set N to 1

Carry not affected.

The HL register pair will contain 3FFF$_{16}$, and BC=0.

## CPDR — COMPARE ACCUMULATOR WITH MEMORY. DECREMENT ADDRESS AND BYTE COUNTER. CONTINUE UNTIL MATCH IS FOUND OR BYTE COUNTER IS ZERO

CPDR<br/>
<br/>
ED B9

This instruction is identical to CPD, except that it is repeated until a match is found or the byte counter is zero. After each data transfer, interrupts will be recognized and two refresh cycles will be executed.

Suppose the HL register pair contains 5000$_{16}$, the BC register pair contains 00FF$_{16}$, the Accumulator contains F9$_{16}$, and memory has contents as follows:

| Location | Contents |
|---|---|
| 5000$_{16}$ | AA$_{16}$ |
| 4FFF$_{16}$ | BC$_{16}$ |
| 4FFE$_{16}$ | 19$_{16}$ |
| 4FFD$_{16}$ | 7A$_{16}$ |
| 4FFC$_{16}$ | F9$_{16}$ |
| 4FFB$_{16}$ | DD$_{16}$ |

After execution of

CPDR

the P/O flag will be 1, the Z flag will be 1, the HL register pair will contain 4FFB$_{16}$, and the BC register pair will contain 00FA$_{16}$.

# CPI — COMPARE ACCUMULATOR WITH MEMORY. DECREMENT BYTE COUNTER. INCREMENT ADDRESS

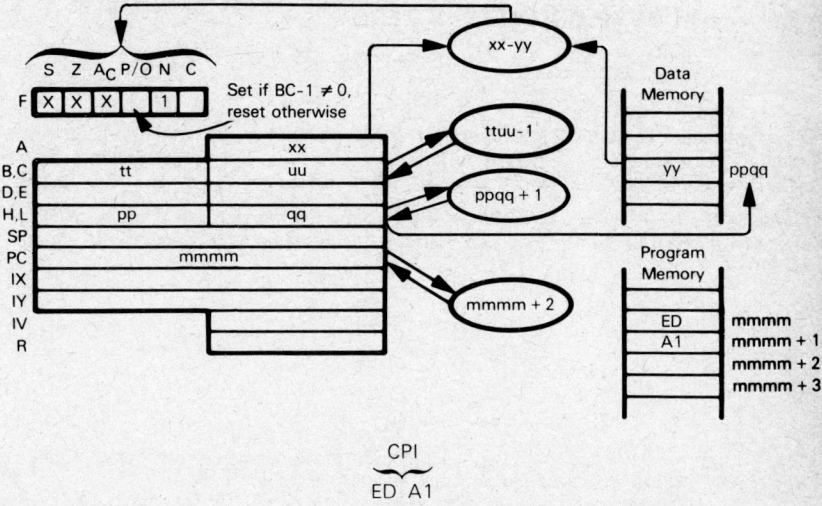

$$\underbrace{\text{CPI}}_{\text{ED A1}}$$

Compare the contents of the Accumulator with the contents of memory location (specified by the HL register pair). If A is equal to memory, set the Z flag. Increment the HL register pair and decrement the BC register pair (BC is used as Byte Counter).

Suppose $xx = E3_{16}$, $ppqq = 4000_{16}$, BC contains $0032_{16}$, and $yy = E3_{16}$. After the instruction

CPI

has executed, the Accumulator will still contain $E3_{16}$, but statuses will be modified as follows:

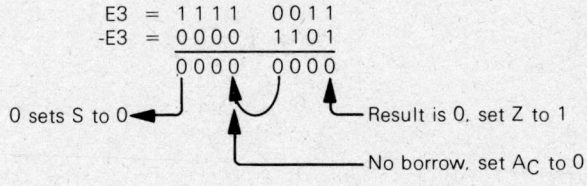

0 sets S to 0

Result is 0, set Z to 1

No borrow, set $A_C$ to 0

The P/O flag will be set because BC−1 ≠ 0.

Subtract instruction involved, set N to 1.

Carry not affected.

The HL register pair will contain $4001_{16}$, and BC will contain $0031_{16}$.

# CPIR — COMPARE ACCUMULATOR WITH MEMORY.
## DECREMENT BYTE COUNTER.
## INCREMENT ADDRESS.
## CONTINUE UNTIL MATCH IS FOUND OR BYTE COUNTER IS ZERO

CPIR

ED B1

This instruction is identical to CPI, except that it is repeated until a match is found or the byte counter is zero. After each data transfer, interrupts will be recognized and two refresh cycles will be executed.

Suppose the HL register pair contains $4500_{16}$, the BC register pair contains $00FF_{16}$, the Accumulator contains $F9_{16}$, and memory has contents as follows:

| Location | Contents |
|---|---|
| $4500_{16}$ | $AA_{16}$ |
| $4501_{16}$ | $15_{16}$ |
| $4502_{16}$ | $F9_{16}$ |

After execution of

CPIR

the P/O flag will be 1, and the Z flag will be 1. The HL register pair will contain $4503_{16}$, and the BC register pair will contain $00FC_{16}$.

## CPL — COMPLEMENT THE ACCUMULATOR

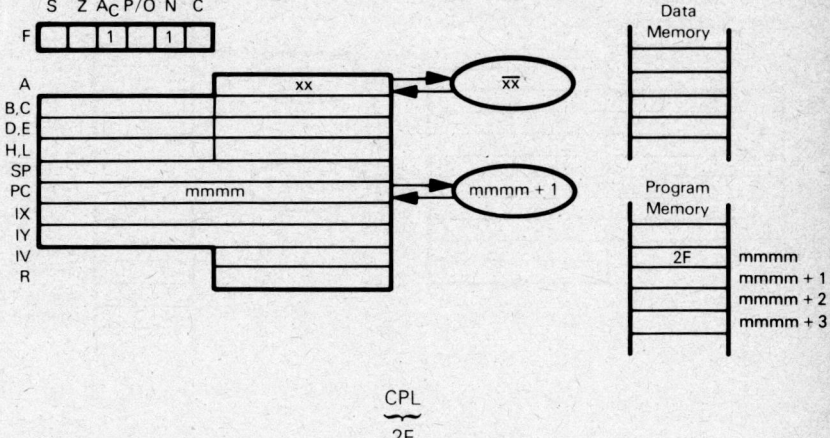

$$\underbrace{\text{CPL}}_{\text{2F}}$$

Complement the contents of the Accumulator. No other register's contents are affected.

Suppose the Accumulator contains $3A_{16}$. After the instruction

CPL

has executed, the Accumulator will contain $C5_{16}$.

$$3A = 0011 \quad 1010$$
$$\text{Complement} = 1100 \quad 0101$$

This is a routine logical instruction. You need not use it for binary subtraction; there are special subtract instructions (SUB, SBC).

## DAA — DECIMAL ADJUST ACCUMULATOR

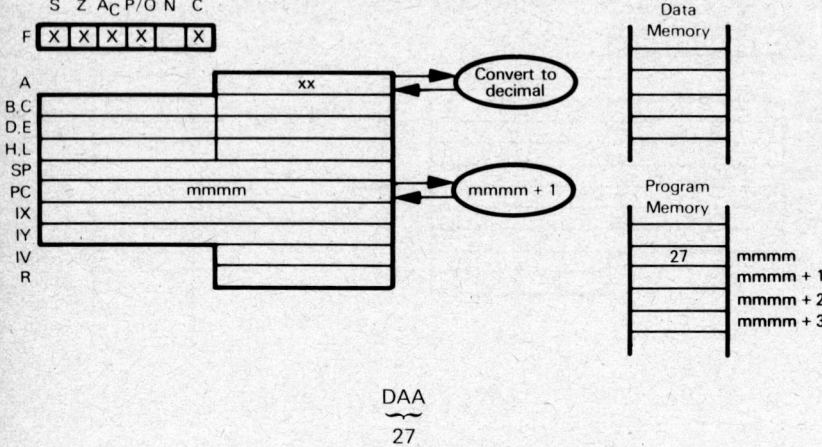

$$\underset{27}{\underbrace{\text{DAA}}}$$

Convert the contents of the Accumulator to binary-coded decimal form. This instruction should only be used after adding or subtracting two BCD numbers; i.e., look upon ADD DAA or ADC DAA or INC DAA or SUB DAA or SBC DAA or DEC DAA or NEG DAA as compound, decimal arithmetic instructions which operate on BCD sources to generate BCD answers.

Suppose the Accumulator contains $39_{16}$ and the B register contains $47_{16}$. After the instructions

                ADD B
                DAA

have executed, the Accumulator will contain $86_{16}$, not $80_{16}$.

Z80 CPU logic uses the values in the Carry and Auxiliary Carry, as well as the Accumulator contents, in the Decimal Adjust operation.

# DEC reg — DECREMENT REGISTER CONTENTS

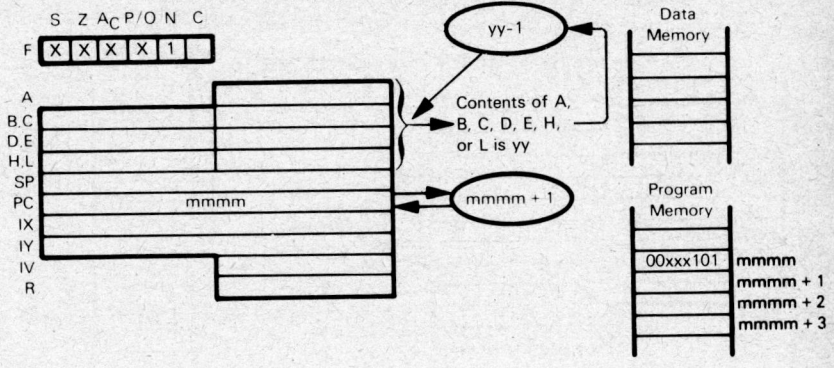

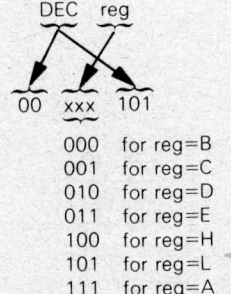

```
000   for reg=B
001   for reg=C
010   for reg=D
011   for reg=E
100   for reg=H
101   for reg=L
111   for reg=A
```

Subtract 1 from the contents of the specified register.

Suppose Register A contains $50_{16}$. After execution of

> DEC A

Register A will contain $4F_{16}$.

# DEC rp — DECREMENT CONTENTS OF SPECIFIED REGISTER PAIR
# DEC IX
# DEC IY

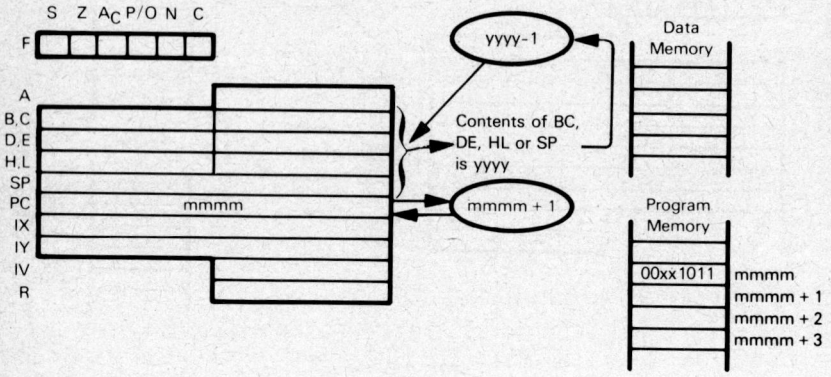

The illustration shows execution of DEC rp:

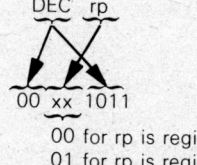

00 for rp is register pair BC
01 for rp is register pair DE
10 for rp is register pair HL
11 for rp is Stack Pointer

Subtract 1 from the 16-bit value contained in the specified register pair. No status flags are affected.

Suppose the H and L registers contain $2F00_{16}$. After the instruction

DEC HL

has executed, the H and L registers will contain $2EFF_{16}$.

DEC IX
DD 2B

Subtract 1 from the 16-bit value contained in the IX register.

DEC IY
FD 2B

Subtract 1 from the 16-bit value contained in the IY register.

Neither DEC rp, DEC IX nor DEC IY affects any of the status flags. This is a defect in the Z80 instruction set, inherited from the 8080. Whereas the DEC reg instruction is used in iterative instruction loops that use a counter with a value of 256 or less, the DEC rp (DEC IX or DEC IY) instruction must be used if the counter value is more than 256. Since the DEC rp instruction sets no status flags, other instructions must be added to simply test for a zero result. This is a typical loop form:

```
       LD    DE,DATA     ;LOAD INITIAL 16-BIT COUNTER VALUE
LOOP   -                 ;FIRST INSTRUCTION OF LOOP
       -
       -
       DEC   DE          ;DECREMENT COUNTER
       LD    A,D         ;TO TEST FOR ZERO, MOVE D TO A
       OR    E           ;THEN OR A WITH E
       JP    NZ,LOOP     ;RETURN IF NOT ZERO
```

# DEC (HL) — DECREMENT MEMORY CONTENTS
# DEC (IX+disp)
# DEC (IY+disp)

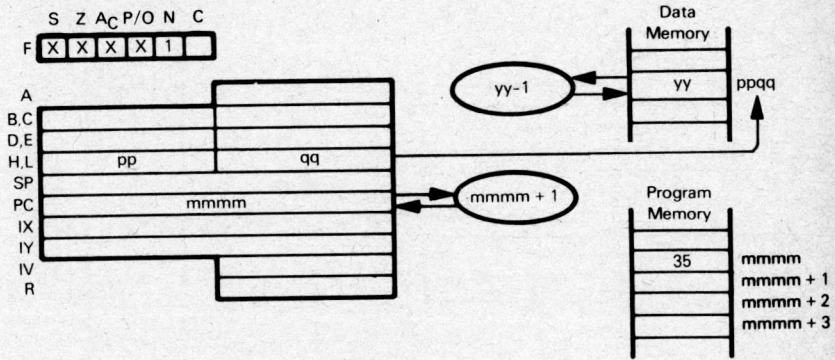

The illustration shows execution of DEC (HL):

Subtract 1 from the contents of memory location (specified by the contents of the HL register pair).

Suppose $ppqq = 4500_{16}$, $yy = 5F_{16}$. After execution of

$$DEC \ (HL)$$

memory location $4500_{16}$ will contain $5E_{16}$.

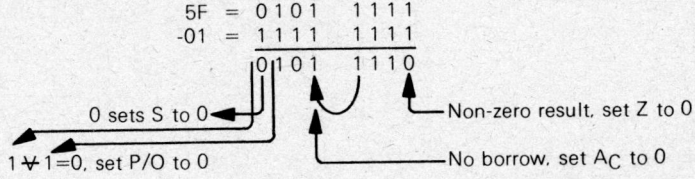

$$\underbrace{\text{DEC}}\ \underbrace{\text{(IX+disp)}}$$
$$\text{DD 35}\quad \text{d}$$

Subtract 1 from the contents of memory location (specified by the sum of the contents of the IX register and the displacement value d).

$$\underbrace{\text{DEC}}\ \underbrace{\text{(IY+disp)}}$$
$$\text{FD 35}\quad \text{d}$$

This instruction is identical to DEC (IX+disp), except that it uses the IY register instead of the IX register.

## DI — DISABLE INTERRUPTS

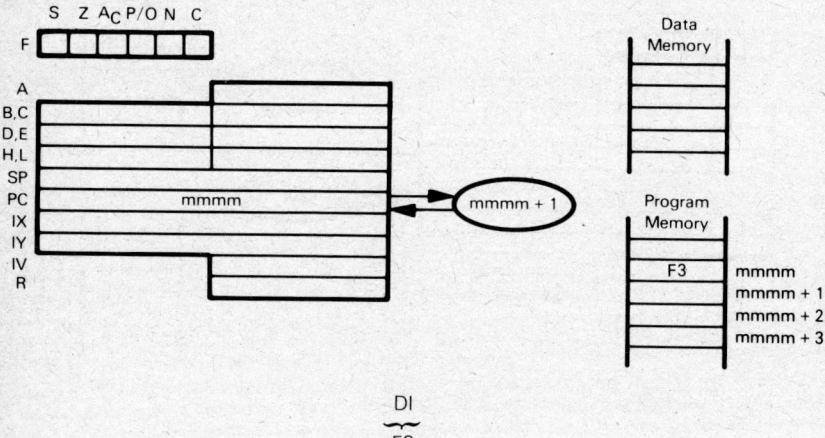

$$\underbrace{\text{DI}}$$
$$\text{F3}$$

When this instruction is executed, the maskable interrupt request is disabled and the $\overline{\text{INT}}$ input to the CPU will be ignored. Remember that when an interrupt is acknowledged, the maskable interrupt is automatically disabled.

The maskable interrupt request remains disabled until it is subsequently enabled by an EI instruction.

No registers or flags are affected by this instruction.

# DJNZ disp — JUMP RELATIVE TO PRESENT CONTENTS OF PROGRAM COUNTER IF REG B IS NOT ZERO

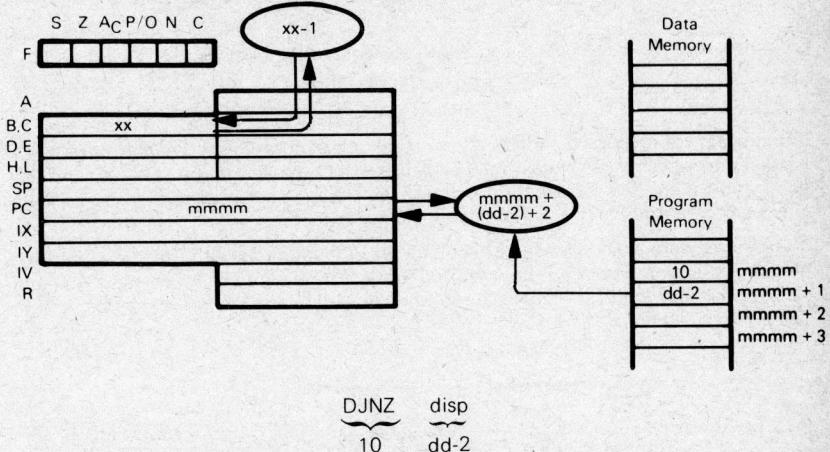

```
DJNZ   disp
 10    dd-2
```

Decrement Register B. If remaining contents are not zero, add the contents of the DJNZ instruction object code second byte and 2 to the Program Counter. The jump is measured from the address of the instruction operation code, and has a range of -126 to +129 bytes. The Assembler automatically adjusts for the twice-incremented PC.

If the contents of B are zero after decrementing, the next sequential instruction is executed.

The DJNZ instruction is extremely useful for any program loop operation, since the one instruction replaces the typical "decrement-then-branch on condition" instruction sequence.

# EI — ENABLE INTERRUPTS

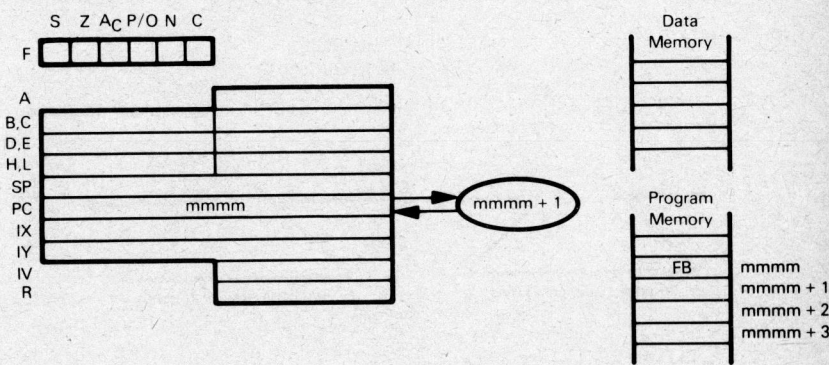

EI
$\overbrace{\text{FB}}$

Execution of this instruction causes interrupts to be enabled, but not until one more instruction executes.

Most interrupt service routines end with the two instructions:

```
EI              ;ENABLE INTERRUPTS
RET             ;RETURN TO INTERRUPTED PROGRAM
```

If interrupts are processed serially, then for the entire duration of the interrupt service routine all maskable interrupts are disabled — which means that in a multi-interrupt application there is a significant possibility for one or more interrupts to be pending when any interrupt service routine completes execution.

If interrupts were acknowledged as soon as the EI instructions had executed, then the Return instruction would not be executed. Under these circumstances, returns would stack up one on top of the other — and unnecessarily consume stack memory space. This may be illustrated as follows:

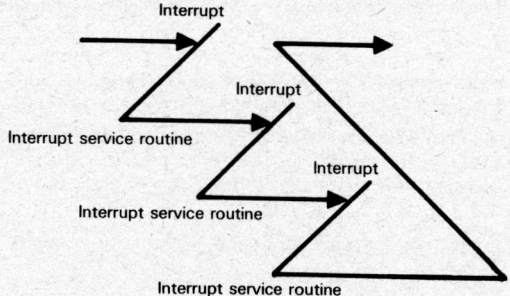

By inhibiting interrupts for one more instruction following execution of EI, the Z80 CPU ensures that the RET instruction gets executed in the sequence:

```
-
-
-
EI              ;ENABLE INTERRUPTS
RET             ;RETURN FROM INTERRUPT
```

It is not uncommon for interrupts to be kept disabled while an interrupt service routine is executing. Interrupts are processed serially:

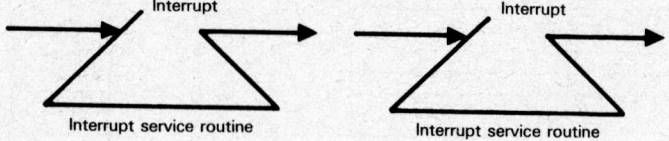

# EX AF,AF' — EXCHANGE PROGRAM STATUS AND ALTERNATE PROGRAM STATUS

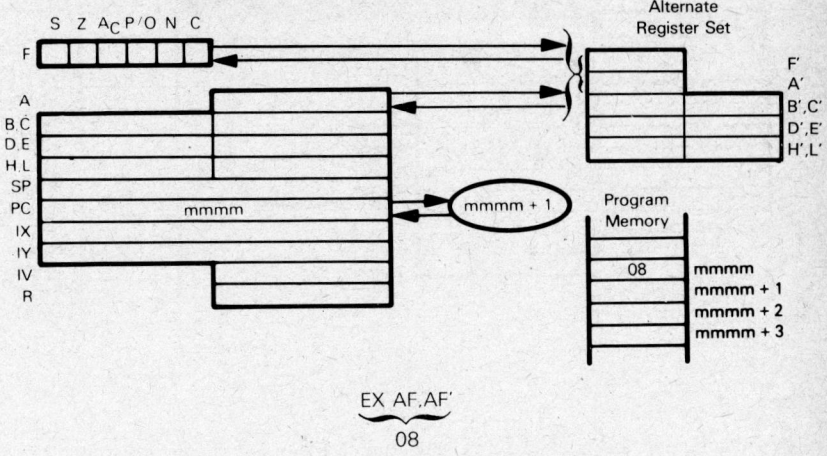

$$\underbrace{\text{EX AF,AF'}}_{08}$$

The two-byte contents of register pairs AF and A'F' are exchanged.

Suppose AF contains $4F99_{16}$ and A'F' contains $10AA_{16}$. After execution of

EX AF,AF'

AF will contain $10AA_{16}$ and AF' will contain $4F99_{16}$.

# EX DE,HL — EXCHANGE DE AND HL CONTENTS

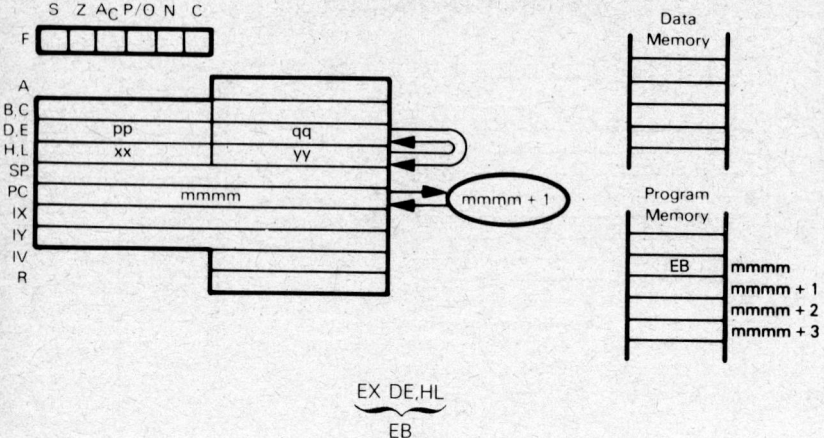

$$\underbrace{\text{EX DE,HL}}_{\text{EB}}$$

The D and E registers' contents are swapped with the H and L registers' contents.
Suppose $pp=03_{16}$, $qq=2A_{16}$, $xx=41_{16}$ and $yy=FC_{16}$. After the instruction

                    EX DE,HL

has executed, H will contain $03_{16}$, L will contain $2A_{16}$, D will contain $41_{16}$ and E will contain $FC_{16}$.

The two instructions:

                    EX DE,HL
                    LD A,(HL)

are equivalent to:

                    LD A,(DE)

but if you want to load data addressed by the D and E register into the B register,

                    EX DE,HL
                    LD B,(HL)

has no single instruction equivalent.

# EX (SP),HL — EXCHANGE CONTENTS OF REGISTER AND TOP OF STACK
# EX (SP),IX
# EX (SP),IY

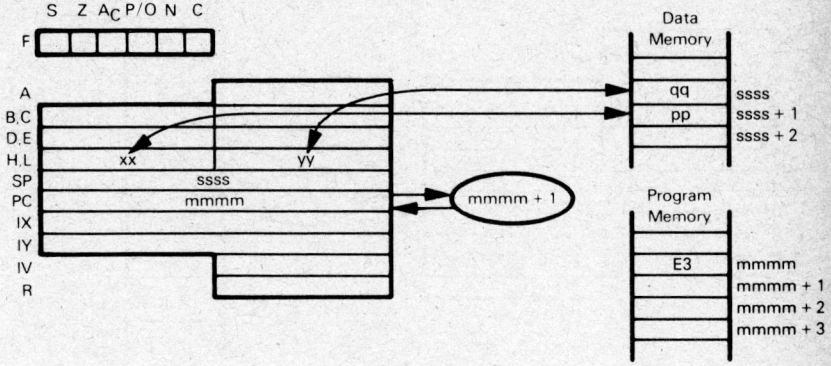

The illustration shows execution of EX (SP),HL.

$$\underbrace{\text{EX (SP),HL}}_{\text{E3}}$$

Exchange the contents of the L register with the top stack byte. Exchange the contents of the H register with the byte below the stack top.

Suppose $xx=21_{16}$, $yy=FA_{16}$, $pp=3A_{16}$, $qq=E2_{16}$. After the instruction

$$\text{EX (SP),HL}$$

has executed, H will contain $3A_{16}$, L will contain $E2_{16}$ and the two top stack bytes will contain $FA_{16}$ and $21_{16}$ respectively.

The EX (SP),HL instruction is used to access and manipulate data at the top of the stack.

$$\underbrace{\text{EX (SP),IX}}_{\text{DD E3}}$$

Exchange the contents of the IX register's low-order byte with the top stack byte. Exchange the IX register's high-order byte with the byte below the stack top.

$$\underbrace{\text{EX (SP),IY}}_{\text{FD E3}}$$

This instruction is identical to EX (SP),IX, but uses the IY register instead of the IX register.

# EXX — EXCHANGE REGISTER PAIRS AND ALTERNATE REGISTER PAIRS

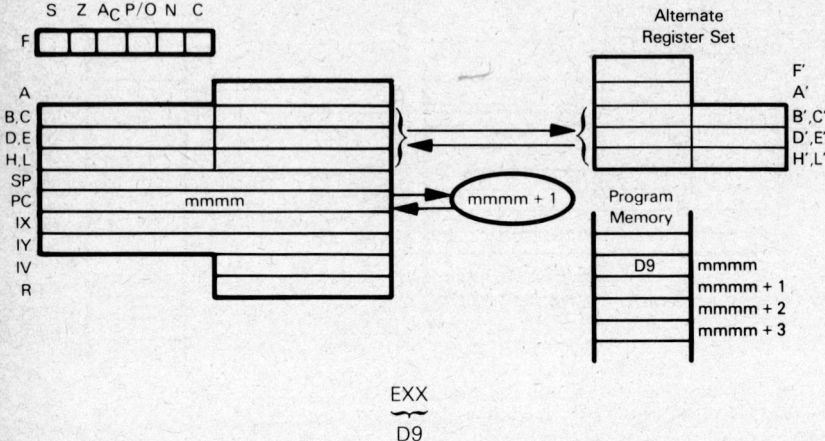

EXX
﹈
D9

The contents of register pairs BC, DE and HL are swapped with the contents of register pairs B'C', D'E', and H'L'.

Suppose register pairs BC, DE and HL contain $4901_{16}$, $5F00_{16}$ and $7251_{16}$ respectively, and register pairs B'C', D'E', H'L' contain $0000_{16}$, $10FF_{16}$ and $3333_{16}$ respectively. After the execution of

EXX

the registers will have the following contents:

BC: $0000_{16}$; DE: $10FF_{16}$; HL: $3333_{16}$;
B'C': $4901_{16}$; D'E': $5F00_{16}$; H'L': $7251_{16}$

This instruction can be used to exchange register banks to provide very fast interrupt response times.

# HALT

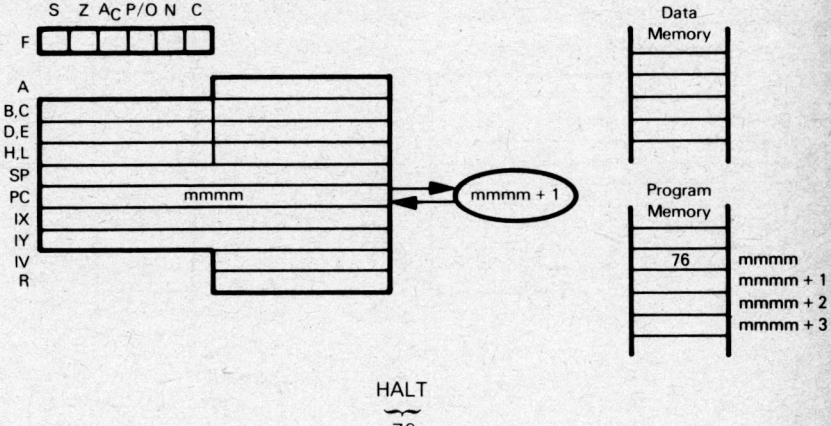

HALT
$\underbrace{\phantom{xx}}$
76

When the HALT instruction is executed, program execution ceases. The CPU requires an interrupt or a reset to restart execution. No registers or statuses are affected; however, memory refresh logic continues to operate.

# IM 0 — INTERRUPT MODE 0

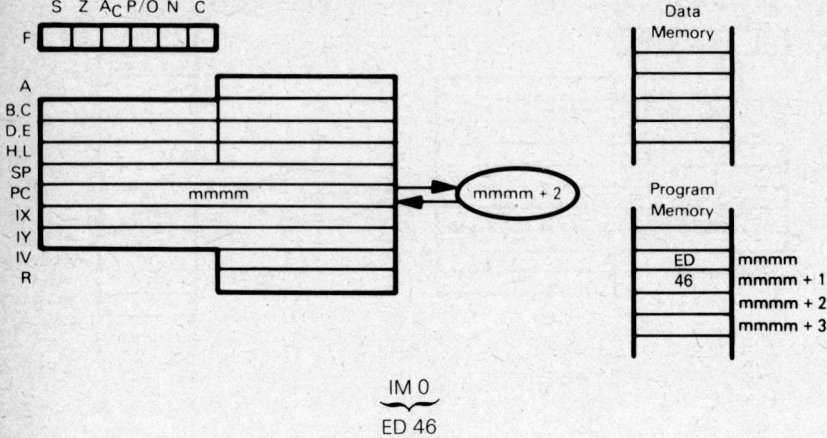

$$\underbrace{\text{IM 0}}_{\text{ED 46}}$$

This instruction places the CPU in interrupt mode 0. In this mode, the interrupting device will place an instruction on the Data Bus and the CPU will then execute that instruction. No registers or statuses are affected.

# IM 1 — INTERRUPT MODE 1

$$\underbrace{\text{IM 1}}_{\text{ED 56}}$$

This instruction places the CPU in interrupt mode 1. In this mode, the CPU responds to an interrupt by executing a restart (RST) to location $0038_{16}$.

# IM 2 — INTERRUPT MODE 2

$$\underbrace{\text{IM 2}}_{\text{ED 5E}}$$

This instruction places the CPU in interrupt mode 2. In this mode, the CPU performs an indirect call to any specified location in memory. A 16-bit address is formed using the contents of the Interrupt Vector (IV) register for the upper eight bits, while the lower eight bits are supplied by the interrupting device. Refer to Chapter 5 for a full description of interrupt modes. No registers or statuses are affected by this instruction.

# IN A,(port) — INPUT TO ACCUMULATOR

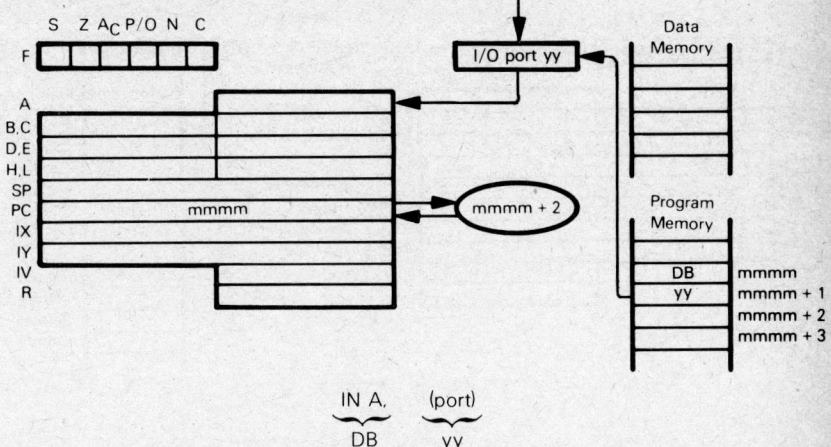

Load a byte of data into the Accumulator from the I/O port (identified by the second IN instruction object code byte).

Suppose $36_{16}$ is held in the buffer of I/O port $1A_{16}$. After the instruction

IN A,(1AH)

has executed, the Accumulator will contain $36_{16}$.

The IN instruction does not affect any statuses.

Use of the IN instruction is very hardware dependent. Valid I/O port addresses are determined by the way in which I/O logic has been implemented. It is also possible to design a microcomputer system that accesses external logic using memory reference instructions with specific memory addresses.

# INC reg — INCREMENT REGISTER CONTENTS

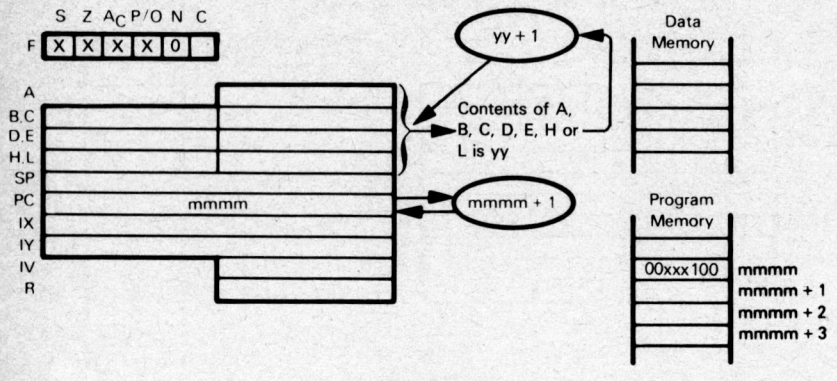

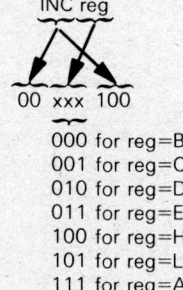

```
00 xxx 100
   ‾‾‾
   000 for reg=B
   001 for reg=C
   010 for reg=D
   011 for reg=E
   100 for reg=H
   101 for reg=L
   111 for reg=A
```

Add 1 to the contents of the specified register.

Suppose Register E contains $A8_{16}$. After execution of

INC E

Register E will contain $A9_{16}$.

# INC rp — INCREMENT CONTENTS OF SPECIFIED REGISTER PAIR
# INC IX
# INC IY

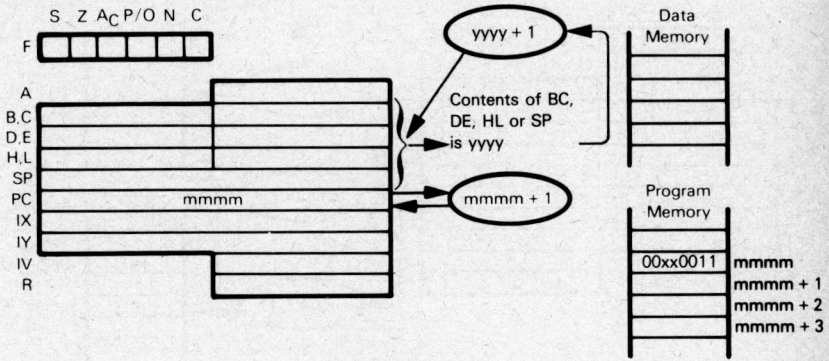

The illustration shows execution of INC rp:

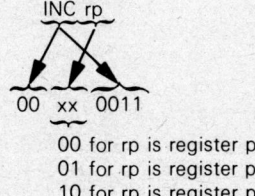

00 for rp is register pair BC
01 for rp is register pair DE
10 for rp is register pair HL
11 for rp is Stack Pointer

Add 1 to the 16-bit value contained in the specified register pair. No status flags are affected.

Suppose the D and E registers contain $2F7A_{16}$. After the instruction

INC DE

has executed, the D and E registers will contain $2F7B_{16}$.

$$\underbrace{\text{INC IX}}_{\text{DD 23}}$$

Add 1 to the 16-bit value contained in the IX register.

$$\underbrace{\text{INC IY}}_{\text{FD 23}}$$

Add 1 to the 16-bit value contained in the IY register.

Just like the DEC rp, DEC IX and DEC IY, neither INC rp, INC IX nor INC IY affects any status flags. This is a defect in the Z80 instruction set inherited from the 8080.

# INC (HL) — INCREMENT MEMORY CONTENTS
# INC (IX+disp)
# INC (IY+disp)

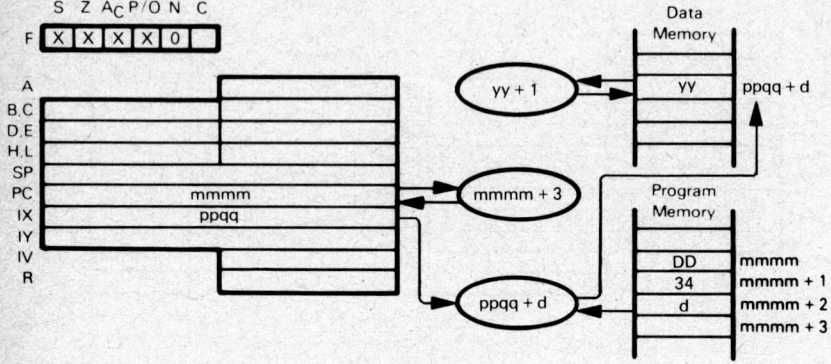

The illustration shows execution of INC (IX+d):

$$\underbrace{\text{INC (IX+disp)}}_{\text{DD 34}} \quad \underbrace{}_{d}$$

Add 1 to the contents of memory location (specified by the sum of the contents of Register IX and the displacement value d).

Suppose ppqq=$4000_{16}$ and memory location $400F_{16}$ contains $36_{16}$. After execution of the instruction

$$\text{INC (IX+0FH)}$$

memory location $400F_{16}$ will contain $37_{16}$.

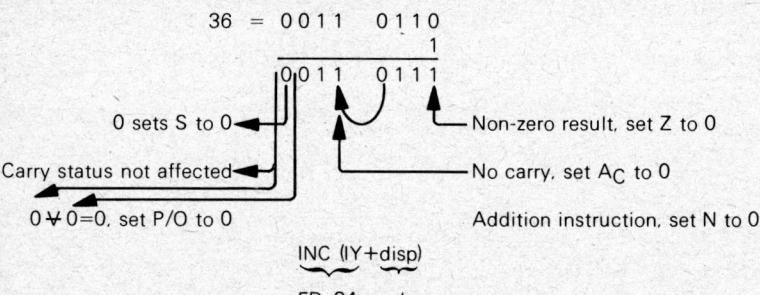

$$\underbrace{\text{INC (IY+disp)}}_{\text{FD 34}} \quad \underbrace{}_{d}$$

This instruction is identical to INC (IX+disp), except that it uses the IY register instead of the IX register.

$$\underbrace{\text{INC (HL)}}_{34}$$

Add 1 to the contents of memory location (specified by the contents of the HL register pair).

# IND — INPUT TO MEMORY AND DECREMENT POINTER

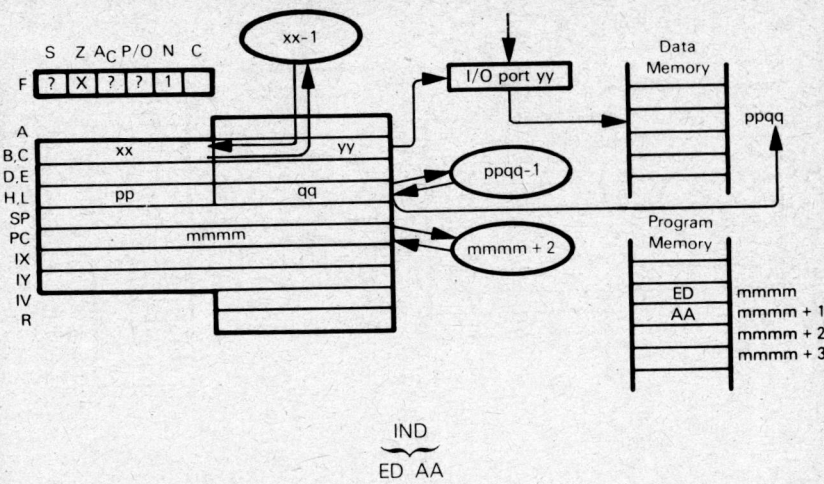

$$\underbrace{\text{IND}}_{\text{ED AA}}$$

Input from I/O port (addressed by Register C) to memory location (specified by HL). Decrement Registers B and HL.

Suppose xx=$05_{16}$, yy=$15_{16}$, ppqq=$2400_{16}$, and $19_{16}$ is held in the buffer of I/O port $15_{16}$. After the instruction

$$\text{IND}$$

has executed, memory location $2400_{16}$ will contain $19_{16}$. The B register will contain $04_{16}$ and the HL register pair $23FF_{16}$.

# INDR — INPUT TO MEMORY AND DECREMENT POINTER UNTIL BYTE COUNTER IS ZERO

$$\underbrace{\text{INDR}}_{\text{ED BA}}$$

INDR is identical to IND, but is repeated until Register B=0.

Suppose Register B contains $03_{16}$, Register C contains $15_{16}$, and HL contains $2400_{16}$. The following sequence of bytes is available at I/O port $15_{16}$:

$$17_{16}, 59_{16} \text{ and } AE_{16}$$

After the execution of

$$\text{INDR}$$

the HL register pair will contain $23FD_{16}$ and Register B will contain zero, and memory locations will have contents as follows:

| Location | Contents |
|----------|----------|
| 2400 | $17_{16}$ |
| 23FF | $59_{16}$ |
| 23FE | $AE_{16}$ |

This instruction is extremely useful for loading blocks of data from an input device into memory.

# INI — INPUT TO MEMORY AND INCREMENT POINTER

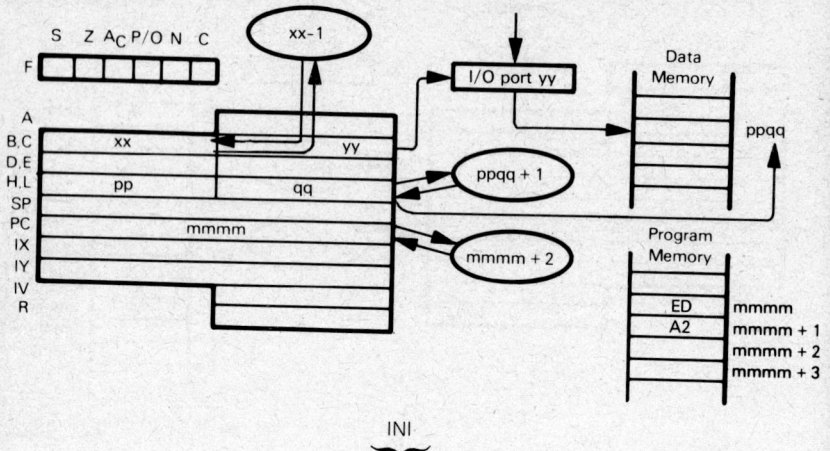

$$\underbrace{\text{INI}}_{\text{ED A2}}$$

Input from I/O port (addressed by Register C) to memory location (specified by HL). Decrement Register B; increment register pair HL.

Suppose $xx = 05_{16}$, $yy = 15_{16}$, $ppqq = 2400_{16}$, and $19_{16}$ is held in the buffer of I/O port $15_{16}$.

After the instruction

$$\text{INI}$$

has executed, memory location $2400_{16}$ will contain $19_{16}$. The B register will contain $04_{16}$ and the HL register pair $2401_{16}$.

# INIR — INPUT TO MEMORY AND INCREMENT POINTER UNTIL BYTE COUNTER IS ZERO

$$\underbrace{\text{INIR}}_{\text{ED B2}}$$

INIR is identical to INI, but is repeated until Register B=0.

Suppose Register B contains $03_{16}$, Register C contains $15_{16}$, and HL contains $2400_{16}$. The following sequence of bytes is available at I/O port $15_{16}$:

$$17_{16}, 59_{16} \text{ and } AE_{16}$$

After the execution of

$$\text{INIR}$$

the HL register pair will contain $2403_{16}$ and Register B will contain zero, and memory locations will have contents as follows:

| Location | Contents |
|----------|----------|
| 2400 | $17_{16}$ |
| 2401 | $59_{16}$ |
| 2402 | $AE_{16}$ |

This instruction is extremely useful for loading blocks of data from a device into memory.

## IN reg,(C) — INPUT TO REGISTER

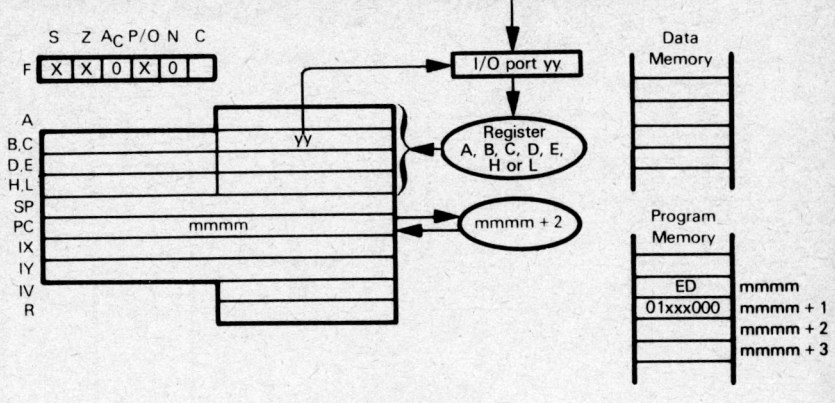

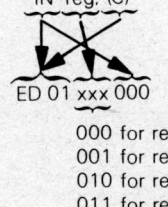

```
ED 01 xxx 000
```

000 for reg=B
001 for reg=C
010 for reg=D
011 for reg=E
100 for reg=H
101 for reg=L
111 for reg=A
110 for setting of status flags without changing registers

Load a byte of data into the specified register (reg) from the I/O port (identified by the contents of the C register).

Suppose $42_{16}$ is held in the buffer of I/O port $36_{16}$, and Register C contains $36_{16}$. After the instruction

                IN D,(C)

has executed, the D register will contain $42_{16}$.

During the execution of the instruction, the contents of Register B are placed on the top half of the Address Bus, making it possible to extend the number of addressable I/O ports.

## JP label — JUMP TO THE INSTRUCTION IDENTIFIED IN THE OPERAND

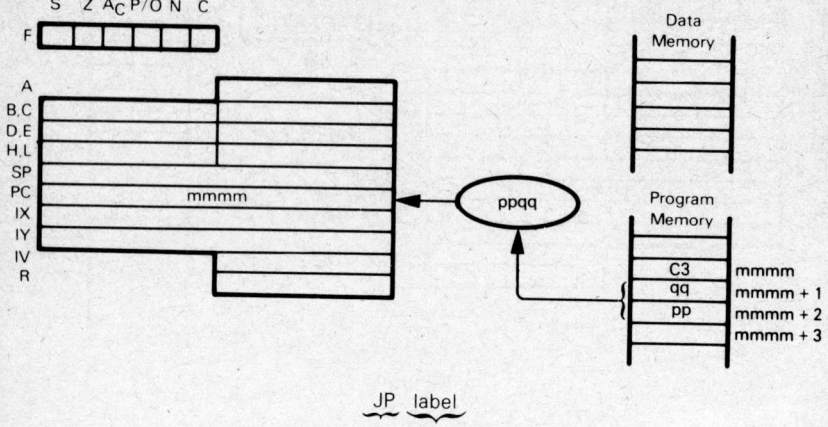

```
    JP   label
    C3   ppqq
```

Load the contents of the Jump instruction object code second and third bytes into the Program Counter; this becomes the memory address for the next instruction to be executed. The previous Program Counter contents are lost.

In the following sequence:

```
        JP      NEXT
        AND     7FH
        -
NEXT    CPL
```

The CPL instruction will be executed after the JP instruction. The AND instruction will never be executed, unless a Jump instruction somewhere else in the instruction sequence jumps to this instruction.

## JP condition,label — JUMP TO ADDRESS IDENTIFIED IN THE OPERAND IF CONDITION IS SATISIFED

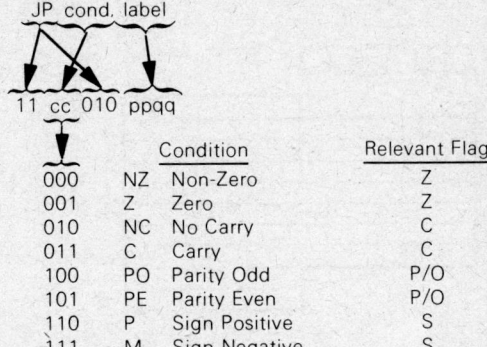

| | | Condition | Relevant Flag |
|---|---|---|---|
| 000 | NZ | Non-Zero | Z |
| 001 | Z | Zero | Z |
| 010 | NC | No Carry | C |
| 011 | C | Carry | C |
| 100 | PO | Parity Odd | P/O |
| 101 | PE | Parity Even | P/O |
| 110 | P | Sign Positive | S |
| 111 | M | Sign Negative | S |

This instruction is identical to the JP instruction, except that the jump will be performed only if the condition is satisfied; otherwise, the instruction sequentially following the JP condition instruction will be executed.

Consider the instruction sequence

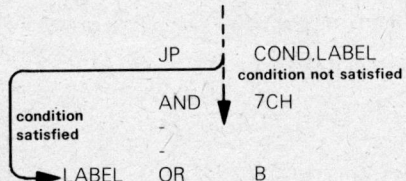

After the JP cond,label instruction has executed, if the condition is satisfied then the OR instruction will be executed. If the condition is not satisfied, the AND instruction, being the next sequential instruction, is executed.

# JP (HL) — JUMP TO ADDRESS SPECIFIED BY CONTENTS
# JP (IX)  OF 16-BIT REGISTER
# JP (IY)

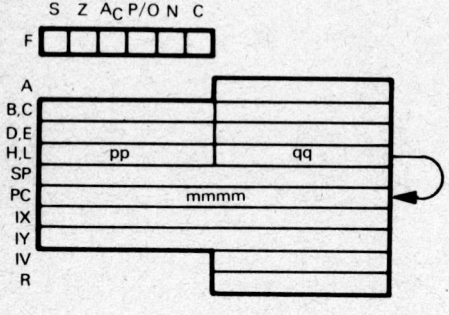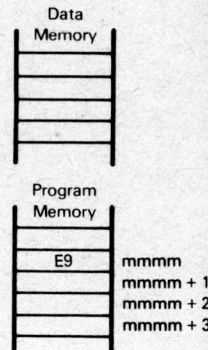

The illustration shows execution of JP (HL):

$$\underbrace{\text{JP (HL)}}_{\text{E9}}$$

The contents of the HL register pair are moved to the Program Counter; therefore, an implied addressing jump is performed.

The instruction sequence

```
LD      H,ADDR
JP      (HL)
```

has exactly the same net effect as the single instruction

```
JP      ADDR
```

Both specify that the instruction with label ADDR is to be executed next.

The JP (HL) instruction is useful when you want to increment a return address for a subroutine that has multiple returns.

Consider the following call to subroutine SUB:

```
CALL    SUB         ;CALL SUBROUTINE
JP      ERR         ;ERROR RETURN
                    ;GOOD RETURN
```

Using RET to return from SUB would return execution of JP ERR; therefore, if SUB executes without detecting error conditions, return as follows:

```
POP     HL          ;POP RETURN ADDRESS TO HL
INC     HL          ;ADD 3 TO RETURN ADDRESS
INC     HL
INC     HL
JP      (HL)        ;RETURN
```

$$\underbrace{\text{JP (IX)}}_{\text{DD E9}}$$

This instruction is identical to the JP (HL) instruction, except that it uses the IX register instead of the HL register pair.

$$\underbrace{\text{JP (IY)}}_{\text{FD E9}}$$

This instruction is identical to the JP (HL) instruction, except that it uses the IY register instead of the HL register pair.

## JR C,disp — JUMP RELATIVE TO CONTENTS OF PROGRAM COUNTER IF CARRY IS SET

$$\underbrace{\text{JR C,}}\ \underbrace{\text{disp}}$$
$$38\quad \text{dd-2}$$

This instruction is identical to the JR disp instruction, except that the jump is only executed if the Carry status equals 1; otherwise, the next instruction is executed.

In the following instruction sequence:

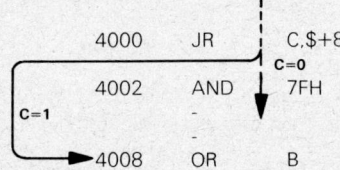

After the JR C,$+8 instruction, the OR instruction is executed if the Carry status equals 1. The AND instruction is executed if the Carry status equals 0.

# JR disp — JUMP RELATIVE TO PRESENT CONTENTS OF PROGRAM COUNTER

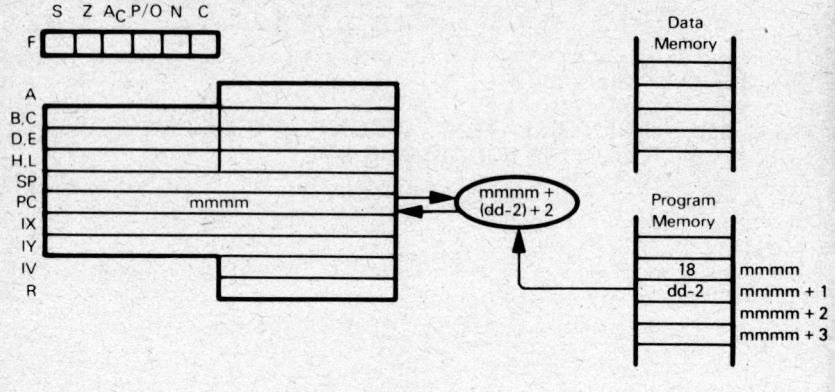

JR disp

18  dd-2

Add the contents of the JR instruction object code second byte, the contents of the Program Counter, and 2. Load the sum into the Program Counter. The jump is measured from the address of the instruction operation code, and has a range of -126 to +129 bytes. The Assembler automatically adjusts for the twice-incremented PC.

The following assembly language statement is used to jump four steps forward from address $4000_{16}$.

JR $+4

Result of this instruction is shown below:

| Location | Instruction |
|----------|-------------|
| 4000 | 18 |
| 4001 | 02 |
| 4002 | - |
| 4003 | - |
| 4004 | - ← new PC value |

## JR NC,disp — JUMP RELATIVE TO CONTENTS OF PROGRAM COUNTER IF CARRY FLAG IS RESET

JR NC,disp

30  dd-2

This instruction is identical to the JR disp instruction, except that the jump is only executed if the Carry status equals 0; otherwise, the next instruction is executed.

In the following instruction sequence:

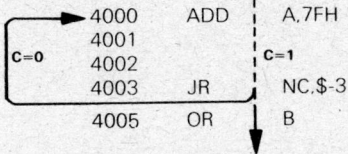

After the JR NC,$-3 instruction, the OR instruction is executed if the Carry status equals 1. The ADD instruction is executed if the Carry status equals 0.

## JR NZ,disp — JUMP RELATIVE TO CONTENTS OF PROGRAM COUNTER IF ZERO FLAG IS RESET

JR NZ,disp

20  dd-2

This instruction is identical to the JR disp instruction, except that the jump is only executed if the Zero status equals 0; otherwise, the next instruction is executed.

In the following instruction sequence:

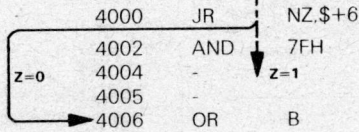

After the JR NZ,$+6 instruction, the OR instruction is executed if the Zero status equals 0. The AND instruction is executed if the Zero status equals 1.

## JR Z,disp — JUMP RELATIVE TO CONTENTS OF PROGRAM COUNTER IF ZERO FLAG IS SET

$$\underbrace{\text{JR Z,disp}}_{\text{28 dd-2}}$$

This instruction is identical to the JR disp instruction, except that the jump is only executed if the Zero status equals 1; otherwise, the next instruction is executed.

In the following instruction sequence:

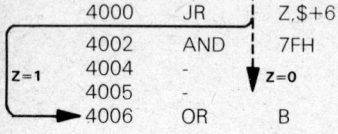

After the JR Z,$+6 instruction, the OR instruction is executed if the Zero status equals 1. The AND instruction is executed if the Zero status equals 0.

## LD A,IV — MOVE CONTENTS OF INTERRUPT VECTOR OR
## LD A,R     REFRESH REGISTER TO ACCUMULATOR

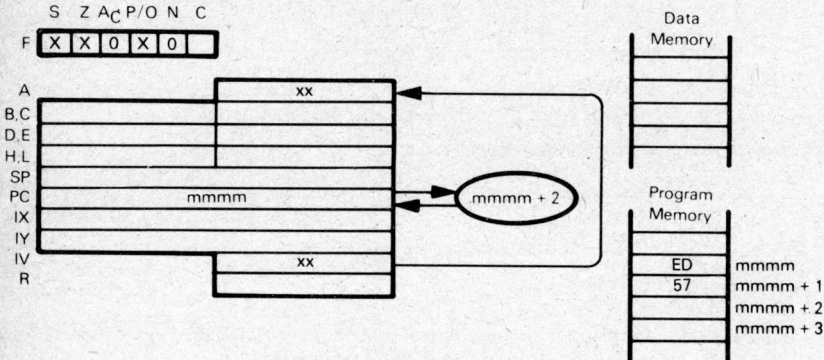

The illustration shows execution of LD A,IV:

$$\underbrace{\text{LD A,IV}}_{\text{ED 57}}$$

Move the contents of the Interrupt Vector register to the Accumulator, and reflect interrupt enable status in Parity/Overflow flag.

Suppose the Interrupt Vector register contains $7F_{16}$, and interrupts are disabled. After execution of

$$\text{LD A,IV}$$

Register A will contain $7F_{16}$, and P/O will be 0.

$$\underbrace{\text{LD A,R}}_{\text{ED 5F}}$$

Move the contents of the Refresh register to the Accumulator. The value of the interrupt flip-flop will appear in the Parity/Overflow flag.

# LD A,(addr) — LOAD ACCUMULATOR FROM MEMORY USING DIRECT ADDRESSING

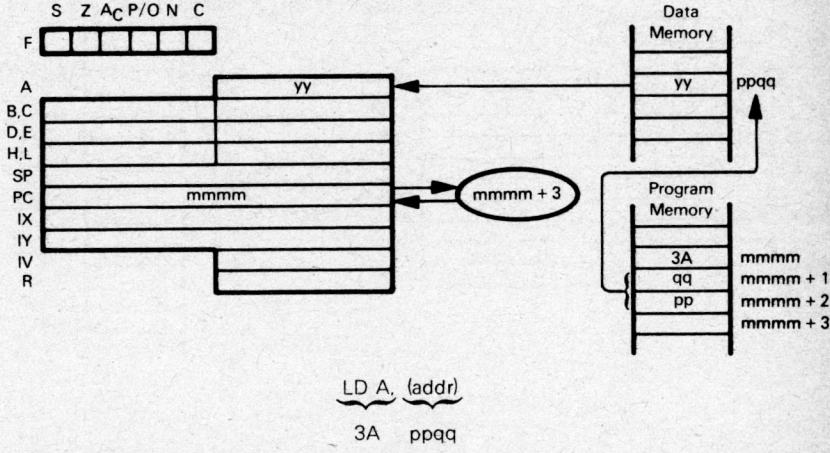

```
   LD A,  (addr)
   ‾‾‾‾   ‾‾‾‾‾‾
    3A     ppqq
```

Load the contents of the memory byte (addressed directly by the second and third bytes of the LD A,(addr) instruction object code) into the Accumulator. Suppose memory byte $084A_{16}$ contains $20_{16}$. After the instruction

      label     EQU     084AH
- 
- 
- 
      LD        A,(label)

has executed, the Accumulator will contain $20_{16}$.

Remember that EQU is an assembler directive rather than an instruction; it tells the Assembler to use the 16-bit value $084A_{16}$ wherever the label appears.

The instruction

      LD        A,(label)

is equivalent to the two instructions

      LD        HL,label
      LD        A,(HL)

When you are loading a single value from memory, the LD A,(label) instruction is preferred; it uses one instruction and three object program bytes to do what the LD HL,label, LD A,(HL) combination does in two instructions and four object program bytes. Also, the LD HL,label, LD A,(HL) combination uses the H and L registers, which LD A,(label) does not.

# LD A,(rp) — LOAD ACCUMULATOR FROM MEMORY LOCATION ADDRESSED BY REGISTER PAIR

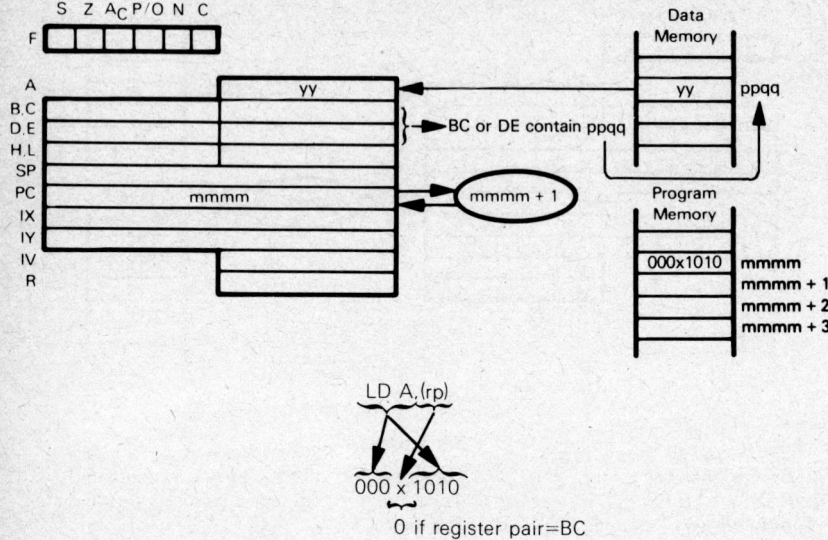

0 if register pair=BC
1 if register pair=DE

Load the contents of the memory byte (addressed by the BC or DE register pair) into the Accumulator.

Suppose the B register contains $08_{16}$, the C register contains $4A_{16}$, and memory byte $084A_{16}$ contains $3A_{16}$. After the instruction

                LD  A,(BC)

has executed, the Accumulator will contain $3A_{16}$.

Normally, the LD A,(rp) and LD rp,data will be used together, since the LD rp,data instruction loads a 16-bit address into the BC or DE registers as follows:

    LD        BC,084AH
    LD        A,(BC)

## LD dst,src — MOVE CONTENTS OF SOURCE REGISTER TO DESTINATION REGISTER

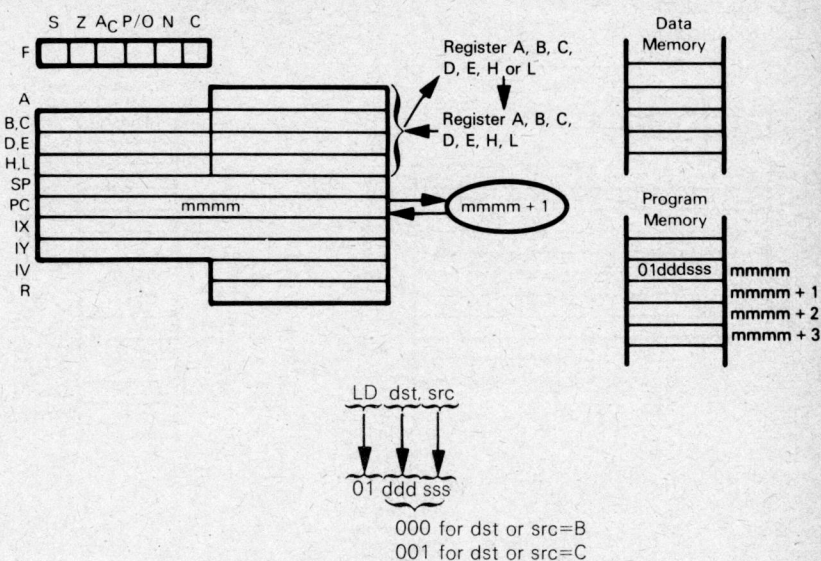

```
    LD dst, src
     ↓   ↓   ↓
  01 ddd sss
```

000 for dst or src=B
001 for dst or src=C
010 for dst or src=D
011 for dst or src=E
100 for dst or src=H
101 for dst or src=L
111 for dst or src=A

The contents of any designated register are loaded into any other register.

For example:

LD A,B

loads the contents of Register B into Register A.

LD L,D

loads the contents of Register D into Register L.

LD C,C

does nothing, since the C register has been specified as both the source and the destination:

# LD HL,(addr) — LOAD REGISTER PAIR OR INDEX REGISTER
# LD rp,(addr)  FROM MEMORY USING DIRECT ADDRESSING
# LD IX,(addr)
# LD IY,(addr)

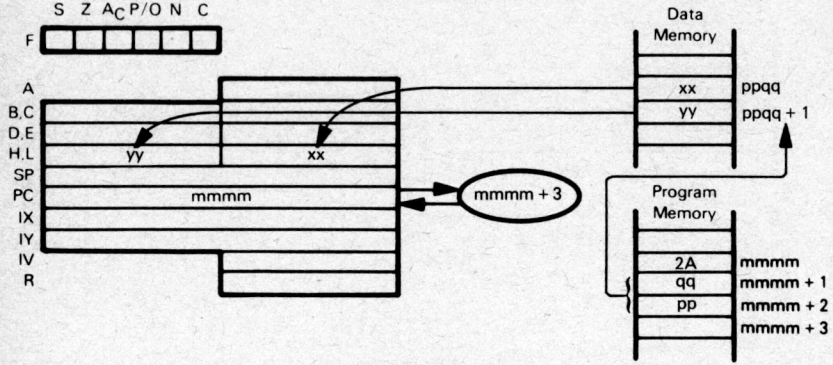

The illustration shows execution of LD HL(ppqq):

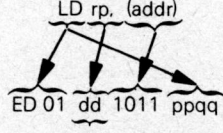

Load the HL register pair from directly addressed memory location.

Suppose memory location $4004_{16}$ contains $AD_{16}$ and memory location $4005_{16}$ contains $12_{16}$. After the instruction

LD HL,(4004H)

has executed, the HL register pair will contain $12AD_{16}$.

```
         LD rp, (addr)
         
ED 01  dd  1011  ppqq
```

00 for rp is register pair BC
01 for rp is register pair DE
10 for rp is register pair HL
11 for rp is Stack Pointer

Load register pair from directly addressed memory.

Suppose memory location $49FF_{16}$ contains $BE_{16}$ and memory location $4A00_{16}$ contains $33_{16}$. After the instruction

LD DE,(49FFH)

has executed, the DE register pair will contain $33BE_{16}$.

```
  LD IX,(addr)
  
  DD 2A  ppqq
```

Load IX register from directly addressed memory.

Suppose memory location $D111_{16}$ contains $FF_{16}$ and memory location $D112_{16}$ contains $56_{16}$. After the instruction

$$\text{LD IX,(D111H)}$$

has executed, the IX register will contain $56FF_{16}$.

$$\underbrace{\text{LD IY,(addr)}}_{\text{FD 2A ppqq}}$$

Load IY register from directly addressed memory.

Affects IY register instead of IX. Otherwise identical to LD IX(addr).

## LD IV,A — LOAD INTERRUPT VECTOR OR REFRESH
## LD R,A  REGISTER FROM ACCUMULATOR

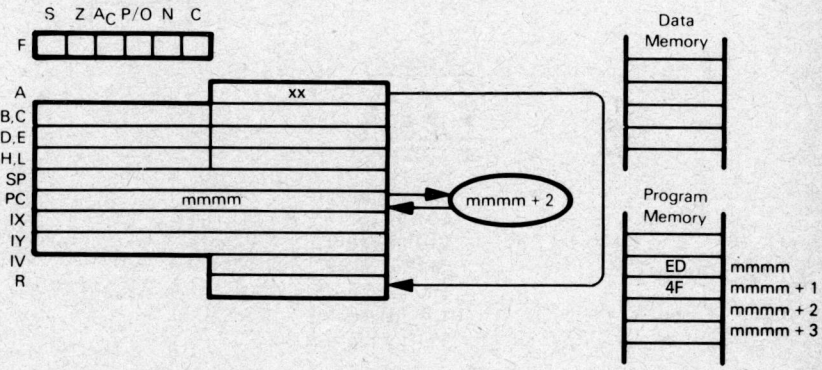

The illustration shows execution of LD R,A:

$$\underbrace{\text{LD R,A}}_{\text{ED 4F}}$$

Load Refresh register from Accumulator.

Suppose the Accumulator contains $7F_{16}$. After the instruction

$$\text{LD R,A}$$

has executed, the Refresh register will contain $7F_{16}$.

$$\underbrace{\text{LD IV,A}}_{\text{ED 47}}$$

Load Interrupt Vector register from Accumulator.

## LD reg,data — LOAD IMMEDIATE INTO REGISTER

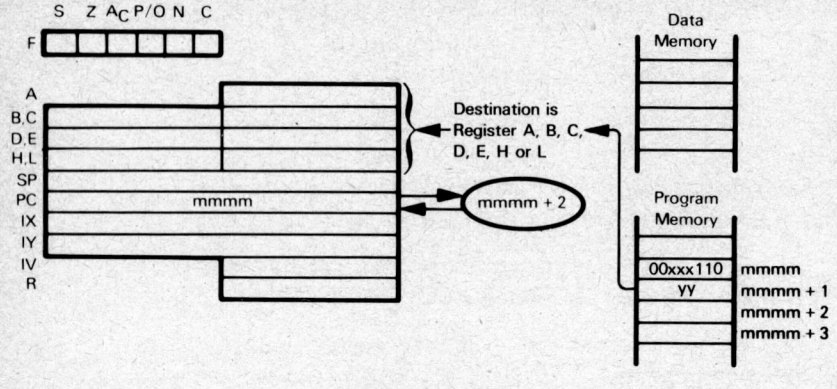

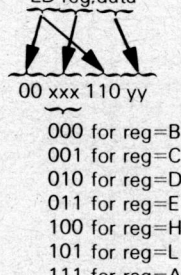

```
00 xxx 110 yy
```

000 for reg=B
001 for reg=C
010 for reg=D
011 for reg=E
100 for reg=H
101 for reg=L
111 for reg=A

Load the contents of the second object code byte into one of the registers. When the instruction

　　　　　　　　LD A,2AH

has executed, $2A_{16}$ is loaded into the Accumulator.

# LD rp,data — LOAD 16 BITS OF DATA IMMEDIATE INTO
# LD IX,data    REGISTER
# LD IY,data

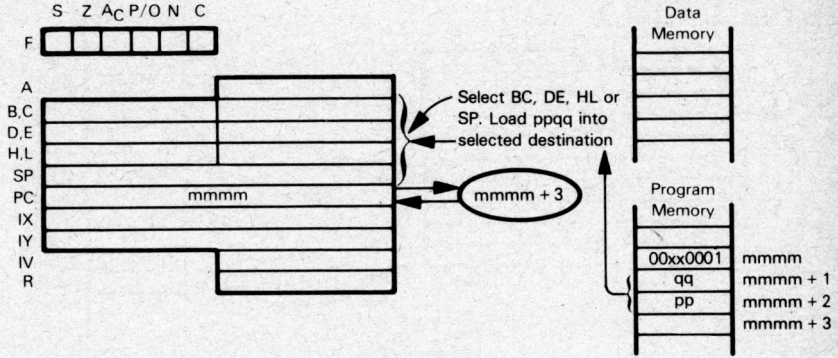

The illustration shows execution of LD rp,data:

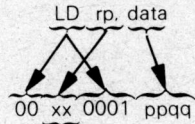

```
00 xx 0001 ppqq
```

00 for rp is register pair BC
01 for rp is register pair DE
10 for rp is register pair HL
11 for rp is Stack Pointer

Load the contents of the second and third object code bytes into the selected register pair. After the instruction

        LD SP,217AH

has executed, the Stack Pointer will contain $217A_{16}$.

        LD IX, data

        DD 21 ppqq

Load the contents of the second and third object code bytes into the Index register IX.

        LD IY, data

        FD 21 ppqq

Load the contents of the second and third object code bytes into the Index Register IY.

Notice that the LD rp,data instruction is equivalent to two LD reg,data instructions. For example:

    LD        HL,032AH

is equivalent to

    LD        H,03H
    LD        L,2AH

# LD reg,(HL) — LOAD REGISTER FROM MEMORY
# LD reg,(IX+disp)
# LD reg,(IY+disp)

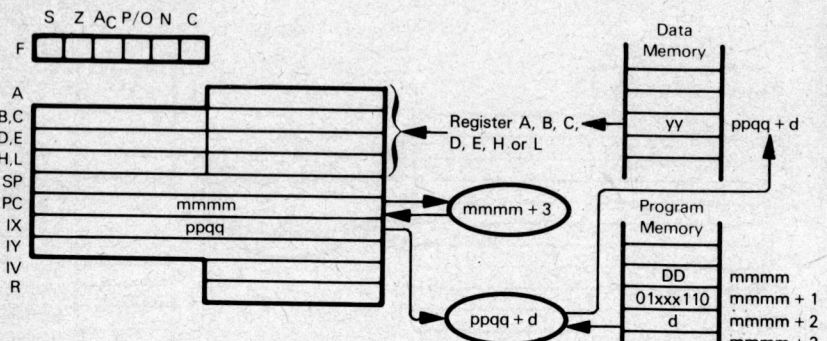

The illustration shows execution of LD reg,(IX+disp):

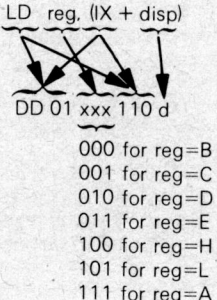

```
000 for reg=B
001 for reg=C
010 for reg=D
011 for reg=E
100 for reg=H
101 for reg=L
111 for reg=A
```

Load specified register from memory location (specified by the sum of the contents of the IX register and the displacement digit d).

Suppose ppqq=$4004_{16}$ and memory location $4010_{16}$ contains $FF_{16}$. After the instruction

LD B(IX+0CH)

has executed, Register B will contain $FF_{16}$.

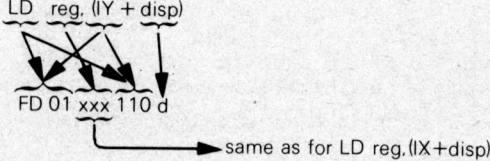

This instruction is identical to LD reg,(IX+disp), except that it uses the IY register instead of the IX register.

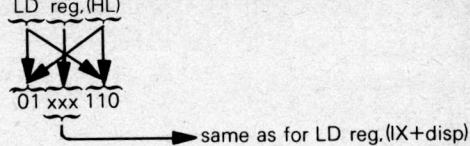

Load specified register from memory location (specified by the contents of the HL register pair).

## LD SP,HL — MOVE CONTENTS OF HL OR INDEX REGISTER
## LD SP,IX     TO STACK POINTER
## LD SP,IY

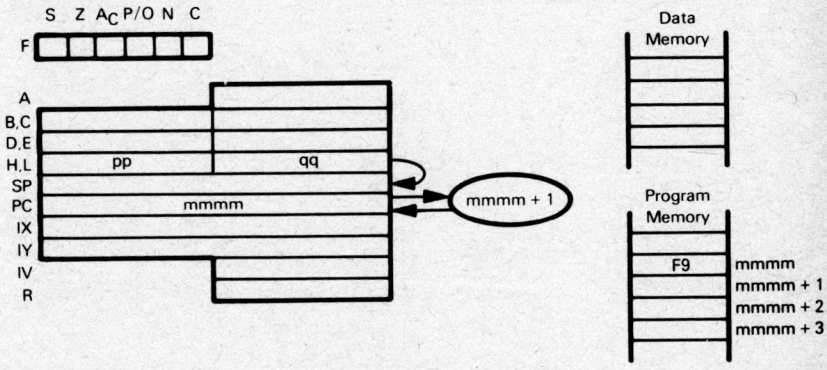

The illustration shows execution of LD SP,HL:

$$\underbrace{\text{LD SP,HL}}_{\text{F9}}$$

Load contents of HL into Stack Pointer.

Suppose $pp=08_{16}$ and $qq=3F_{16}$. After the instruction

$$\text{LD SP,HL}$$

has executed, the Stack Pointer will contain $083F_{16}$.

$$\underbrace{\text{LD SP,IX}}_{\text{DD F9}}$$

Load contents of Index Register IX into Stack Pointer.

$$\underbrace{\text{LD SP,IY}}_{\text{FD F9}}$$

Load contents of Index Register IY into Stack Pointer.

# LD (addr),A — STORE ACCUMULATOR IN MEMORY USING DIRECT ADDRESSING

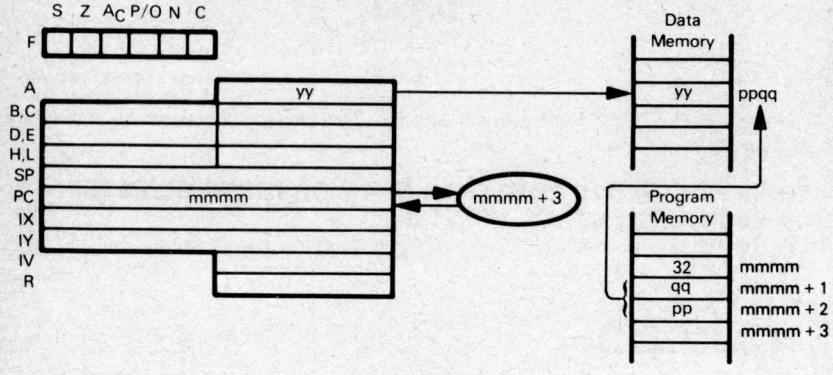

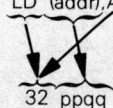

Store the Accumulator contents in the memory byte addressed directly by the second and third bytes of the LD (addr),A instruction object code.

Suppose the Accumulator contains $3A_{16}$. After the instruction

```
label    EQU      084AH
  -
  -
         LD       (label),A
```

has executed, memory byte $084A_{16}$ will contain $3A_{16}$.

Remember that EQU is an assembler directive rather than an instruction; it tells the Assembler to use the 16-bit value 084AH whenever the word "label" appears.

The instruction

>       LD (addr),A

is equivalent to the two instructions

>       LD H,label
>       LD (HL),A

When you are storing a single data value in memory, the LD (label),A instruction is preferred because it uses one instruction and three object program bytes to do what the LD H(label), LD (HL),A combination does in two instructions and four object program bytes. Also, the LD H(label), LD (HL),A combination uses the H and L registers, while the LD (label),A instruction does not.

# LD (addr),HL — STORE REGISTER PAIR OR INDEX
# LD (addr),rp   REGISTER IN MEMORY USING DIRECT
# LD (addr),xy   ADDRESSING

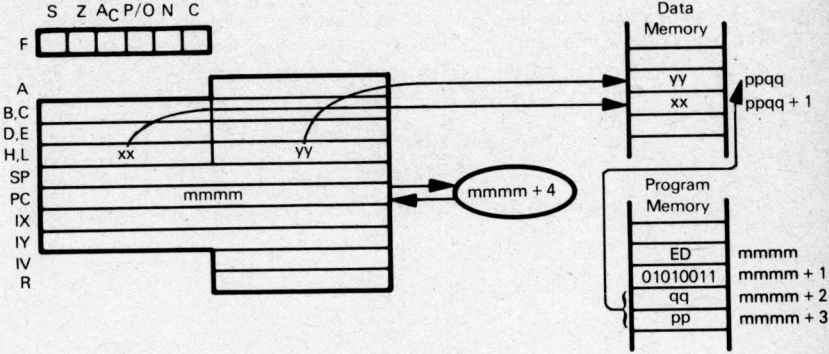

The illustration shows execution of LD (ppqq),DE:

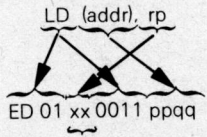

ED 01 xx 0011 ppqq

    00 for rp is register pair BC
    01 for rp is register pair DE
    10 for rp is register pair HL
    11 for rp is Stack Pointer

Store the contents of the specified register pair in memory. The third and fourth object code bytes give the address of the memory location where the low-order byte is to be written. The high-order byte is written into the next sequential memory location.

Suppose the BC register pair contains $3C2A_{16}$. After the instruction

    label  EQU  084AH
    -
    -
    -
    LD   (label),BC

has executed, memory byte $084A_{16}$ will contain $2A_{16}$. Memory byte $084B_{16}$ will contain $3C_{16}$.

Remember that EQU is an assembler directive rather than an instruction; it tells the Assembler to use the 16-bit value $084A_{16}$ whenever the word "label" appears.

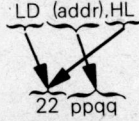

This is a three-byte version of LD (addr),rp which directly specifies HL as the source register pair.

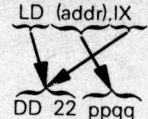

Store the contents of Index register IX in memory. The third and fourth object code bytes give the address of the memory location where the low-order byte is to be written. The high-order byte is written into the next sequential memory location.

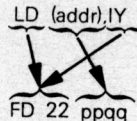

This instruction is identical to the LD (addr),IX instruction, except that it uses the IY register instead of the IX register.

# LD (HL),data — LOAD IMMEDIATE INTO MEMORY
# LD (IX+disp),data
# LD (IY+disp),data

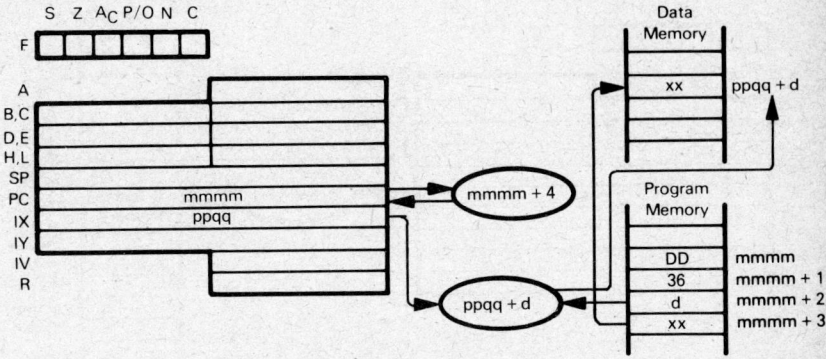

The illustration shows execution of LD (IX+d),xx:

$$\underbrace{\text{LD}}\ \underbrace{\text{(IX+disp)}},\underbrace{\text{data}}$$
$$\text{DD}\ 36\quad\ \ d\quad\ \ xx$$

Load Immediate into the Memory location designated by base relative addressing.

Suppose ppqq=$5400_{16}$. After the instruction

LD (IX+9),FAH

has executed, memory location $5409_{16}$ will contain $FA_{16}$

$$\underbrace{\text{LD}}\ \underbrace{\text{(IY+disp)}},\underbrace{\text{data}}$$
$$\text{FD}\ 36\quad\ \ d\quad\ \ xx$$

This instruction is identical to LD (IX+disp),data, but uses the IY register instead of the IX register.

$$\underbrace{\text{LD}}\ \underbrace{\text{(HL)}},\underbrace{\text{data}}$$
$$36\quad\ \ xx$$

Load Immediate into the Memory location (specified by the contents of the HL register pair).

The Load Immediate into Memory instructions are used much less than the Load Immediate into Register instructions.

# LD (HL),reg — LOAD MEMORY FROM REGISTER
# LD (IX+disp),reg
# LD (IY+disp),reg

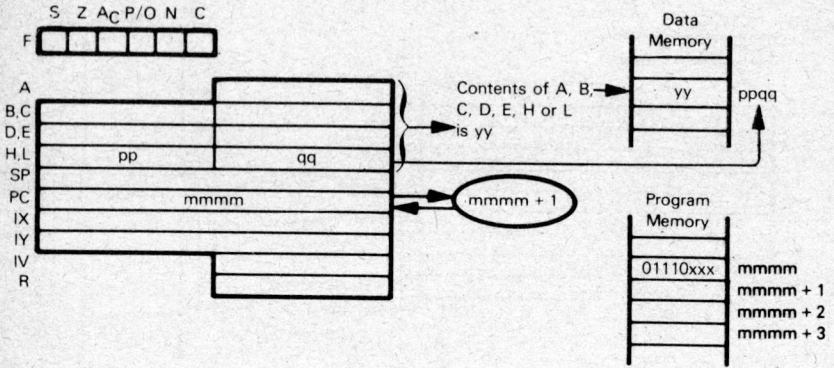

The illustration shows execution of LD (HL),reg:

LD (HL),reg

01110 xxx

000 for reg=B
001 for reg=C
010 for reg=D
011 for reg=E
100 for reg=H
101 for reg=L
111 for reg=A

Load memory location (specified by the contents of the HL register pair) from specified register.

Suppose ppqq=$4500_{16}$ and Register C contains $F9_{16}$. After the instruction

LD (HL),C

has executed, memory location $4500_{16}$ will contain $F9_{16}$.

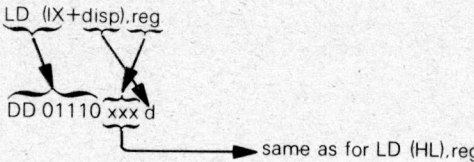

DD 01110 xxx d

→ same as for LD (HL),reg

Load memory location (specified by the sum of the contents of the IX register and the displacement value d) from specified register.

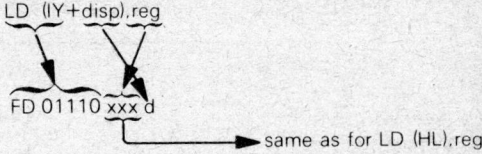

This instruction is identical to LD (IX+disp),reg, except that it uses the IY register instead of the IX register.

## LD (rp),A — LOAD ACCUMULATOR INTO THE MEMORY LOCATION ADDRESSED BY REGISTER PAIR

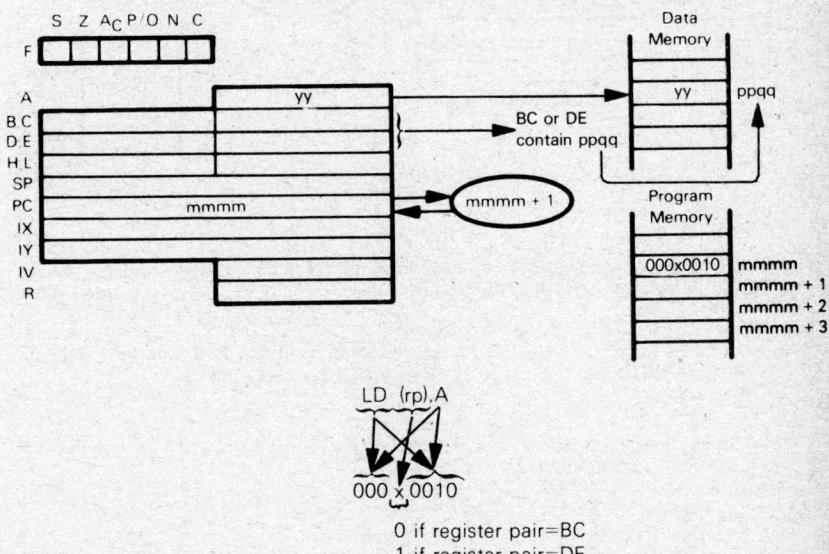

0 if register pair=BC
1 if register pair=DE

Store the Accumulator in the memory byte addressed by the BC or DE register pair.

Suppose the BC register pair contains $084A_{16}$ and the Accumulator contains $3A_{16}$. After the instruction

LD (BC),A

has executed, memory byte $084A_{16}$ will contain $3A_{16}$.

The LD (rp),A and LD rp,data will normally be used together, since the LD rp,data instruction loads a 16-bit address into the BC or DE registers as follows:

LD BC,084AH
LD (BC),A

# LDD — TRANSFER DATA BETWEEN MEMORY LOCATIONS, DECREMENT DESTINATION AND SOURCE ADDRESSES

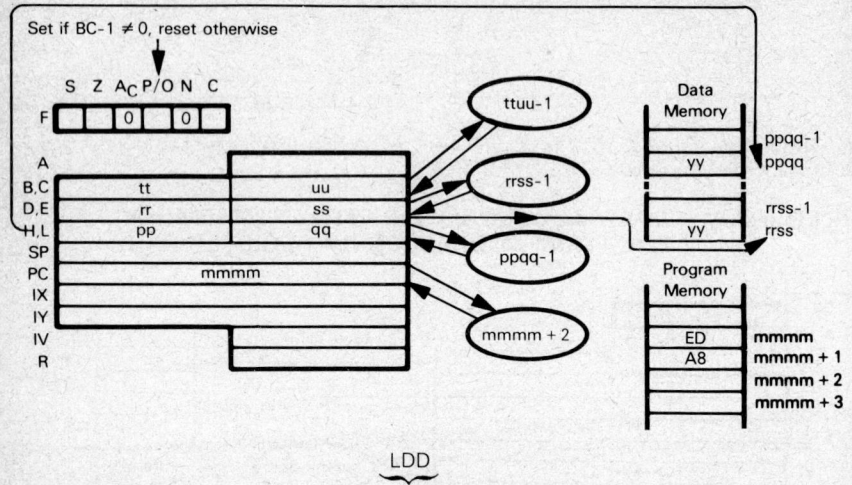

$$\underbrace{\text{LDD}}_{\text{ED A8}}$$

Transfer a byte of data from memory location addressed by the HL register pair to memory location addressed by the DE register pair. Decrement contents of register pairs BC, DE, and HL.

Suppose register pair BC contains $004F_{16}$, DE contains $4545_{16}$, HL contains $2012_{16}$, and memory location $2012_{16}$ contains $18_{16}$. After the instruction

LDD

has executed, memory location $4545_{16}$ will contain $18_{16}$, register pair BC will contain $004E_{16}$, DE will contain $4544_{16}$, and HL will contain $2011_{16}$.

## LDDR — TRANSFER DATA BETWEEN MEMORY LOCATIONS UNTIL BYTE COUNTER IS ZERO, DECREMENT DESTINATION AND SOURCE ADDRESSES

$$\underbrace{\text{LDDR}}_{\text{ED B8}}$$

This instruction is identical to LDD, except that it is repeated until the BC register pair contains zero. After each data transfer, interrupts will be recognized and two refresh cycles will be executed.

Suppose we have the following contents in memory and register pairs:

| Register/Contents | Location/Contents |
|---|---|
| HL $2012_{16}$ | $2012_{16}$ $18_{16}$ |
| DE $4545_{16}$ | $2011_{16}$ $AA_{16}$ |
| BC $0003_{16}$ | $2010_{16}$ $25_{16}$ |

After execution of

LDDR

register pairs and memory locations will have the following contents:

| Register/Contents | Location/Contents | Location/Contents |
|---|---|---|
| HL $2009_{16}$ | $2012_{16}$ $18_{16}$ | $4545_{16}$ $18_{16}$ |
| DE $4542_{16}$ | $2011_{16}$ $AA_{16}$ | $4544_{16}$ $AA_{16}$ |
| BC $0000_{16}$ | $2010_{16}$ $25_{16}$ | $4543_{16}$ $25_{16}$ |

This instruction is extremely useful for transferring blocks of data from one area of memory to another.

# LDI — TRANSFER DATA BETWEEN MEMORY LOCATIONS, INCREMENT DESTINATION AND SOURCE ADDRESSES

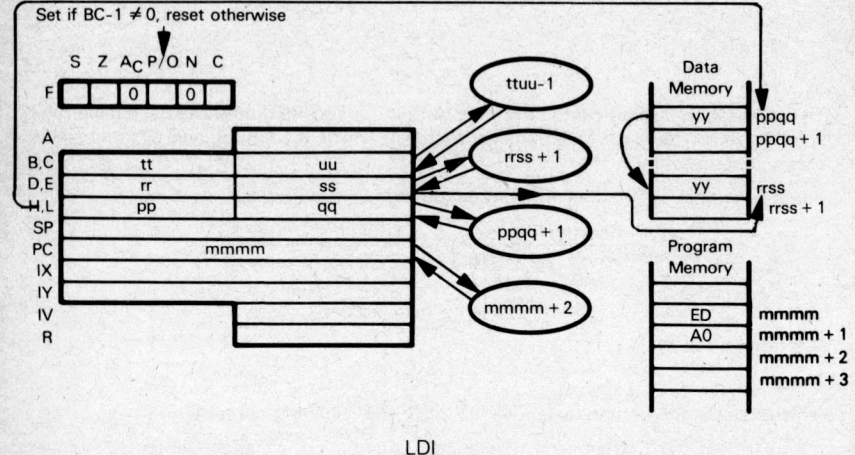

$$\underbrace{\text{LDI}}_{\text{ED A0}}$$

Transfer a byte of data from memory location addressed by the HL register pair to memory location addressed by the DE register pair. Increment contents of register pairs HL and DE. Decrement contents of the BC register pair.

Suppose register pair BC contains $004F_{16}$, DE contains $4545_{16}$, HL contains $2012_{16}$, and memory location $2012_{16}$ contains $18_{16}$. After the instruction

LDI

has executed, memory location $4545_{16}$ will contain $18_{16}$, register pair BC will contain $004E_{16}$, DE will contain $4546_{16}$, and HL will contain $2013_{16}$.

# LDIR — TRANSFER DATA BETWEEN MEMORY LOCATIONS UNTIL BYTE COUNTER IS ZERO. INCREMENT DESTINATION AND SOURCE ADDRESSES

$$\underbrace{\text{LDIR}}_{\text{ED B0}}$$

This instruction is identical to LDI, except that it is repeated until the BC register pair contains zero. After each data transfer, interrupts will be recognized and two refresh cycles will be executed.

Suppose we have the following contents in memory and register pairs:

| Register/Contents | | Location/Contents | |
|---|---|---|---|
| HL | $2012_{16}$ | $2012_{16}$ | $18_{16}$ |
| DE | $4545_{16}$ | $2013_{16}$ | $CD_{16}$ |
| BC | $0003_{16}$ | $2014_{16}$ | $F0_{16}$ |

After execution of

LDIR

register pairs and memory will have the following contents:

| Register/Contents | | Location/Contents | | Location/Contents | |
|---|---|---|---|---|---|
| HL | $2015_{16}$ | $2012_{16}$ | $18_{16}$ | $4545_{16}$ | $18_{16}$ |
| DE | $4548_{16}$ | $2013_{16}$ | $CD_{16}$ | $4546_{16}$ | $CD_{16}$ |
| BC | $0000_{16}$ | $2014_{16}$ | $F0_{16}$ | $4547_{16}$ | $F0_{16}$ |

This instruction is extremely useful for transferring blocks of data from one area of memory to another.

# NEG — NEGATE CONTENTS OF ACCUMULATOR

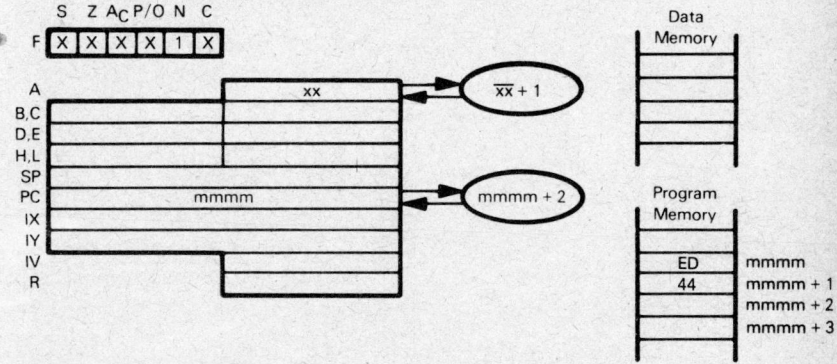

Negate contents of Accumulator. This is the same as subtracting contents of the Accumulator from zero. The result is the two's complement. 80H will be left unchanged.

Suppose xx=$5A_{16}$. After the instruction

NEG

has executed, the Accumulator will contain $A6_{16}$.

```
      5A          = 0101  1010
Two's complement  = 1010  0110
```

# NOP — NO OPERATION

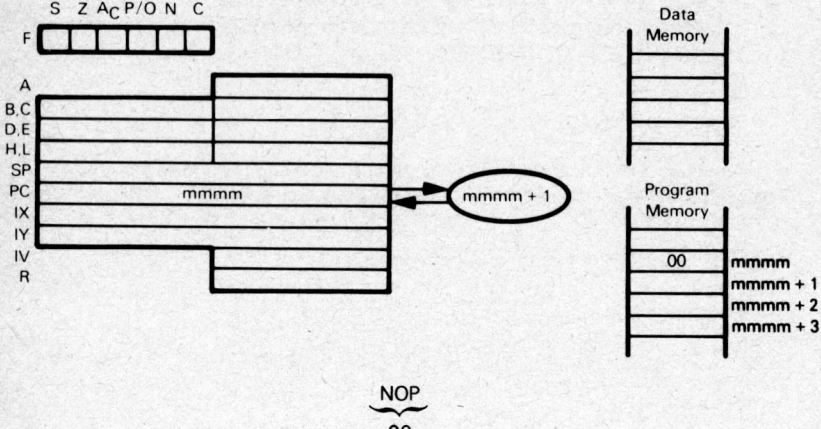

$$\underbrace{\text{NOP}}_{00}$$

This is a one-byte instruction which performs no operation, except that the Program Counter is incremented and memory refresh continues. This instruction is present for several reasons:

1) A program error that fetches an object code from non-existent memory will fetch 00. It is a good idea to ensure that the most common program error will do nothing.

2) The NOP instruction allows you to give a label to an object program byte:
   HERE   NOP

3) To fine-tune delay times. Each NOP instruction adds four clock cycles to a delay.

NOP is not a very useful or frequently used instruction.

## OR data — OR IMMEDIATE WITH ACCUMULATOR

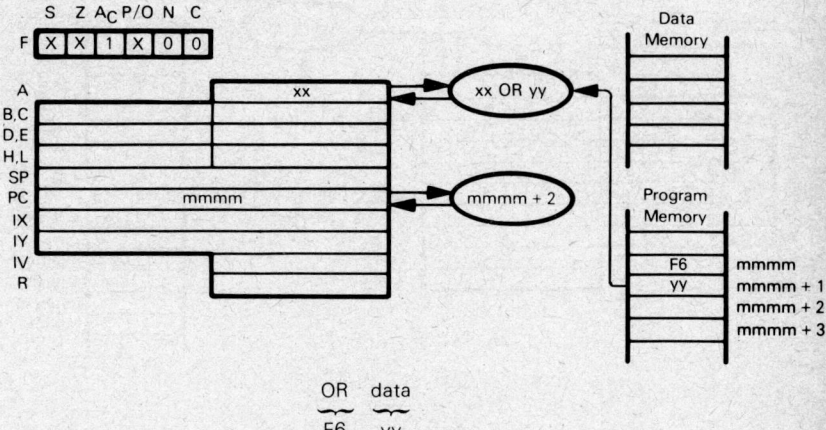

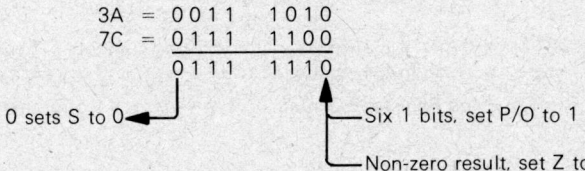

OR the Accumulator with the contents of the second instruction object code byte.

Suppose xx=$3A_{16}$. After the instruction

OR 7CH

has executed, the Accumulator will contain $7E_{16}$.

```
3A =  0011 1010
7C =  0111 1100
      0111 1110
```

0 sets S to 0

Six 1 bits, set P/O to 1

Non-zero result, set Z to 0

This is a routine logical instruction; it is often used to turn bits "on". For example, the instruction

OR 80H

will unconditionally set the high-order Accumulator bit to 1.

## OR reg — OR REGISTER WITH ACCUMULATOR

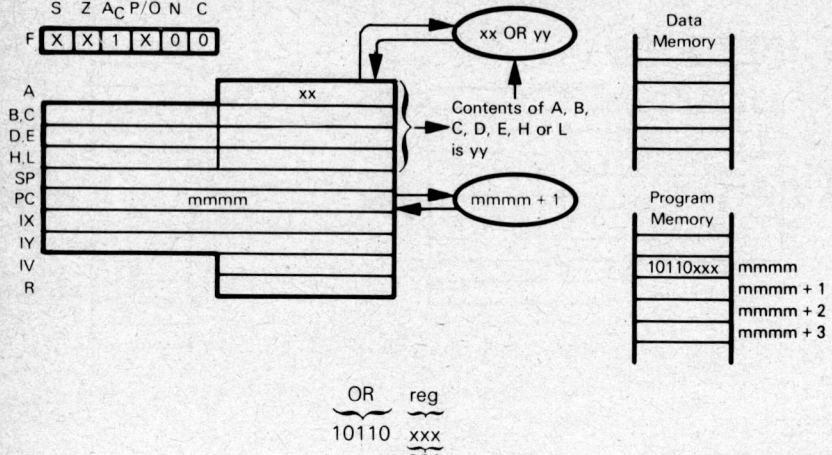

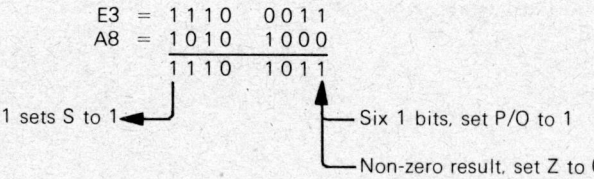

```
 OR    reg
10110  xxx
       000 for reg=B
       001 for reg=C
       010 for reg=D
       011 for reg=E
       100 for reg=H
       101 for reg=L
       111 for reg=A
```

Logically OR the contents of the Accumulator with the contents of Register A, B, C, D, E, H or L. Store the result in the Accumulator.

Suppose xx=$E3_{16}$ and Register E contains $A8_{16}$. After the instruction

                OR E

has executed, the Accumulator will contain $EB_{16}$.

```
E3 = 1110 0011
A8 = 1010 1000
     1110 1011
```

1 sets S to 1

Six 1 bits, set P/O to 1

Non-zero result, set Z to 0

# OR (HL) — OR MEMORY WITH ACCUMULATOR
# OR (IX+disp)
# OR (IY+disp)

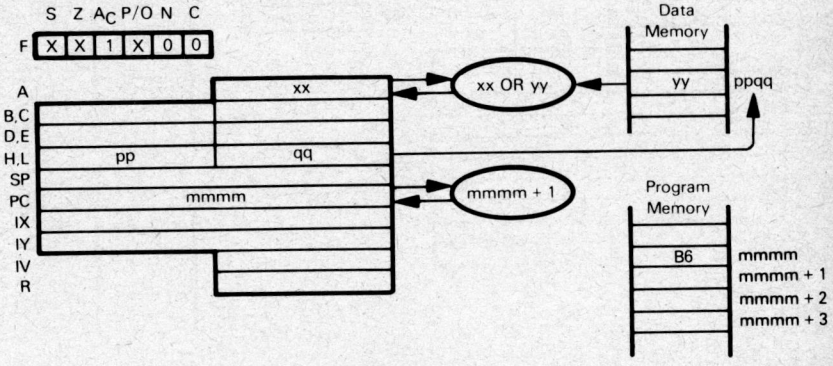

The illustration shows execution of OR (HL):

$$\underbrace{\text{OR (HL)}}_{\text{B6}}$$

OR contents of memory location (specified by the contents of the HL register pair) with the Accumulator.

Suppose $xx = E3_{16}$, $ppqq = 4000_{16}$, and memory location $4000_{16}$ contains $A8_{16}$. After the instruction

$$\text{OR (HL)}$$

has executed, the Accumulator will contain $EB_{16}$.

```
E3 = 1110 0011
A8 = 1010 1000
     1110 1011
```

1 sets S to 1 ←┘            └→ Six 1 bits, set P/O to 1
                             └→ Non-zero result, set Z to 0

$$\underbrace{\text{OR (IX+disp)}}_{\text{DD B6 d}}$$

OR contents of memory location (specified by the sum of the contents of the IX register and the displacement value d) with the Accumulator.

$$\underbrace{\text{OR (IY+disp)}}_{\text{FD B6 d}}$$

This instruction is identical to OR (IX+disp), except that it uses the IY register instead of the IX register.

# OUT (C),reg — OUTPUT FROM REGISTER

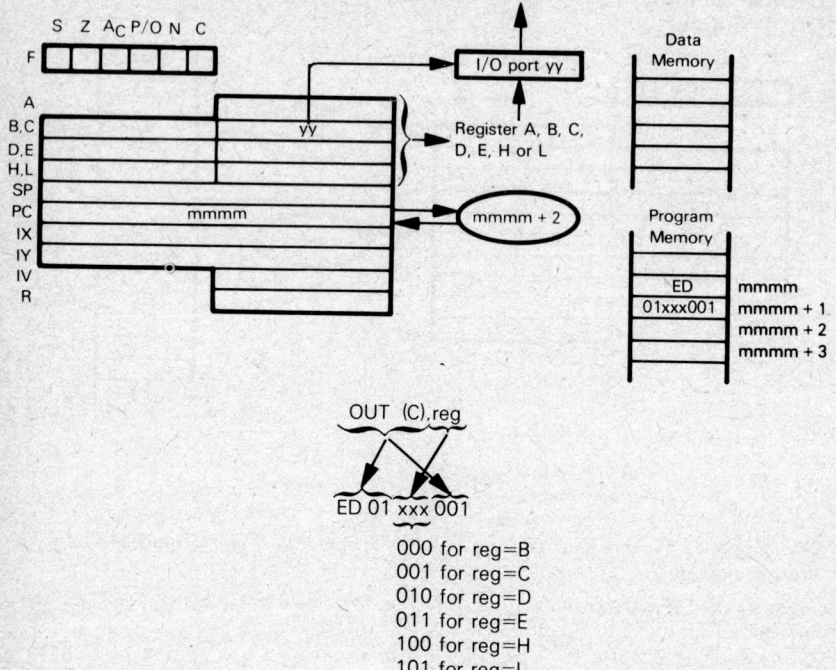

```
        OUT (C),reg
        ╱╲    ╱╲
       ╱  ╲  ╱  ╲
      ▼    ╲╱    ▼
    ED 01 xxx 001
```

000 for reg=B
001 for reg=C
010 for reg=D
011 for reg=E
100 for reg=H
101 for reg=L
111 for reg=A

Suppose yy=$1F_{16}$ and the contents of H are $AA_{16}$. After the execution of

OUT (C),H

$AA_{16}$ will be in the buffer of I/O port $1F_{16}$.

# OUTD — OUTPUT FROM MEMORY. DECREMENT ADDRESS

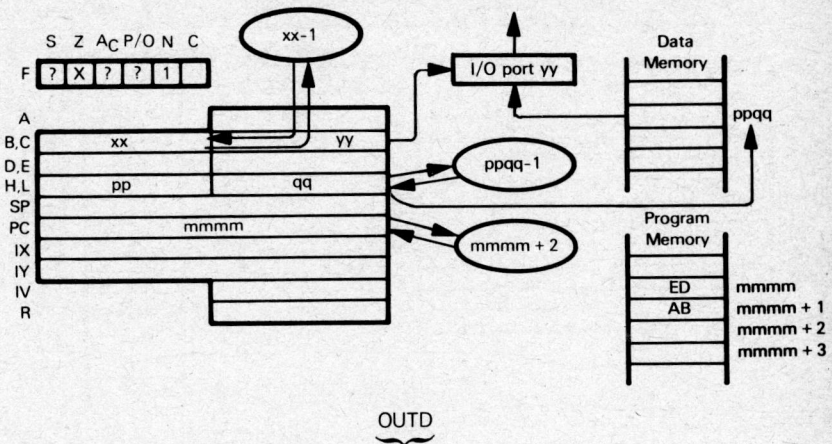

$$\underbrace{\text{OUTD}}_{\text{ED AB}}$$

Output from memory location specified by HL to I/O port addressed by Register C. Registers B and HL are decremented.

Suppose xx=0A$_{16}$, yy=FF$_{16}$, ppqq=5000$_{16}$, and memory location 5000$_{16}$ contains 77$_{16}$. After the instruction

OUTD

has executed, 77$_{16}$ will be held in the buffer of I/O port FF$_{16}$. The B register will contain 09$_{16}$, and the HL register pair 4FFF$_{16}$.

# OTDR — OUTPUT FROM MEMORY. DECREMENT ADDRESS, CONTINUE UNTIL REGISTER B=0

$$\underbrace{\text{OTDR}}_{\text{ED BB}}$$

OTDR is identical to OUTD, but is repeated until Register B contains 0.

Suppose Register B contains 03$_{16}$, Register C contains FF$_{16}$, and HL contains 5000$_{16}$. Memory locations 4FFE$_{16}$ through 5000$_{16}$ contain:

| Location/Contents | |
|---|---|
| 4FFE$_{16}$ | CA$_{16}$ |
| 4FFF$_{16}$ | 1B$_{16}$ |
| 5000$_{16}$ | F1$_{16}$ |

After execution of

OTDR

register pair HL will contain 4FFD$_{16}$, Register B will contain zero, and the sequence F1$_{16}$, 1B$_{16}$, CA$_{16}$ will have been written to I/O port FF$_{16}$.

This instruction is very useful for transferring blocks of data from memory to output devices.

# OUTI — OUTPUT FROM MEMORY. INCREMENT ADDRESS

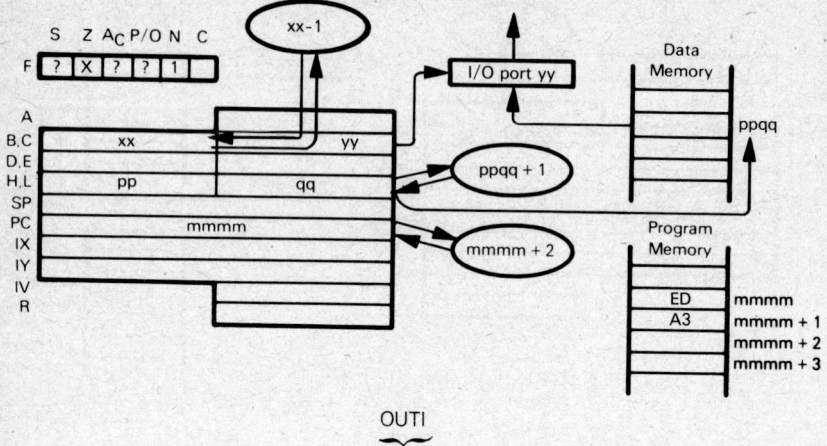

$$\underbrace{\text{OUTI}}_{\text{ED A3}}$$

Output from memory location specified by HL to I/O port addressed by Register C. Register B is decremented and the HL register pair is incremented.

Suppose $xx=0A_{16}$, $yy=FF_{16}$, $ppqq=5000_{16}$, and memory location $5000_{16}$ contains $77_{16}$. After the instruction

OUTI

has executed, $77_{16}$ will be held in the buffer of I/O port $FF_{16}$. The B register will contain $09_{16}$ and the HL register pair will contain $5001_{16}$.

# OTIR — OUTPUT FROM MEMORY. INCREMENT ADDRESS, CONTINUE UNTIL REGISTER B=0

$$\underbrace{\text{OTIR}}_{\text{ED B3}}$$

OTIR is identical to OUTI, except that it is repeated until Register B contains 0.

Suppose Register B contains $04_{16}$, Register C contains $FF_{16}$, and HL contains $5000_{16}$. Memory locations $5000_{16}$ through $5003_{16}$ contain:

| Location/Contents | |
|---|---|
| $5000_{16}$ | $CA_{16}$ |
| $5001_{16}$ | $1B_{16}$ |
| $5002_{16}$ | $B1_{16}$ |
| $5003_{16}$ | $AD_{16}$ |

After execution of

OTIR

register pair HL will contain $5004_{16}$. Register B will contain zero and the sequence $CA_{16}$, $1B_{16}$, $B1_{16}$ and $AD_{16}$ will have been written to I/O port $FF_{16}$.

This instruction is very useful for transferring blocks of data from memory to an output device.

## OUT (port),A — OUTPUT FROM ACCUMULATOR

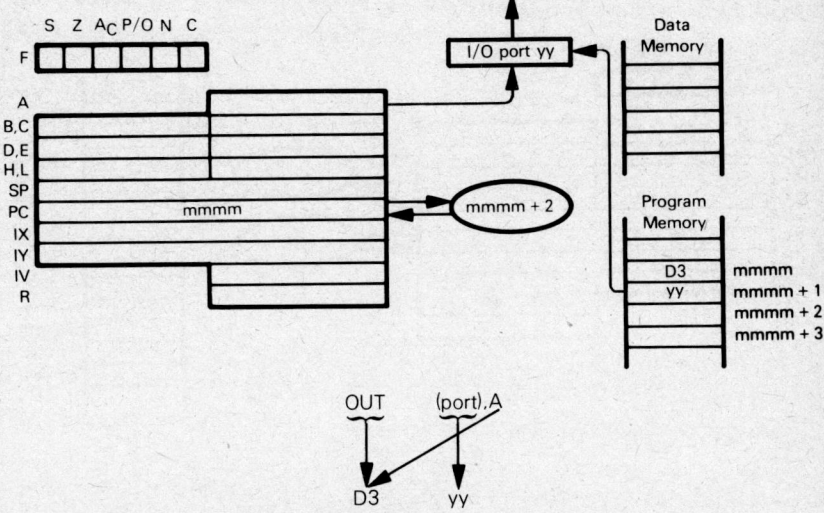

Output the contents of the Accumulator to the I/O port identified by the second OUT instruction object code byte.

Suppose $36_{16}$ is held in the Accumulator. After the instruction

OUT (1AH),A

has executed, $36_{16}$ will be in the buffer of I/O port $1A_{16}$.

The OUT instruction does not affect any statuses. Use of the OUT instruction is very hardware-dependent. Valid I/O port addresses are determined by the way in which I/O logic has been implemented. It is also possible to design a microcomputer system that accesses external logic using memory reference instructions with specific memory addresses. OUT instructions are frequently used in special ways to control microcomputer logic external to the CPU.

# POP rp — READ FROM THE TOP OF THE STACK
# POP IX
# POP IY

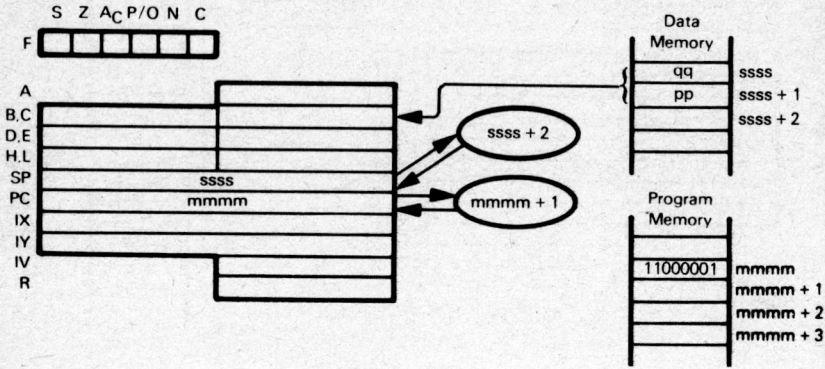

The illustration shows execution of POP BC:

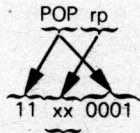

```
11 xx 0001
```

    00 for rp is register pair BC
    01 for rp is register pair DE
    10 for rp is register pair HL
    11 for rp is register pair A and F

POP the two top stack bytes into the designated register pair.

Suppose $qq=01_{16}$ and $pp=2A_{16}$. Execution of

    POP HL

loads $01_{16}$ into the L register and $2A_{16}$ into the H register. Execution of the instruction

    POP AF

loads 01 into the status flags and $2A_{16}$ into the Accumulator. Thus, the Carry status will be set to 1 and other statuses will be cleared.

$$\underbrace{\text{POP IX}}_{\text{DD E1}}$$

POP the two top stack bytes into the IX register.

$$\underbrace{\text{POP IY}}_{\text{FD E1}}$$

POP the two top stack bytes into the IY register.

The POP instruction is most frequently used to restore register and status contents which have been saved on the stack; for example, while servicing an interrupt.

# PUSH rp — WRITE TO THE TOP OF THE STACK
# PUSH IX
# PUSH IY

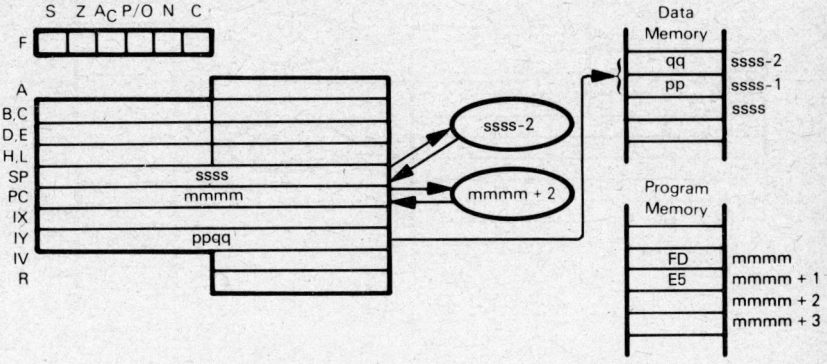

The illustration shows execution of PUSH IY:

PUSH IY
FD E5

PUSH the contents of the IY register onto the top of the stack.

Suppose the IY register contains $45FF_{16}$. Execution of the instruction

PUSH IY

loads $45_{16}$, then $FF_{16}$ onto the top of the stack.

PUSH IX
DD E5

PUSH the contents of the IX register onto the top of the stack.

PUSH rp

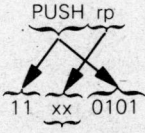

11 xx 0101

00 for rp is register pair BC
01 for rp is register pair DE
10 for rp is register pair HL
11 for rp is register pair A and F

PUSH contents of designated register pair onto the top of the stack.

Execution of the instruction

PUSH AF

loads the Accumulator and then the status flags onto the top of the stack.

The PUSH instruction is most frequently used to save register and status contents; for example, before servicing an interrupt.

# RES b,reg — RESET INDICATED REGISTER BIT

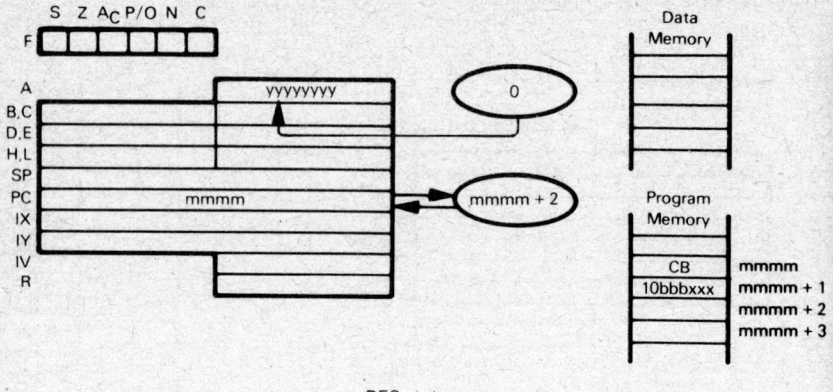

```
       RES    b,reg
        ↓      ↓  ↓
   CB  10   bbb xxx
       Bit  bbb xxx   Register
        0   000 000      B
        1   001 001      C
        2   010 010      D
        3   011 011      E
        4   100 100      H
        5   101 101      L
        6   110 111      A
        7   111
```

Reset indicated bit within specified register.

After the instruction

RES 6,H

has executed, bit 6 in Register H will be reset. (Bit 0 is the least significant bit.)

# RES b,(HL) — RESET BIT b OF INDICATED MEMORY POSITION
# RES b,(IX+disp)
# RES b,(IY+disp)

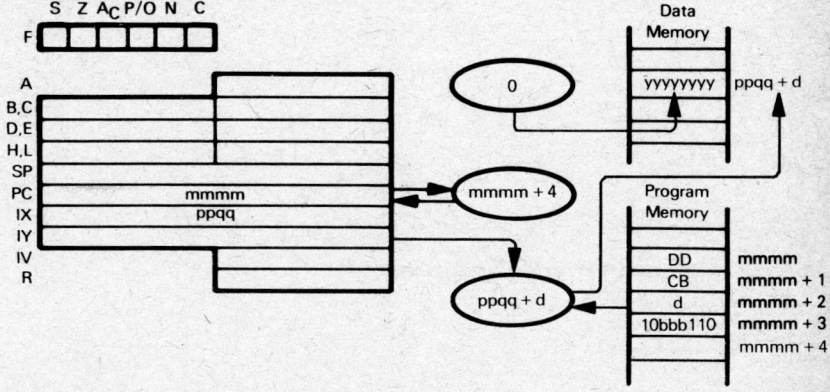

The illustration shows execution of SET b,(IX+disp). Bit 0 is execution of SET b,(IX+disp). Bit 0 is the least significant bit.

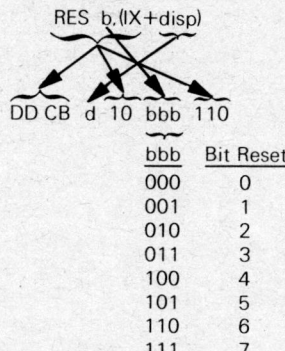

| bbb | Bit Reset |
|-----|-----------|
| 000 | 0 |
| 001 | 1 |
| 010 | 2 |
| 011 | 3 |
| 100 | 4 |
| 101 | 5 |
| 110 | 6 |
| 111 | 7 |

Reset indicated bit within memory location indicated by the sum of Index Register IX and d.

Suppose IX contains $4110_{16}$. After the instruction

$$\text{RES 0,(IX+7)}$$

has executed, bit 0 in memory location $4117_{16}$ will be 0.

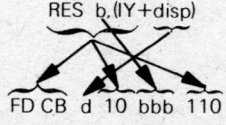

bbb is the same as in RES b,(IX+disp)

This instruction is identical to RES b,(IX+disp), except that it uses the IY register instead of the IX register.

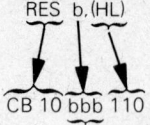

bbb is the same as in RES b,(IX+disp)

Reset indicated bit within memory location indicated by HL.

Suppose HL contains $4444_{16}$. After execution of

RES 7,(HL)

bit 7 in memory location $4444_{16}$ will be 0.

## RET — RETURN FROM SUBROUTINE

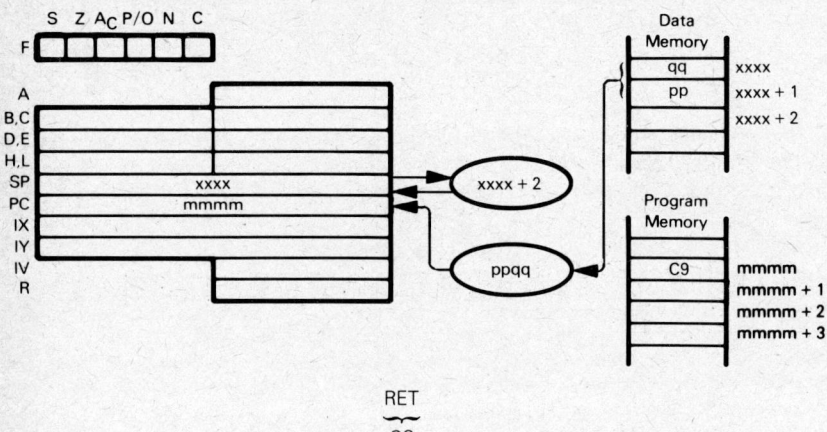

RET
C9

Move the contents of the top two stack bytes to the Program Counter; these two bytes provide the address of the next instruction to be executed. Previous Program Counter contents are lost. Increment the Stack Pointer by 2, to address the new top of stack.

Every subroutine must contain at least one Return (or conditional Return) instruction; this is the last instruction executed within the subroutine, and causes execution to return to the calling program.

# RET cond — RETURN FROM SUBROUTINE IF CONDITION IS SATISFIED

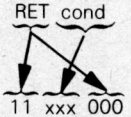

|     |    | Condition     | Relevant Flag |
|-----|----|---------------|---------------|
| 000 | NZ | Non-Zero      | Z             |
| 001 | Z  | Zero          | Z             |
| 010 | NC | Non-Carry     | C             |
| 011 | C  | Carry         | C             |
| 100 | PO | Parity Odd    | P/O           |
| 101 | PE | Parity Even   | P/O           |
| 110 | P  | Sign Positive | S             |
| 111 | M  | Sign Negative | S             |

This instruction is identical to the RET instruction, except that the return is not executed unless the condition is satisfied; otherwise, the instruction sequentially following the RET cond instruction will be executed.

Consider the instruction sequence:

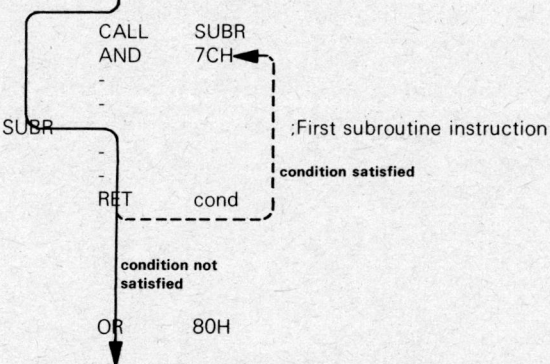

After the RET cond is executed, if the condition is satisfied then execution returns to the AND instruction which follows the CALL. If the condition is not satisfied, the OR instruction, being the next sequential instruction, is executed.

# RETI — RETURN FROM INTERRUPT

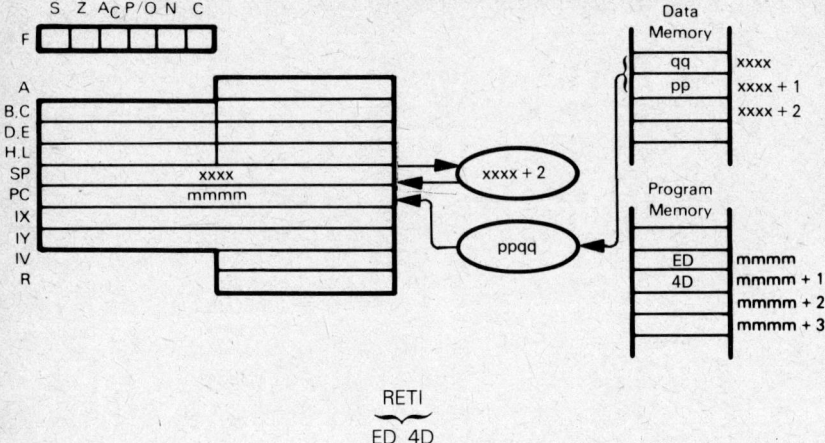

$$\underbrace{\text{RETI}}_{\text{ED 4D}}$$

Move the contents of the top two stack bytes to the Program Counter; these two bytes provide the address of the next instruction to be executed. Previous Program Counter contents are lost. Increment the Stack Pointer by 2, and address the new top of stack.

This instruction is used at the end of an interrupt service routine, and, in addition to returning control to the interrupted program, it is used to signal an I/O device that the interrupt routine has been completed. The I/O device must provide the logic necessary to sense the instruction operation code: refer to Chapter 7 of An Introduction to Microcomputers: Volume II for a description of how the RETI instruction operates with the Z80 family of devices.

# RETN — RETURN FROM NON-MASKABLE INTERRUPT

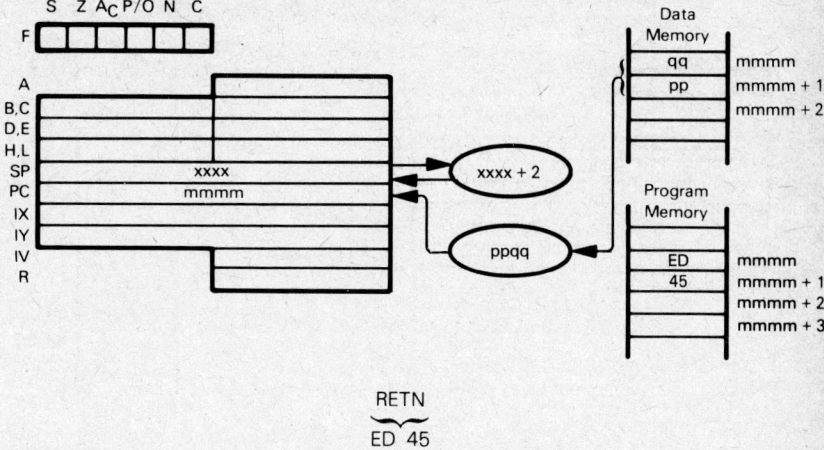

$$\underbrace{\text{RETN}}_{\text{ED 45}}$$

Move the contents of the top two stack bytes to the Program Counter; these two bytes provide the address of the next instruction to be executed. Previous Program Counter contents are lost. Increment the Stack Pointer by 2 to address the new top of stack. Restore the interrupt enable logic to the state it had prior to the occurrence of the non-maskable interrupt.

This instruction is used at the end of a service routine for a non-maskable interrupt, and causes execution to return to the program that was interrupted.

## RL reg — ROTATE CONTENTS OF REGISTER LEFT THROUGH CARRY

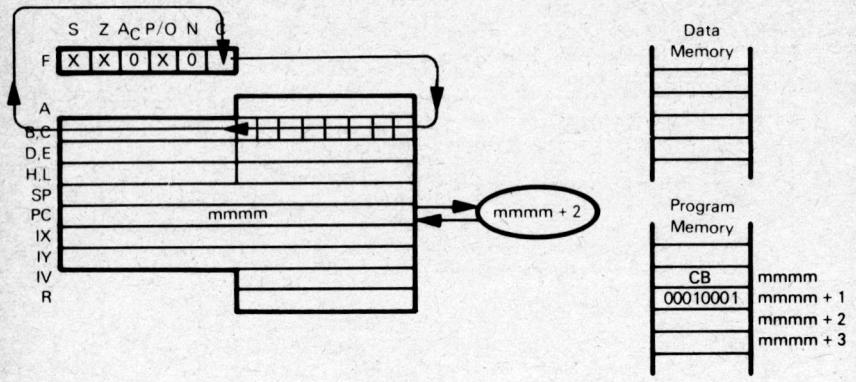

The illustration shows execution of RL C:

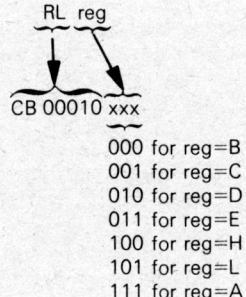

000 for reg=B
001 for reg=C
010 for reg=D
011 for reg=E
100 for reg=H
101 for reg=L
111 for reg=A

Rotate contents of specified register left one bit through Carry.

Suppose D contains $A9_{16}$ and Carry=0. After the instruction

RL D

has executed, D will contain $52_{16}$ and Carry will be 1:

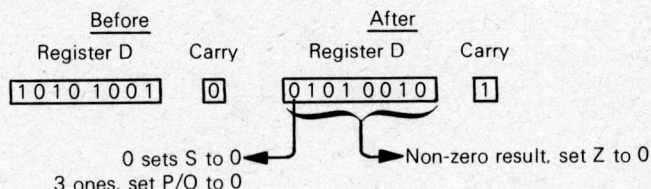

# RL (HL) — ROTATE CONTENTS OF MEMORY LOCATION
# RL (IX+disp) LEFT THROUGH CARRY
# RL (IY+disp)

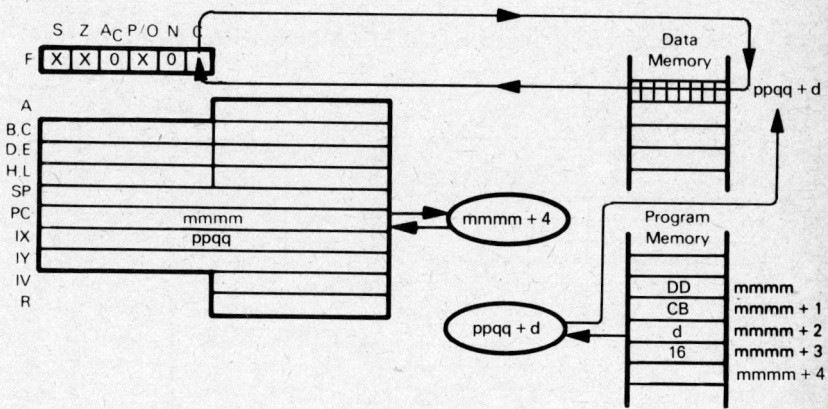

The illustration shows execution of RL (IX+disp):

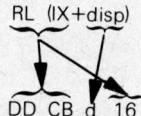

Rotate contents of memory location (specified by the sum of the contents of Index Register IX and displacement integer d) left one bit through Carry.

Suppose the IX register contains $4000_{16}$, memory location $4007_{16}$ contains $2F_{16}$, and Carry is set to 1. After execution of the instruction

RL (IX+7)

memory location $4007_{16}$ will contain $5F_{16}$, and Carry is 0:

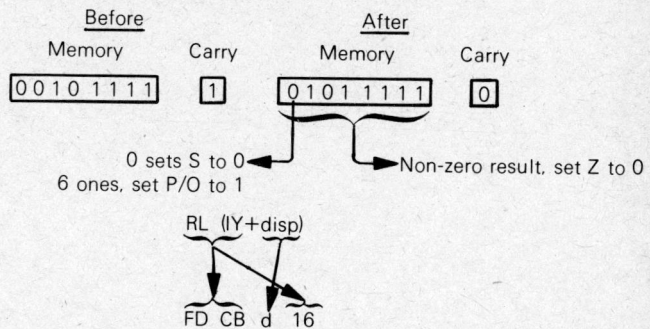

This instruction is identical to RL (IX+disp), but uses the IY register instead of the IX register.

$$\underbrace{\text{RL (HL)}}_{\text{CB 16}}$$

Rotate contents of memory location (specified by the contents of the HL register pair) left one bit through Carry.

## RLA — ROTATE ACCUMULATOR LEFT THROUGH CARRY

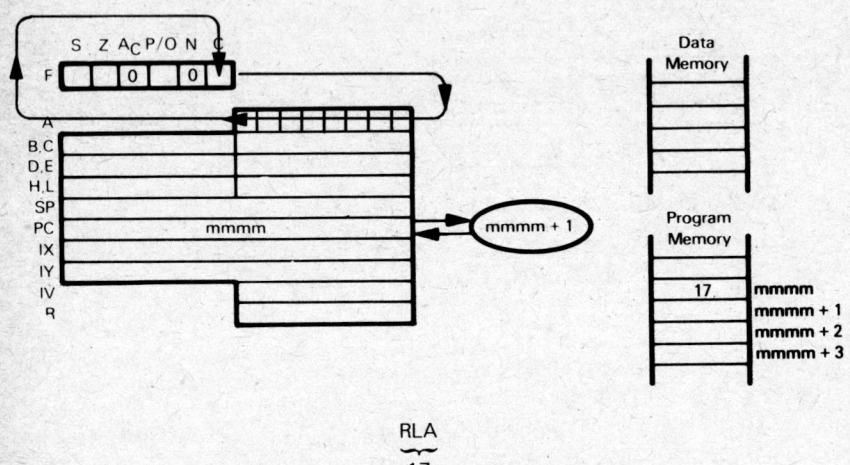

$$\underbrace{\text{RLA}}_{17}$$

Rotate Accumulator contents left one bit through Carry status.

Suppose the Accumulator contains $2A_{16}$ and the Carry status is set to 1. After the instruction

RLA

has executed, the Accumulator will contain $F5_{16}$ and the Carry status will be reset to 0:

| Before | | After | |
|---|---|---|---|
| Accumulator | Carry | Accumulator | Carry |
| 0 1 1 1 1 0 1 0 | 1 | 1 1 1 1 0 1 0 1 | 0 |

# RLC reg — ROTATE CONTENTS OF REGISTER LEFT CIRCULAR

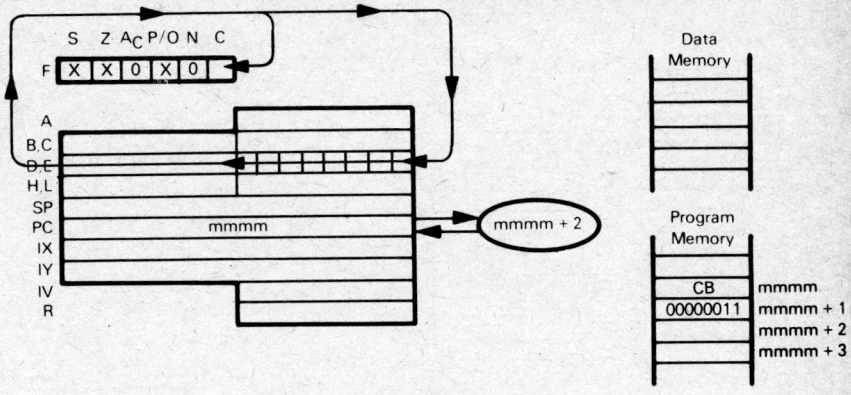

The illustration shows execution of RLC E:

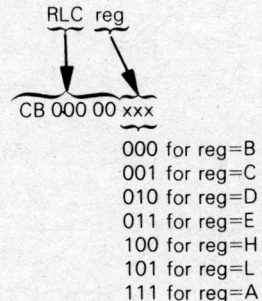

000 for reg=B
001 for reg=C
010 for reg=D
011 for reg=E
100 for reg=H
101 for reg=L
111 for reg=A

Rotate contents of specified register left one bit, copying bit 7 into Carry.

Suppose Register D contains $A9_{16}$ and Carry is 1. After execution of

        RLC D

Register D will contain $53_{16}$ and Carry will be 1:

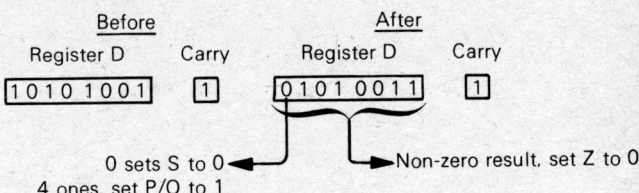

**RLC (HL) —**
**RLC (IX+disp)** **ROTATE CONTENTS OF MEMORY LOCATION**
**RLC (IY+disp)** **LEFT CIRCULAR**

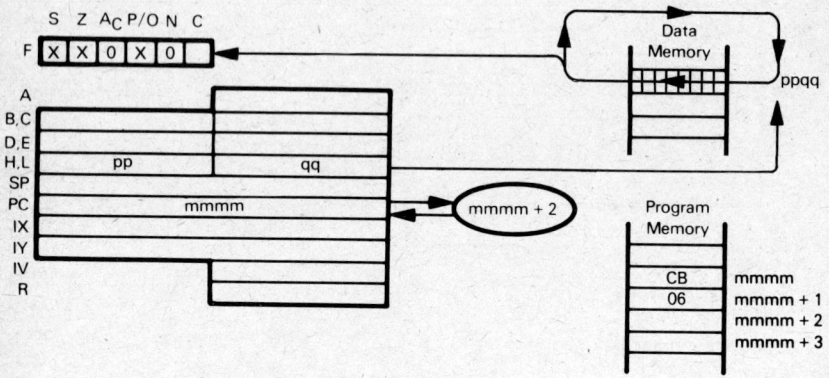

The illustration shows execution of RLC (HL):

$$\underbrace{\text{RLC (HL)}}_{\text{CB 06}}$$

Rotate contents of memory location (specified by the contents of the HL register pair) left one bit, copying bit 7 into Carry.

Suppose register pair HL contains $54FF_{16}$. Memory location $54FF_{16}$ contains $A5_{16}$, and Carry is 0. After execution of

$$\text{RLC (HL)}$$

memory location $54FF_{16}$ will contain $4B_{16}$, and Carry will be 1:

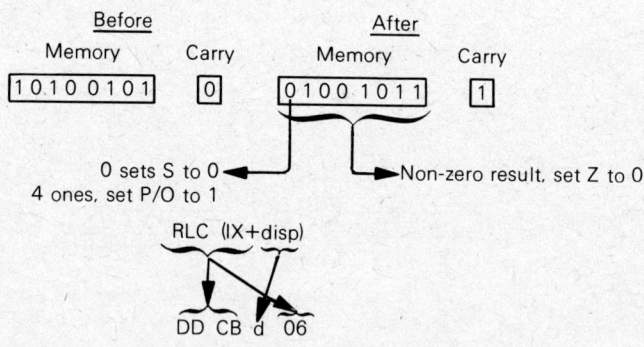

Rotate memory location (specified by the sum of the contents of Index register IX and displacement integer d) left one bit, copying bit 7 into Carry.

Suppose the IX register contains $4000_{16}$. Carry is 1, and memory location $4007_{16}$ contains $2F_{16}$. After the instruction

$$\text{RLC (IX+7)}$$

has executed, memory location $4007_{16}$ will contain $5E_{16}$, and Carry will be 0:

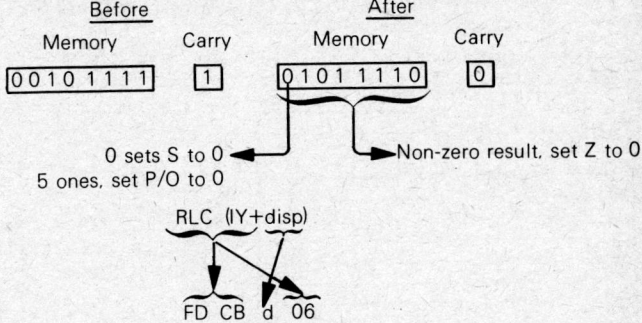

This instruction is identical to RLC (IX+disp), but uses the IY register instead of the IX register.

## RLCA — ROTATE ACCUMULATOR LEFT CIRCULAR

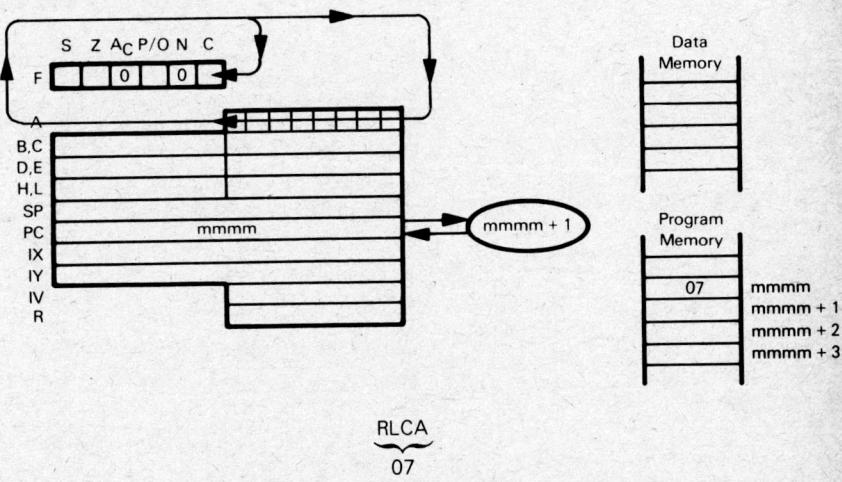

RLCA
07

Rotate Accumulator contents left one bit, copying bit 7 into Carry.

Suppose the Accumulator contains $7A_{16}$ and the Carry status is set to 1. After the instruction

RLCA

has executed, the Accumulator will contain $F4_{16}$ and the Carry status will be reset to 0:

| Before | | After | |
|---|---|---|---|
| Accumulator | Carry | Accumulator | Carry |
| 0 1 1 1 1 0 1 0 | 1 | 1 1 1 1 0 1 0 0 | 0 |

RLCA should be used as a logical instruction.

# RLD — ROTATE ONE BCD DIGIT LEFT BETWEEN THE ACCUMULATOR AND MEMORY LOCATION

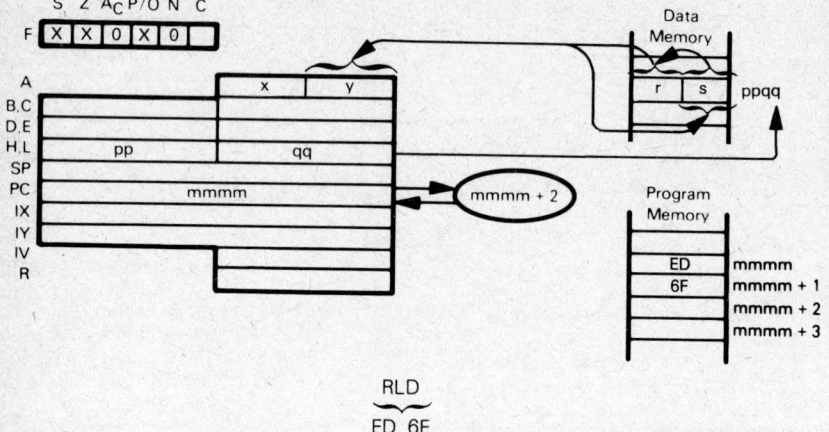

RLD
$\underbrace{\text{ED 6F}}$

The four low-order bits of a memory location (specified by the contents of register pair HL) are copied into the four high-order bits of the same memory location. The previous contents of the four high-order bits of that memory location are copied into the four low-order bits of the Accumulator. The previous four low-order bits of the Accumulator are copied into the four low-order bits of the specified memory location.

Suppose the Accumulator contains $7F_{16}$, HL register pair contains $4000_{16}$, and memory location $4000_{16}$ contains $12_{16}$. After execution of the instruction

RLD

the Accumulator will contain $71_{16}$ and memory location $4000_{16}$ will contain $2F_{16}$:

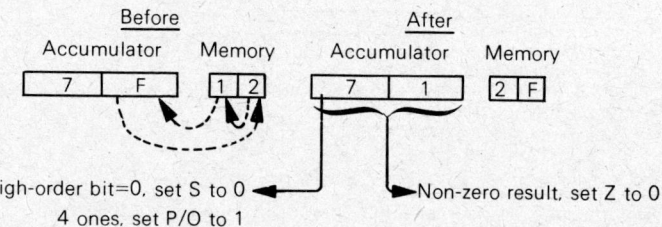

# RR reg — ROTATE CONTENTS OF REGISTER RIGHT THROUGH CARRY

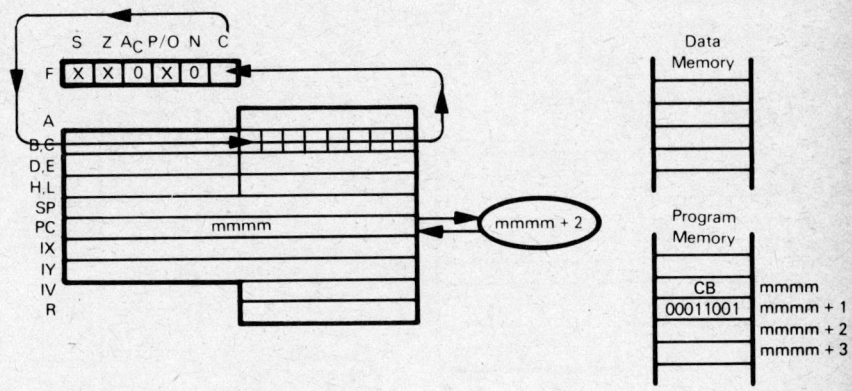

The illustration shows execution of RR C:

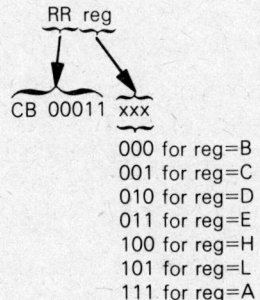

000 for reg=B
001 for reg=C
010 for reg=D
011 for reg=E
100 for reg=H
101 for reg=L
111 for reg=A

Rotate contents of specified register right one bit through Carry.

Suppose Register H contains $0F_{16}$ and Carry is set to 1. After the instruction

RR H

has executed, Register H will contain $87_{16}$, and Carry will be 1:

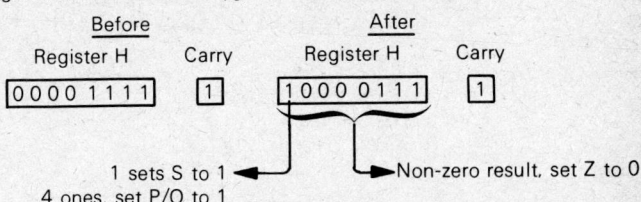

# RR (HL) — ROTATE CONTENTS OF MEMORY LOCATION RIGHT THROUGH CARRY
# RR (IX+disp)
# RR (IY+disp)

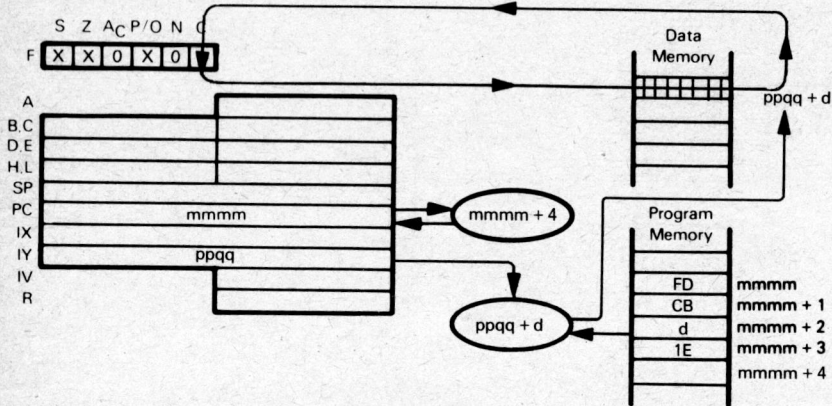

The illustration shows execution of RR (IY+disp):

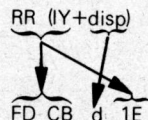

Rotate contents of memory location (specified by the sum of the contents of the IY register and the displacement value d) right one bit through Carry.

Suppose the IY register contains $4500_{16}$, memory location $450F_{16}$ contains $1D_{16}$, and Carry is set to 0. After execution of the instruction

RR (IY+0FH)

memory location $450F_{16}$ will contain $0E_{16}$, and Carry will be 1:

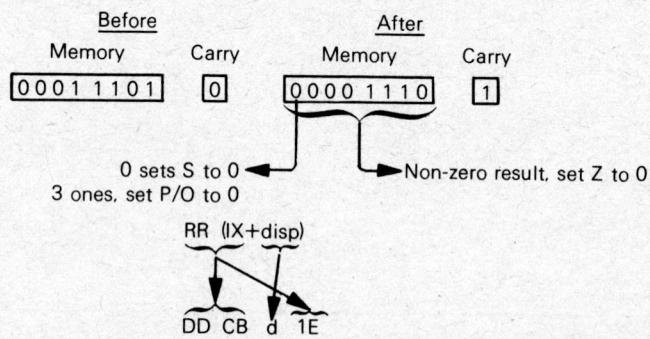

This instruction is identical to RR (IY+disp), but uses the IX register instead of the IY register.

$$\underbrace{\text{RR (HL)}}_{\text{CB 1E}}$$

Rotate contents of memory location (specified by the contents of the HL register pair) right one bit through Carry.

## RRA — ROTATE ACCUMULATOR RIGHT THROUGH CARRY

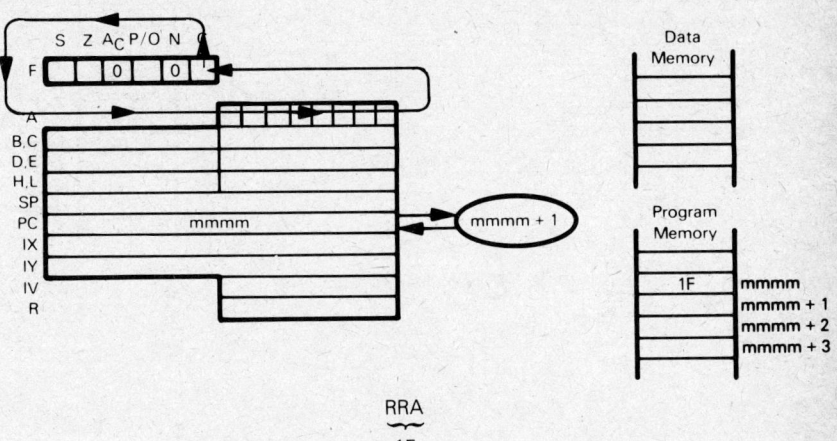

$$\underbrace{\text{RRA}}_{\text{1F}}$$

Rotate Accumulator contents right one bit through Carry status.

Suppose the Accumulator contains $7A_{16}$ and the Carry status is set to 1. After the instruction

RRA

has executed, the Accumulator will contain $BD_{16}$ and the Carry status will be reset to 0:

| Before | | After | |
|---|---|---|---|
| Accumulator | Carry | Accumulator | Carry |
| 0 1 1 1 1 0 1 0 | 1 | 1 0 1 1 1 1 0 1 | 0 |

## RRC reg — ROTATE CONTENTS OF REGISTER RIGHT CIRCULAR

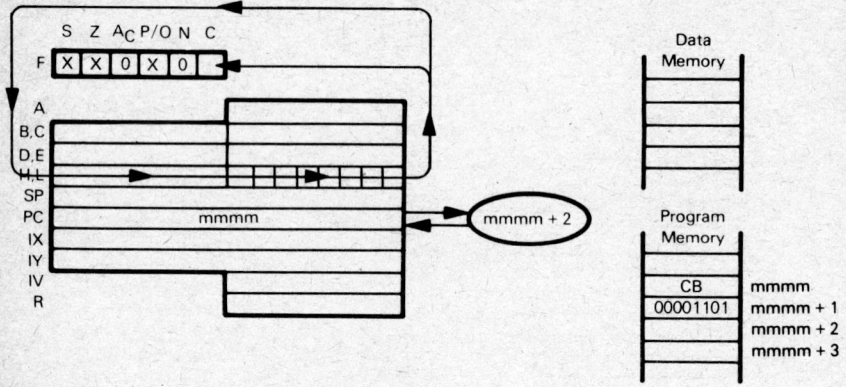

The illustration shows execution of RRC L:

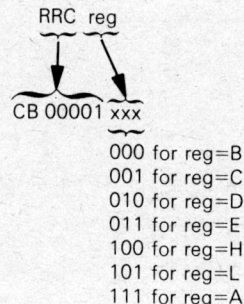

000 for reg=B
001 for reg=C
010 for reg=D
011 for reg=E
100 for reg=H
101 for reg=L
111 for reg=A

Rotate contents of specified register right one bit circularly, copying bit 0 into the Carry status.

Suppose Register D contains $A9_{16}$ and Carry is 0. After execution of

RRC D

Register D will contain $D4_{16}$, and Carry will be 1:

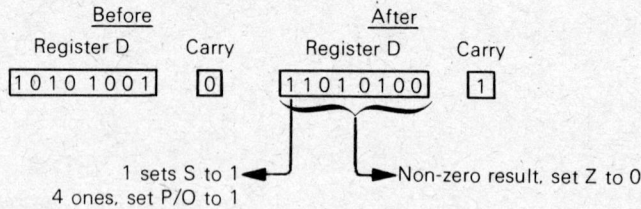

# RRC (HL) — ROTATE CONTENTS OF MEMORY LOCATION
# RRC (IX+disp)   RIGHT CIRCULAR
# RRC (IY+disp)

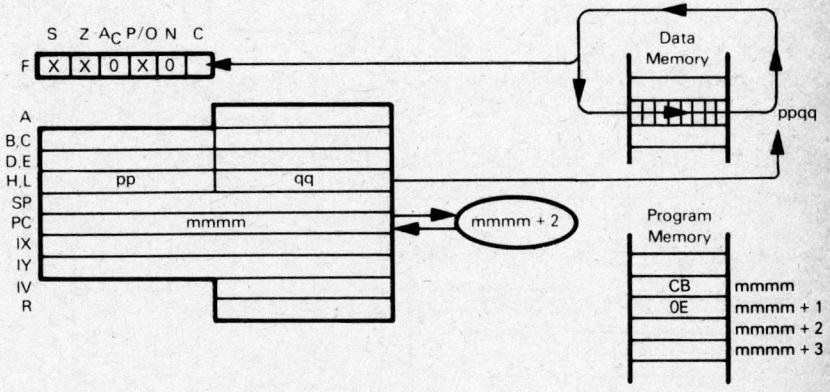

The illustration shows execution of RRC (HL):

$$\underbrace{\text{RRC (HL)}}_{\text{CB 0E}}$$

Rotate contents of memory location (specified by the contents of the HL register pair) right one bit circularly, copying bit 0 into the Carry status.

Suppose the HL register pair contains $4500_{16}$, memory location $4500_{16}$ contains $34_{16}$, and Carry is set to 1. After execution of

RRC (HL)

memory location $4500_{16}$ will contain $1A_{16}$, and Carry will be 0:

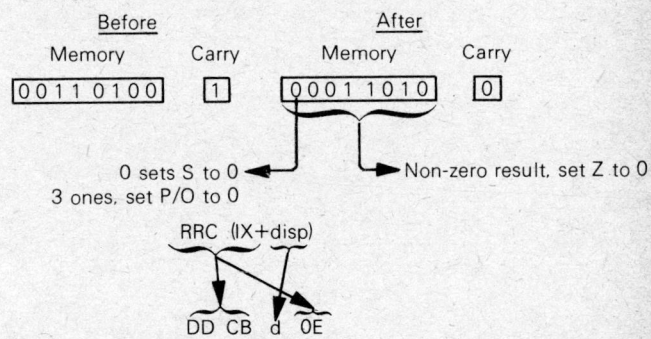

Rotate contents of memory location (specified by the sum of the contents of the IX register and the displacement value d) right one bit circularly, copying bit 0 into the Carry status.

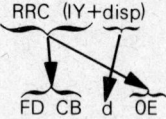

This instruction is identical to the RRC (IX+disp) instruction, but uses the IY register instead of the IX register.

## RRCA — ROTATE ACCUMULATOR RIGHT CIRCULAR

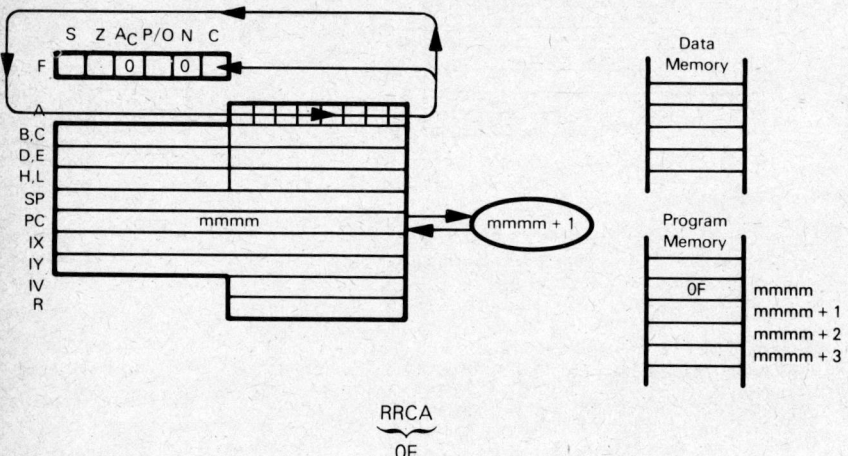

RRCA
0F

Rotate Accumulator contents right one bit circularly, copying bit 0 into the Carry status.

Suppose the Accumulator contains $7A_{16}$ and the Carry status is set to 1. After the instruction

RRCA

has executed, the Accumulator will contain $3D_{16}$ and the Carry status will be reset to 0:

| Before | | After | |
|---|---|---|---|
| Accumulator | Carry | Accumulator | Carry |
| 0 1 1 1 1 0 1 0 | 1 | 0 0 1 1 1 1 0 1 | 0 |

RRCA should be used as a logical instruction.

# RRD — ROTATE ONE BCD DIGIT RIGHT BETWEEN THE ACCUMULATOR AND MEMORY LOCATION

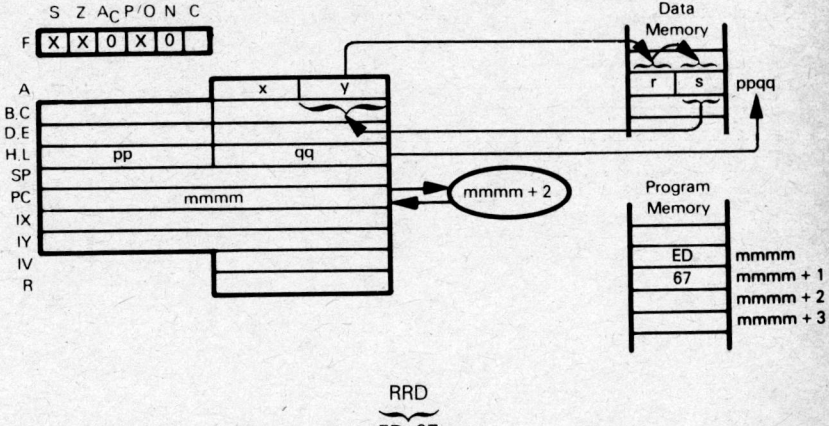

RRD
ED 67

The four high-order bits of a memory location (specified by the contents of register pair HL) are copied into the four low-order bits of the same memory location. The previous contents of the four low-order bits are copied into the four low-order bits of the Accumulator. The previous four low-order bits of the Accumulator are copied into the four high-order bits of the specified memory location.

Suppose the Accumulator contains $7F_{16}$, HL register pair contains $4000_{16}$, and memory location $4000_{16}$ contains $12_{16}$. After execution of the instruction

RRD

the Accumulator will contain $72_{16}$ and memory location $4000_{16}$ will contain $F1_{16}$:

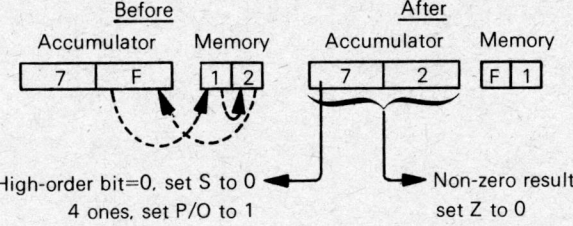

High-order bit=0, set S to 0
4 ones, set P/O to 1

Non-zero result, set Z to 0

# RST n — RESTART

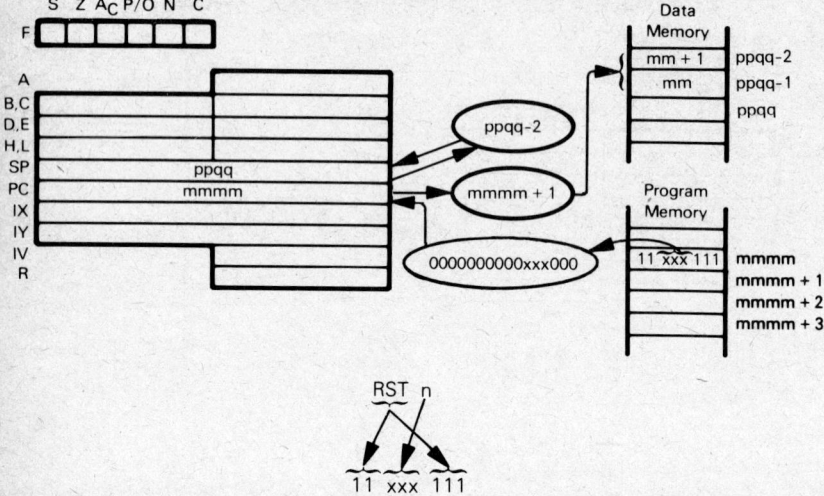

Call the subroutine origined at the low memory address specified by n.

When the instruction

### RST 18H

has executed, the subroutine origined at memory location $0018_{16}$ is called. The previous Program Counter contents are pushed to the top of the stack.

Usually, the RST instruction is used in conjunction with interrupt processing, as described in Chapter 5.

If your application does not use all RST instruction codes to service interrupts, do not overlook the possibility of calling subroutines using RST instructions. Origin frequently used subroutines at appropriate RST addresses, and these subroutines can be called with a single-byte RST instruction instead of a three-byte CALL instruction.

**SUBROUTINE CALL USING RST**

# SBC A,data — SUBTRACT IMMEDIATE DATA FROM ACCUMULATOR WITH BORROW

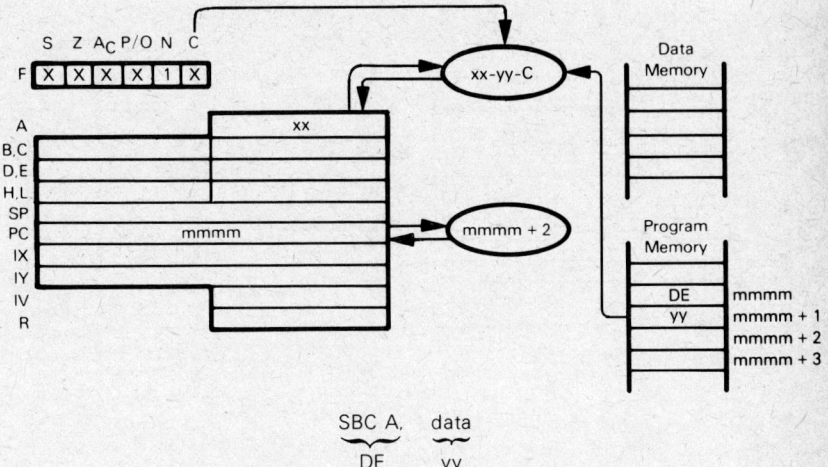

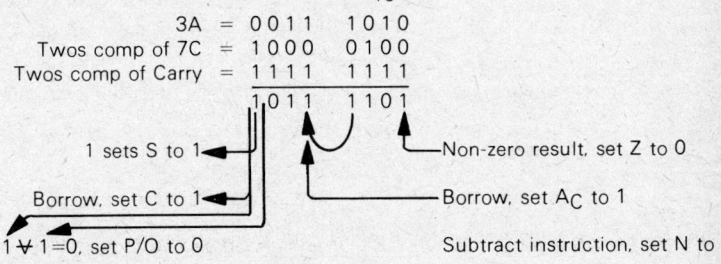

Subtract the contents of the second object code byte and the Carry status from the Accumulator.

Suppose $xx=3A_{16}$ and Carry=1. After the instruction

    SBC A,7CH

has executed, the Accumulator will contain $BD_{16}$.

```
               3A = 0011  1010
    Twos comp of 7C = 1000  0100
 Twos comp of Carry = 1111  1111
                    1,011  1101
```

1 sets S to 1 ◄──────

Borrow, set C to 1 ◄──────

$1 ⩔ 1=0$, set P/O to 0

Non-zero result, set Z to 0

Borrow, set $A_C$ to 1

Subtract instruction, set N to 1

Notice that the resulting carry is complemented.

# SBC A,reg — SUBTRACT REGISTER WITH BORROW FROM ACCUMULATOR

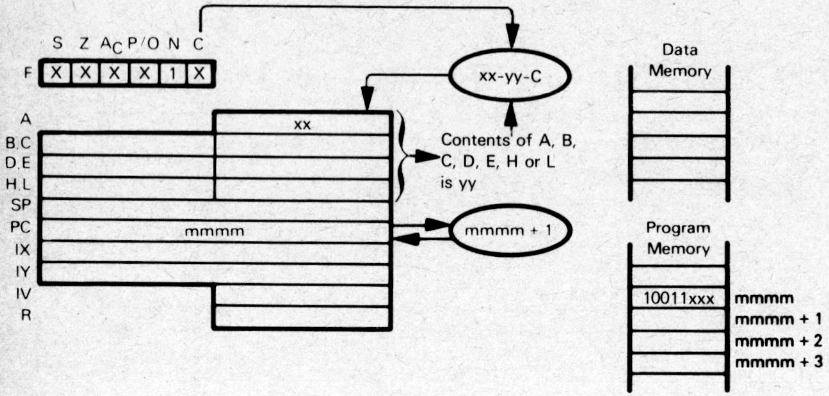

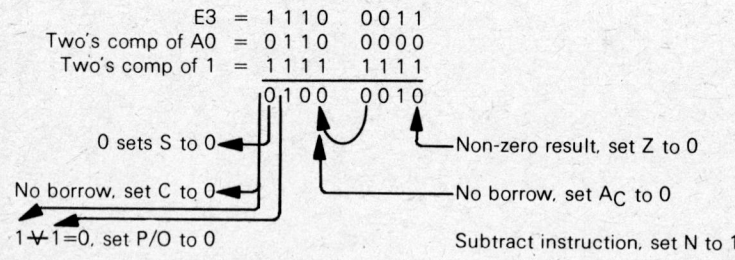

```
SBC A,   reg
‾‾‾‾‾    ‾‾‾
10011    xxx
         ‾‾‾
         000  for reg=B
         001  for reg=C
         010  for reg=D
         011  for reg=E
         100  for reg=H
         101  for reg=L
         111  for reg=A
```

Subtract the contents of the specified register and the Carry status from the Accumulator.

Suppose xx=$E3_{16}$, Register E contains $A0_{16}$, and Carry=1. After the instruction

SBC A,E

has executed, the Accumulator will contain $42_{16}$.

```
          E3 = 1110 0011
Two's comp of A0 = 0110 0000
Two's comp of 1  = 1111 1111
                  0,100 0010
```

0 sets S to 0 ◄—— ——► Non-zero result, set Z to 0

No borrow, set C to 0 ◄—— ——► No borrow, set $A_C$ to 0

$1 ⊻ 1 = 0$, set P/O to 0          Subtract instruction, set N to 1

Notice that the resulting carry is complemented.

**SBC A,(HL) —**
**SBC A,(IX+disp)**
**SBC A,(IY+disp)**

## SUBTRACT MEMORY AND CARRY FROM ACCUMULATOR

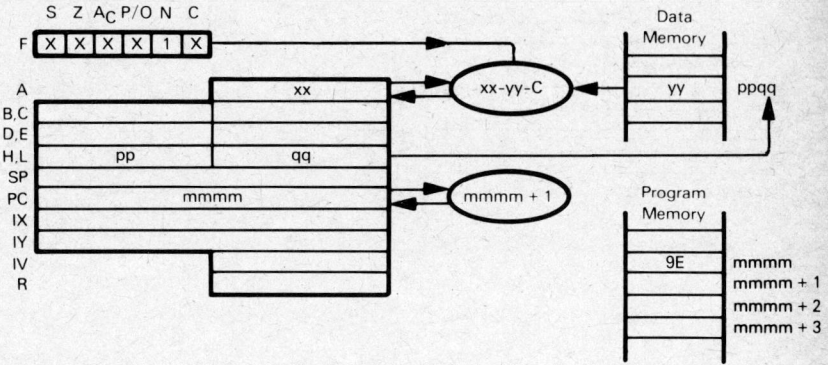

The illustration shows execution of SBC A,(HL):

$$\underbrace{\text{SBC A,(HL)}}_{9E}$$

Subtract the contents of memory location (specified by the contents of the HL register pair) and the Carry from the Accumulator.

Suppose Carry=0, ppqq=$4000_{16}$, xx=$3A_{16}$, and memory location $4000_{16}$ contains $7C_{16}$. After execution of the instruction

SBC A,(HL)

the Accumulator will contain $BE_{16}$.

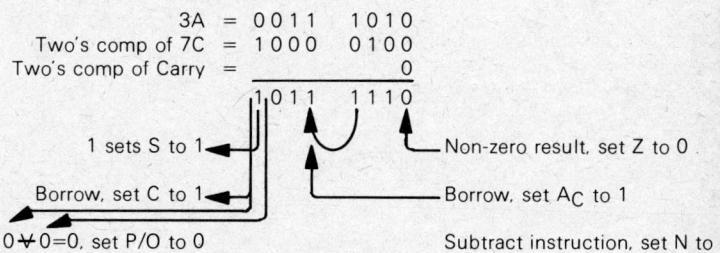

Notice that the resulting carry is complemented.

$$\underbrace{\text{SBC A,}}_{\text{DD 9E}}\underbrace{\text{(IX+disp)}}_{\text{d}}$$

Subtract the contents of memory location (specified by the sum of the contents of the IX register and the displacement value d) and the Carry from the Accumulator.

$$\underbrace{\text{SBC A,}}_{\text{FD 9E}}\underbrace{\text{(IY+disp)}}_{\text{d}}$$

This instruction is identical to the SBC A,(IX+disp) instruction, except that it uses the IY register instead of the IX register.

# SBC HL,rp — SUBTRACT REGISTER PAIR WITH CARRY FROM H AND L

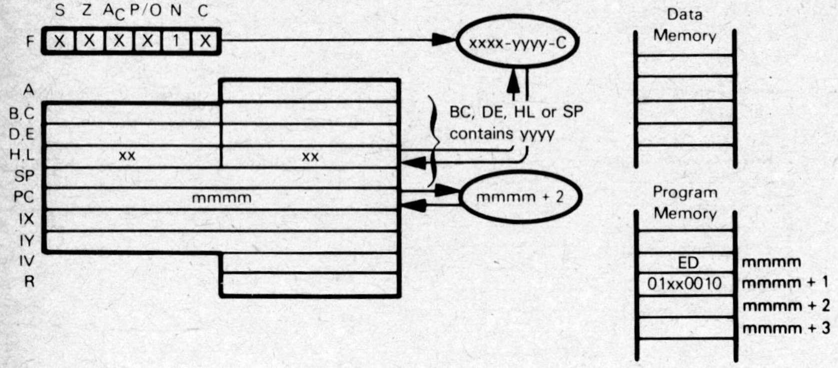

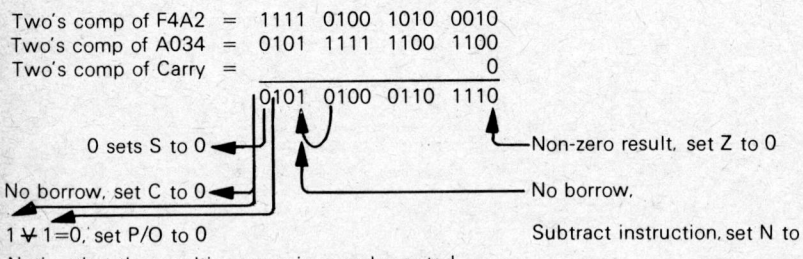

```
        SBC HL, rp
        ↙  ↘ ↙
      01  xx  0010
```

00 for rp is register pair BC
01 for rp is register pair DE
10 for rp is register pair HL
11 for rp is Stack Pointer

Subtract the contents of the designated register pair and the Carry status from the HL register pair.

Suppose HL contains $F4A2_{16}$, BC contains $A034_{16}$, and Carry=0. After the instruction

SBC HL,BC

has executed, the HL register pair will contain $546E_{16}$:

```
Two's comp of F4A2  =   1111 0100 1010 0010
Two's comp of A034  =   0101 1111 1100 1100
Two's comp of Carry =                      0
                       0 0101 0100 0110 1110
```

0 sets S to 0 ◀—                    —Non-zero result, set Z to 0

No borrow, set C to 0 ◀—        —No borrow,

$1 \veebar 1=0$, set P/O to 0                Subtract instruction, set N to 1

Notice that the resulting carry is complemented.

## SCF — SET CARRY FLAG

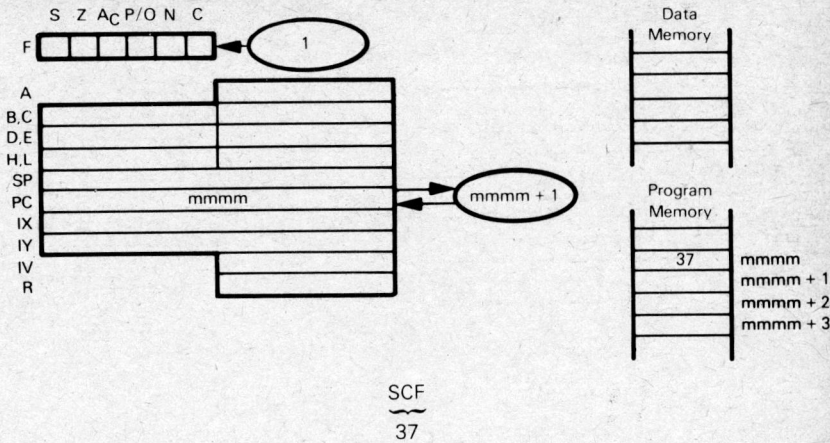

$$\underset{37}{\underbrace{\text{SCF}}}$$

When the SCF instruction is executed, the Carry status is set to 1 regardless of its previous value. No other statuses or register contents are affected.

## SET b,reg — SET INDICATED REGISTER BIT

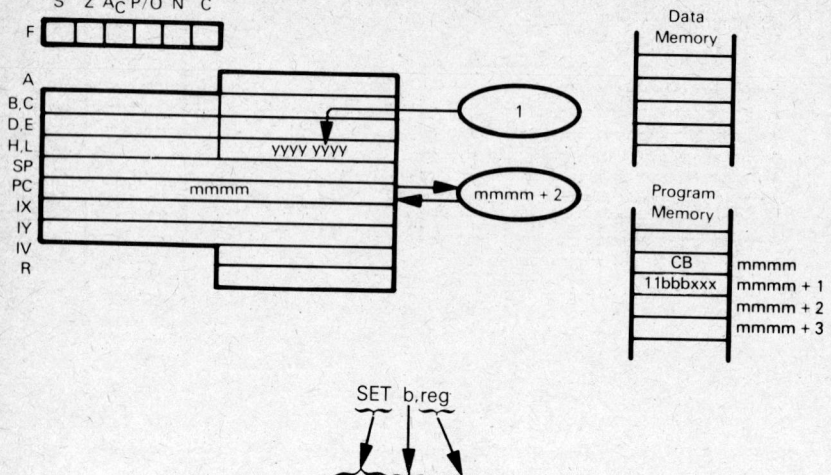

```
        SET b,reg

      CB  11bbb  xxx
```

| Bit | bbb | xxx | Register |
|-----|-----|-----|----------|
| 0   | 000 | 000 | B        |
| 1   | 001 | 001 | C        |
| 2   | 010 | 010 | D        |
| 3   | 011 | 011 | E        |
| 4   | 100 | 100 | H        |
| 5   | 101 | 101 | L        |
| 6   | 110 | 111 | A        |
| 7   | 111 |     |          |

SET indicated bit within specified register. After the instruction

SET 2,L

has executed, bit 2 in Register L will be set. (Bit 0 is the least significant bit.)

# SET b,(HL) — SET BIT b OF INDICATED MEMORY POSITION
# SET b,(IX+disp)
# SET b,(IY+disp)

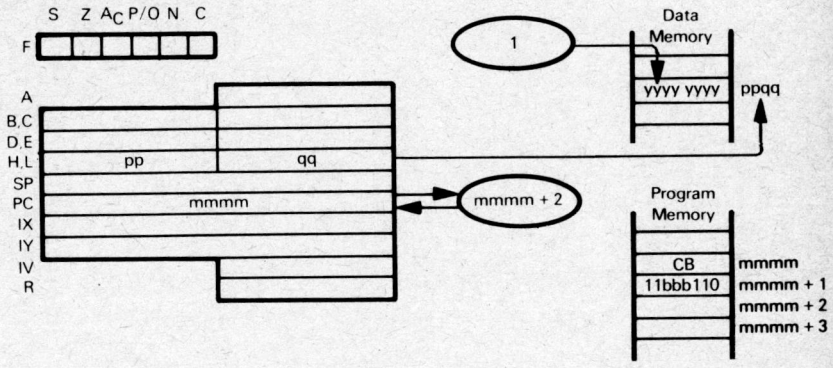

The illustration shows execution of SET b,(HL). Bit 0 is the least significant bit.

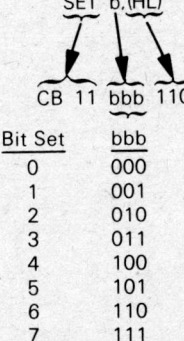

| Bit Set | bbb |
|---------|-----|
| 0 | 000 |
| 1 | 001 |
| 2 | 010 |
| 3 | 011 |
| 4 | 100 |
| 5 | 101 |
| 6 | 110 |
| 7 | 111 |

Set indicated bit within memory location indicated by HL.

Suppose HL contains $4000_{16}$. After the instruction

SET 5,(HL)

has executed, bit 5 in memory position $4000_{16}$ will be 1.

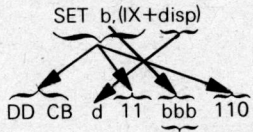

bbb is the same as in SET b,(HL)

Set indicated bit within memory location indicated by the sum of Index Register IX and displacement.

Suppose Index Register IX contains $4000_{16}$. After execution of

SET 6,(IX+5H)

bit 6 in memory location $4005_{16}$ will be 1.

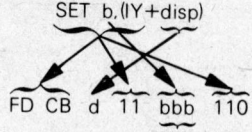

bbb is the same as in SET b,(HL)

This instruction is identical to SET b,(IX+disp), except that it uses the IY register instead of the IX register.

## SLA reg — SHIFT CONTENTS OF REGISTER LEFT ARITHMETIC

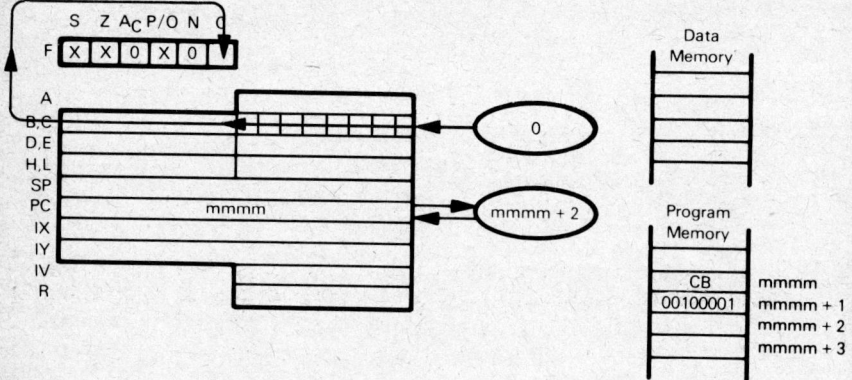

The illustration shows execution of SLA C:

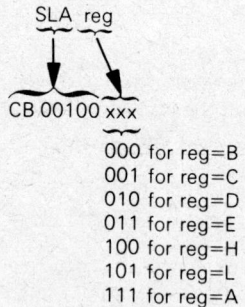

```
000 for reg=B
001 for reg=C
010 for reg=D
011 for reg=E
100 for reg=H
101 for reg=L
111 for reg=A
```

Shift contents of specified register left one bit, resetting the least significant bit to 0.

Suppose Register B contains $1F_{16}$, and Carry=1. After execution of

SLA B

Register B will contain $3E_{16}$ and Carry will be zero.

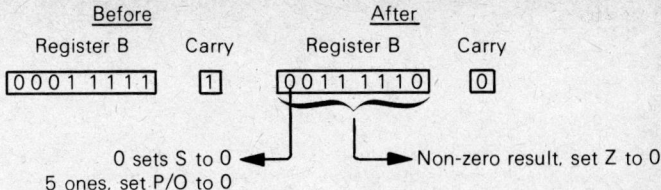

## SLA (HL) — SHIFT CONTENTS OF MEMORY LOCATION
## SLA (IX+disp) LEFT ARITHMETIC
## SLA (IY+disp)

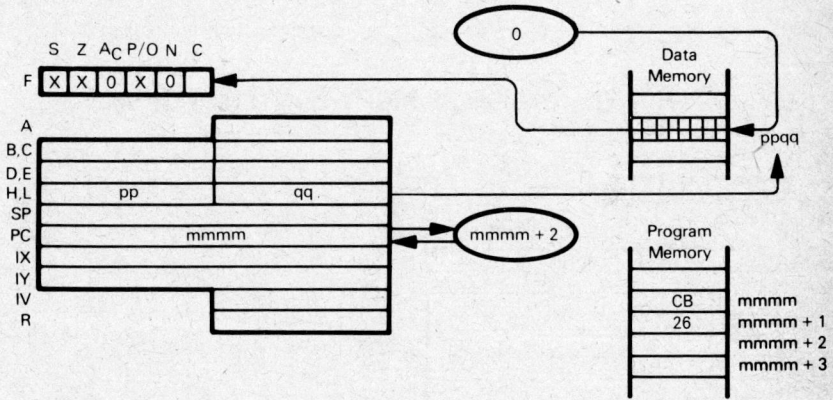

The illustration shows execution of SLA (HL):

$$\underbrace{\text{SLA (HL)}}_{\text{CB 26}}$$

Shift contents of memory location (specified by the contents of the HL register pair) left one bit, resetting the least significant bit to 0.

Suppose the HL register pair contains $4500_{16}$, memory location $4500_{16}$ contains $84_{16}$, and Carry=0. After execution of

$$\text{SLA (HL)}$$

memory location $4500_{16}$ will contain $08_{16}$, and Carry will be 1.

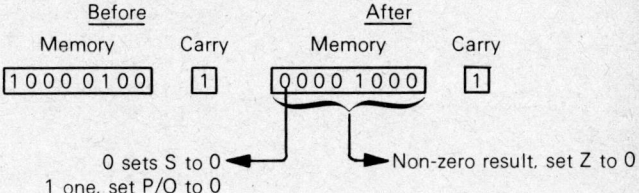

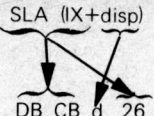

Shift contents of memory location (specified by the sum of the contents of the IX register and the displacement value d) left one bit arithmetically, resetting least significant bit to 0.

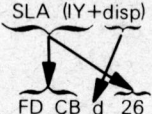

This instruction is identical to SLA (IX+disp), but uses the IY register instead of the IX register.

## SRA reg — ARITHMETIC SHIFT RIGHT CONTENTS OF REGISTER

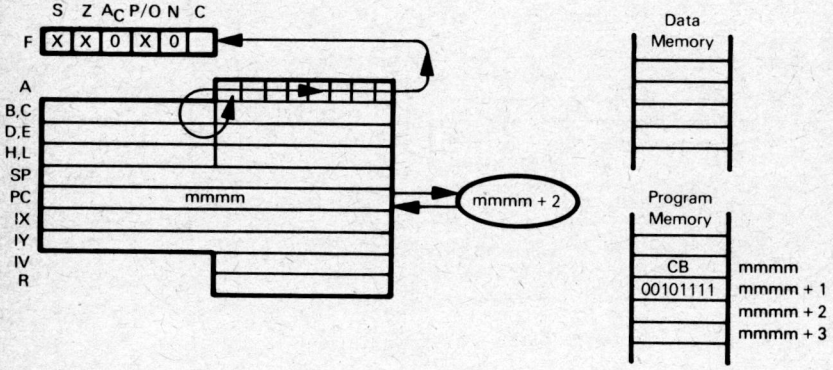

The illustration shows execution of SRA A:

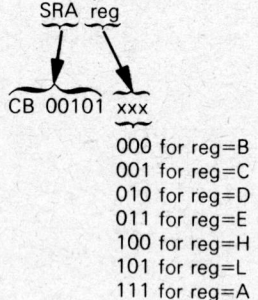

```
000 for reg=B
001 for reg=C
010 for reg=D
011 for reg=E
100 for reg=H
101 for reg=L
111 for reg=A
```

Shift specified register right one bit. Most significant bit is unchanged.

Suppose Register H contains $59_{16}$, and Carry=0. After the instruction

SRA H

has executed, Register H will contain $2C_{16}$ and Carry will be 1.

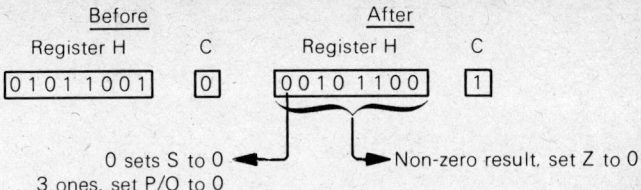

0 sets S to 0
3 ones, set P/O to 0
Non-zero result, set Z to 0

## SRA (HL) — ARITHMETIC SHIFT RIGHT CONTENTS OF
## SRA (IX+disp) MEMORY POSITION
## SRA (IY+disp)

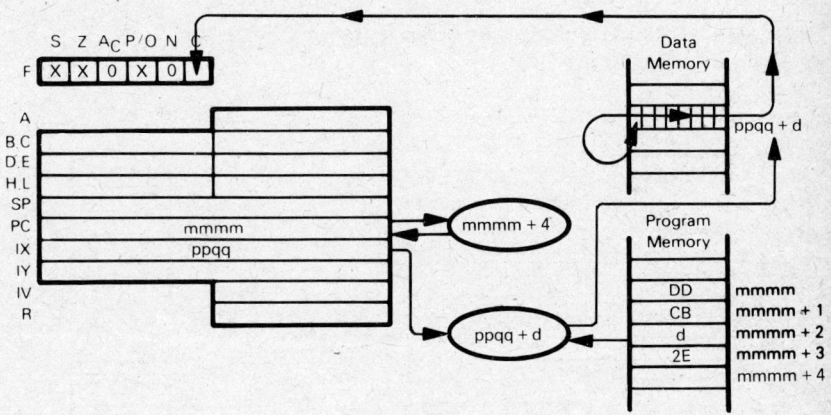

The illustration shows execution of SRA (IX+disp):

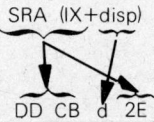

Shift contents of memory location (specified by the sum of the contents of Register IX and the displacement value d) right. Most significant bit is unchanged.

Suppose Register IX contains $3400_{16}$, memory location $34AA_{16}$ contains $27_{16}$, and Carry=1. After execution of

SRA (IX+0AAH)

memory location $34AA_{16}$ will contain $13_{16}$, and Carry will be 1.

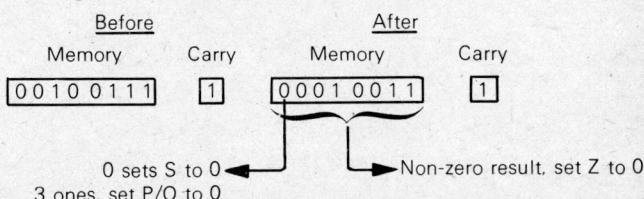

0 sets S to 0
3 ones, set P/O to 0
Non-zero result, set Z to 0

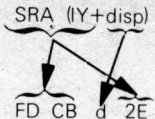

This instruction is identical to SRA (IX+disp), but uses the IY register instead of the IX register.

$$\underbrace{\text{SRA (HL)}}_{\text{CB 2E}}$$

Shift contents of memory location (specified by the contents of the HL register pair) right one bit. Most significant bit is unchanged.

# SRL reg — SHIFT CONTENTS OF REGISTER RIGHT LOGICAL

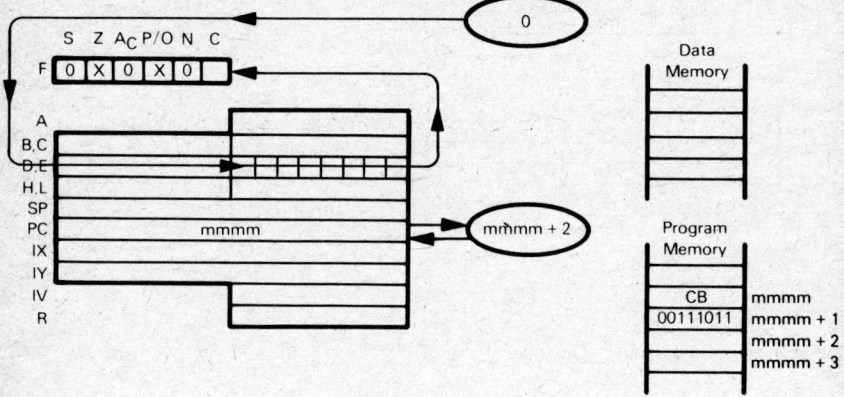

The illustration shows execution of SRL E:

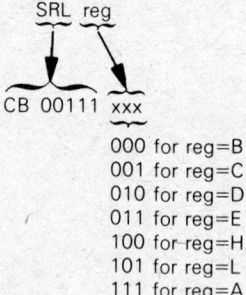

000 for reg=B
001 for reg=C
010 for reg=D
011 for reg=E
100 for reg=H
101 for reg=L
111 for reg=A

Shift contents of specified register right one bit. Most significant bit is reset to 0.

Suppose Register D contains $1F_{10}$, and Carry=0. After execution of

SRL D

Register D will contain $0F_{16}$, and Carry will be 1.

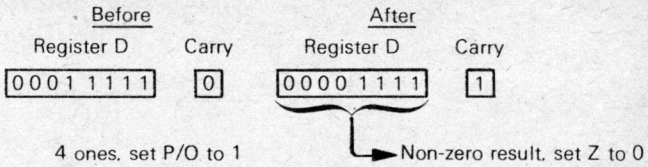

## SRL (HL) — SHIFT CONTENTS OF MEMORY LOCATION
## SRL (IX+disp) RIGHT LOGICAL
## SRL (IY+disp)

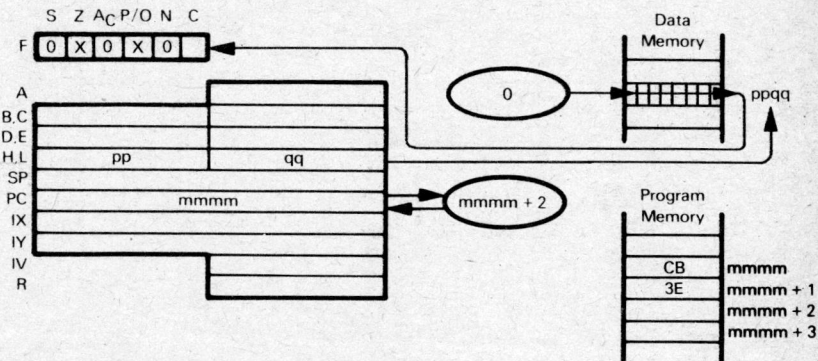

The illustration shows execution of SRL (HL):

$$\underbrace{\text{SRL (HL)}}_{\text{CB 3E}}$$

Shift contents of memory location (specified by the contents of the HL register pair) right one bit. Most significant bit is reset to 0.

Suppose the HL register pair contains $2000_{16}$, memory location $2000_{16}$ contains $8F_{16}$, and Carry=0. After execution of

SRL (HL)

memory location $2000_{16}$ will contain $47_{16}$, and Carry will be 1.

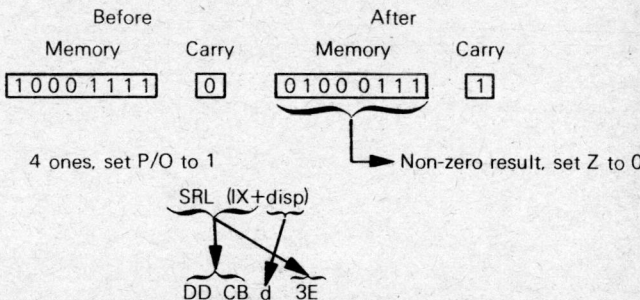

Shift contents of memory location (specified by the sum of the contents of the IX register and the displacement value d) right one bit. Most significant bit is reset to 0.

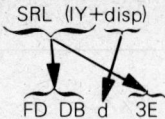

This instruction is identical to SRL (IX+disp), but uses the IY register instead of the IX register.

## SUB data — SUBTRACT IMMEDIATE FROM ACCUMULATOR

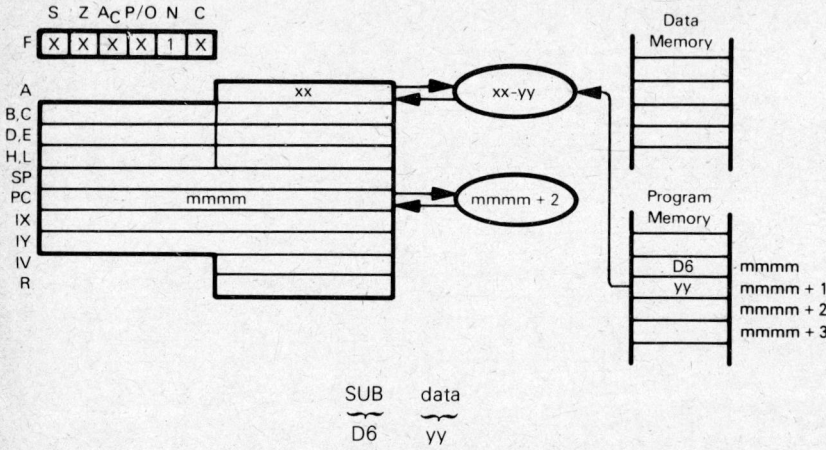

$$\frac{\text{SUB}}{\text{D6}} \quad \frac{\text{data}}{\text{yy}}$$

Subtract the contents of the second object code byte from the Accumulator.

Suppose $xx = 3A_{16}$. After the instruction

SUB 7CH

has executed, the Accumulator will contain $BE_{16}$.

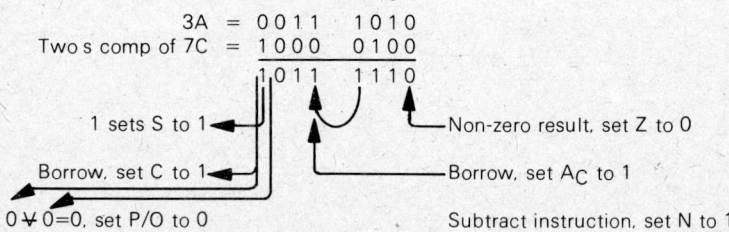

Notice that the resulting carry is complemented.

## SUB reg — SUBTRACT REGISTER FROM ACCUMULATOR

S Z AC P/O N C
F X X X X 1 X
A
B,C
D,E
H,L
SP
PC
IX
IY
IV
R
xx
mmmm
xx-yy
Contents of A, B, C, D, E, H or L is yy
mmmm + 1
Data Memory
Program Memory
10010xxx
mmmm
mmmm + 1
mmmm + 2
mmmm + 3

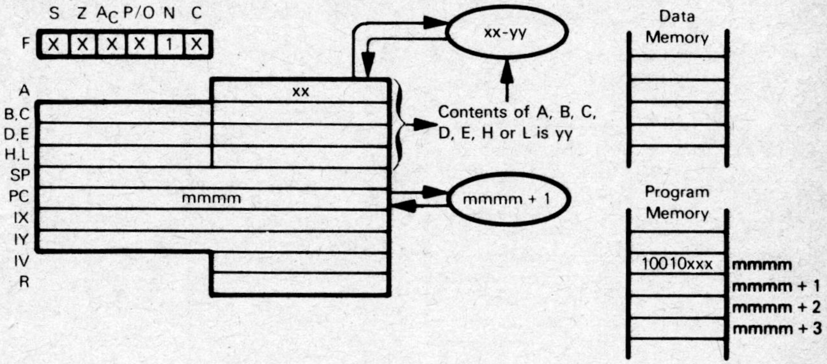

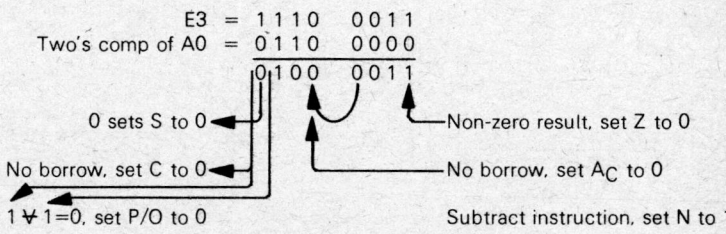

| SUB | reg | |
|---|---|---|
| 10010 | xxx | |
| | 000 | for reg=B |
| | 001 | for reg=C |
| | 010 | for reg=D |
| | 011 | for reg=E |
| | 100 | for reg=H |
| | 101 | for reg=L |
| | 111 | for reg=A |

Subtract the contents of the specified register from the Accumulator.

Suppose xx=E3 and Register H contains $A0_{16}$. After execution of

SUB H

the Accumulator will contain $43_{16}$.

E3 = 1 1 1 0 0 0 1 1
Two's comp of A0 = 0 1 1 0 0 0 0 0
0 1 0 0 0 0 1 1
0 sets S to 0
No borrow, set C to 0
1 ⊻ 1=0, set P/O to 0
Non-zero result, set Z to 0
No borrow, set AC to 0
Subtract instruction, set N to 1

Notice that the resulting carry is complemented.

# SUB (HL) — SUBTRACT MEMORY FROM ACCUMULATOR
# SUB (IX+disp)
# SUB (IY+disp)

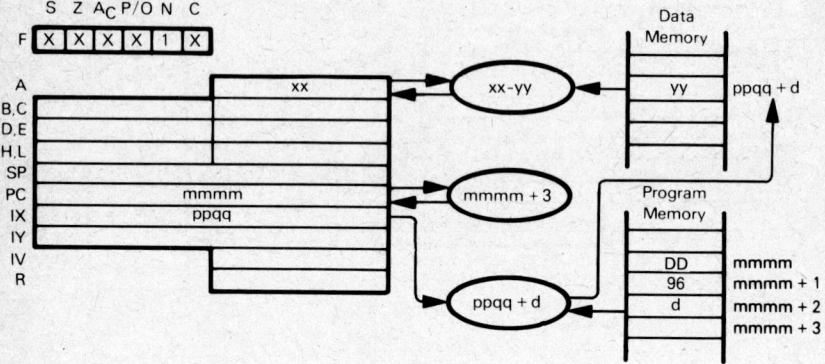

The illustration shows execution of SUB (IX+d):

SUB (IX+disp)

DD 96 d

Subtract contents of memory location (specified by the sum of the contents of the IX register and the displacement value d) from the Accumulator.

Suppose ppqq=$4000_{16}$, xx=$FF_{16}$, and memory location $40FF_{16}$ contains $50_{16}$. After execution of

SUB (IX+0FFH)

the Accumulator will contain $AF_{16}$.

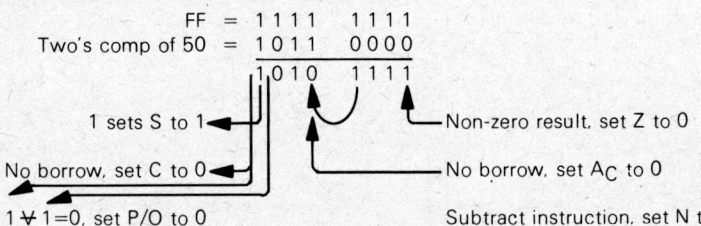

Notice that the resulting carry is complemented.

SUB (IY+disp)

FD 96 d

This instruction is identical to SUB (IX+disp), except that it uses the IY register instead of the IX register.

SUB (HL)

96

Subtract contents of memory location (specified by the contents of the HL register pair) from the Accumulator.

## XOR data — EXCLUSIVE-OR IMMEDIATE WITH ACCUMULATOR

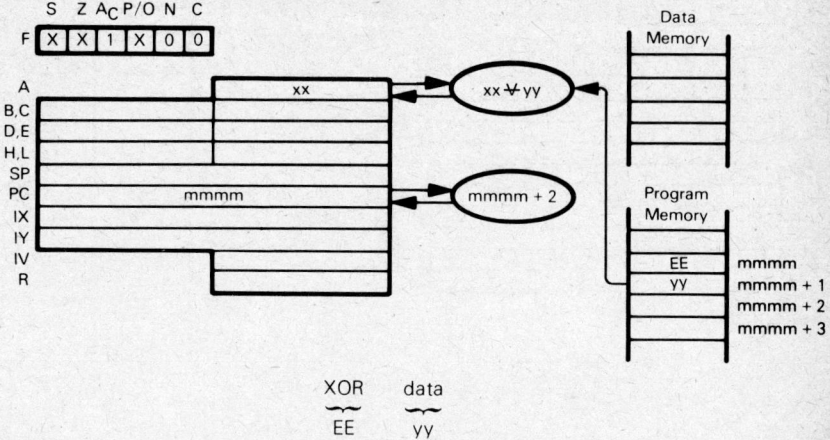

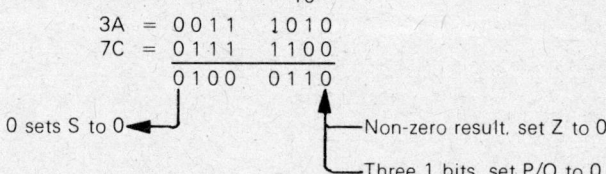

Exclusive-OR the contents of the second object code byte with the Accumulator.

Suppose $xx=3A_{16}$. After the instruction

$$\text{XOR 7CH}$$

has executed, the Accumulator will contain $46_{16}$.

```
 3A =  0011  1010
 7C =  0111  1100
       0100  0110
```

0 sets S to 0

Non-zero result, set Z to 0

Three 1 bits, set P/O to 0

The Exclusive-OR instruction is used to test for changes in bit status.

# XOR reg — EXCLUSIVE-OR REGISTER WITH ACCUMULATOR

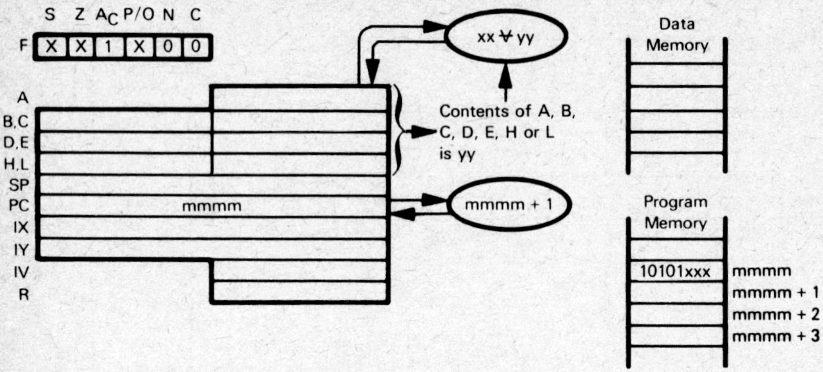

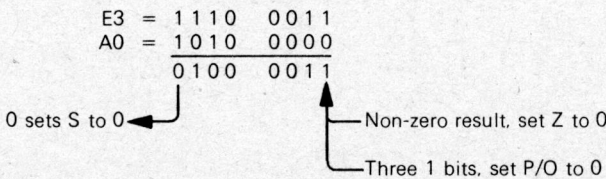

```
XOR    reg
~~~    ~~~
10101  xxx
       000 for reg=B
       001 for reg=C
       010 for reg=D
       011 for reg=E
       100 for reg=H
       101 for reg=L
       111 for reg=A
```

Exclusive-OR the contents of the specified register with the Accumulator.

Suppose xx=$E3_{16}$ and Register E contains $A0_{16}$. After the instruction

$$\text{XOR E}$$

has executed, the Accumulator will contain $43_{16}$.

```
E3 =   1 1 1 0   0 0 1 1
A0 =   1 0 1 0   0 0 0 0
       ---------------
       0 1 0 0   0 0 1 1
```

0 sets S to 0 ◄────┘

Non-zero result, set Z to 0

Three 1 bits, set P/O to 0

The Exclusive-OR instruction is used to test for changes in bit status.

## XOR (HL) — EXCLUSIVE-OR MEMORY WITH ACCUMULATOR
## XOR (IX+disp)
## XOR (IY+disp)

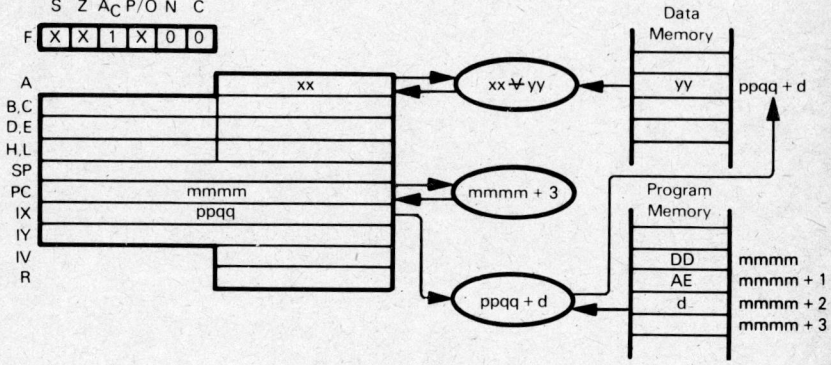

The illustration shows execution of XOR (IX+disp):

$$\underbrace{\text{XOR (IX+}}_{\text{DD AE}}\underbrace{\text{disp)}}_{\text{d}}$$

Exclusive-OR contents of memory location (specified by the sum of the contents of the IX register and the displacement value d) with the Accumulator.

Suppose xx=$E3_{16}$, ppqq=$4500_{16}$, and memory location $45FF_{16}$ contains $A0_{16}$. After the instruction

$$\text{XOR (IX+0FFH)}$$

has executed, the Accumulator will contain $43_{16}$.

```
E3 = 1110  0011
A0 = 1010  0000
     0100  0011
```

0 sets S to 0 ◄─┘           └─ Non-zero result, set Z to 0
                             └─ Three 1 bits, set P/O to 0

$$\underbrace{\text{XOR (IY+}}_{\text{FD AE}}\underbrace{\text{disp)}}_{\text{d}}$$

This instruction is identical to XOR (IX+disp), except that it uses the IY register instead of the IX register.

$$\underbrace{\text{XOR (HL)}}_{\text{AE}}$$

Exclusive-OR contents of memory location (specified by the contents of the HL register pair) with the Accumulator.

# Chapter 7
# SOME COMMONLY USED SUBROUTINES

There are several operations which occur in many microcomputer programs irrespective of the application. This chapter will provide a number of frequently used instruction sequences.

To make the most effective use of this chapter, you should study each subroutine until you know it well enough to modify it. As a simple exercise, you should attempt to rewrite the subroutine so that it does the same job using fewer execution cycles, or fewer instructions, or both. Next, rewrite the programs to implement variations. For example, binary multiplication of 16-bit numbers is illustrated; how about a routine to multiply 32-bit numbers? Look upon each example as a typical illustrative instruction sequence which you will likely modify to meet your immediate needs.

Simple programs at the level covered in this chapter fall into one of four categories:

1) **Memory addressing**
2) **Data movement**
3) **Arithmetic**
4) **Program execution sequence logic**

We will describe programs in the above category sequence.

## MEMORY ADDRESSING

The Z80 has an unusually large variety of memory referencing instructions: direct, indexed, implied, and auto-increment/decrement addressing are all available on the Z80. We are going to show how two other addressing modes — indirect addressing and indirect addressing with post-indexing — may be implemented through simple instruction sequences. Both of these modes are described and illustrated in An Introduction to Microcomputers: Volume I — Basic Concepts.

## INDIRECT ADDRESSING

The Z80 CPU provides register indirect addressing where a register pair (such as HL) serves as a pointer to a location in memory. However, the true indirect memory addressing specifies that the memory address you require be stored in two memory bytes:

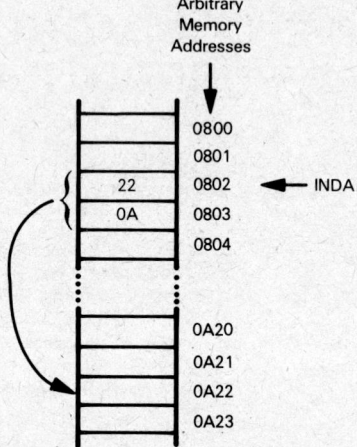

In the illustration above, memory bytes $0802_{16}$ and $0803_{16}$ hold the required memory address: $0A22_{16}$. In keeping with the way the Z80 itself handles 16-bit addresses, the low-order address byte is shown preceding the high-order address byte.

**All that is required to simulate indirect addressing as shown above is the following instruction sequence:**

```
LD    HL,INDA    ;LOAD ADDRESS INTO HL
LD    A,(HL)     ;LOAD DATA INTO A
```

The first instruction moves the address $0A22_{16}$ into HL. The second instruction demonstrates how to access memory location $0A22_{16}$.

## INDIRECT, POST-INDEXED ADDRESSING

In some applications, it is necessary or certainly preferable to perform **indirect post-indexed addressing.** Using Z80 indexed addressing, post-indexed addressing **can be performed in the following manner:**

```
LD    BC,(INDA)  ;LOAD INDIRECT ADDRESS INTO BC
ADD   IX,BC      ;ADD INDIRECT ADDRESS TO INDEX
```

At the beginning of this instruction sequence, we assume that the index is in the Index Register IX.

The index is then added to the indirect address, and the result is placed in the Index register; any memory operation can now be performed using the Index register as the address.

# DATA MOVEMENT

We will now examine some instruction sequences that locate and move contiguous blocks of data bytes — data buffers of any length.

## MOVING SIMPLE DATA BLOCKS

**Beginning with a very simple program, consider moving the contents of a contiguous block of data memory bytes from one area of memory to another.** This operation is made extremely simple by the unique block transfer instruction provided by the Z80 CPU. The block transfer instructions operate with three register pairs:

    HL  addresses the source location
    DE  addresses the destination location
    BC  is a byte counter

The following memory map illustrates the data movement operation:

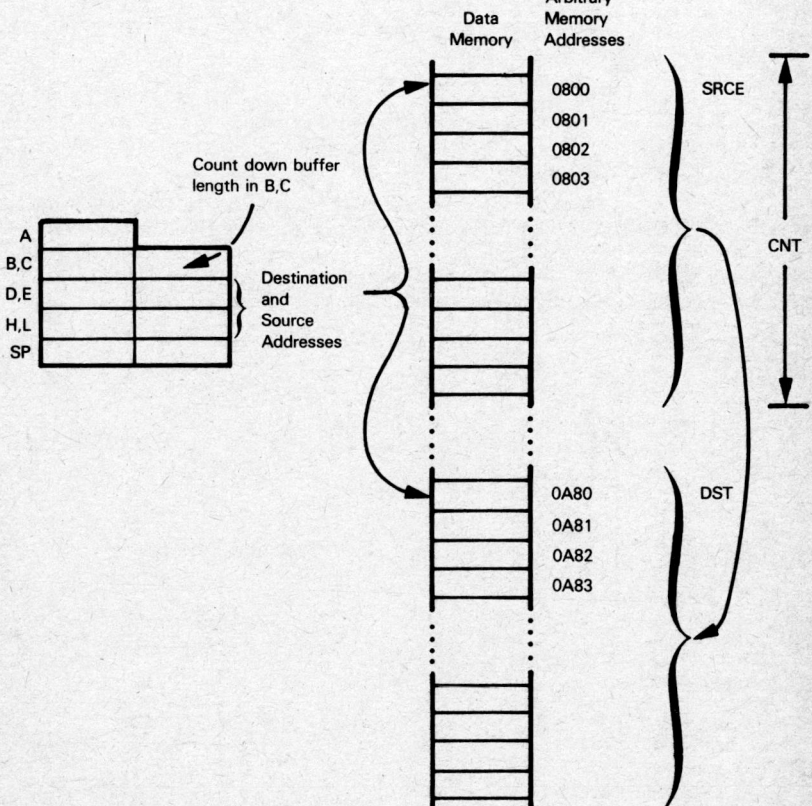

**This is the data move program:**

```
LD    HL,SRCE    ;LOAD SOURCE ADDRESS INTO HL
LD    DE,DST     ;LOAD DESTINATION ADDRESS INTO DE
LD    BC,CNT     ;LOAD BYTE COUNT INTO BC
LDIR             ;TRANSFER DATA
```

The single LDIR instruction does all the work for us — it transfers the byte of data pointed to by HL to the location pointed to by DE, then increments HL and DE to point to the next byte, decrements the count in BC, and repeats the process until the count = 0.

## MULTIPLE TABLE LOOKUPS

**Next, consider a multiple table lookup.** This is a more complex variation of the data move which we just described.

The starting addresses of an indefinite number of data tables are stored in an index table. The index table's starting address is given by the label TABX:

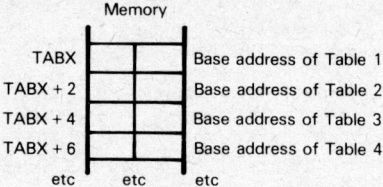

Several data bytes are in temporary storage, starting at a memory location identified by the label CBASE. The actual number of data bytes can be found in a memory location identified by the label CNT. This source buffer is equivalent to the source buffer in the data move program we have just described.

The destination for the block of data is one of the data tables. The table number is identified by the symbol TBNO, which is loaded as immediate data. The first two bytes of every table identify the displacement to the first free byte of the table; in other words, we assume that every table is partially filled and that the block of data is to be moved into the unoccupied end of the selected table. The required data movement may be illustrated as follows:

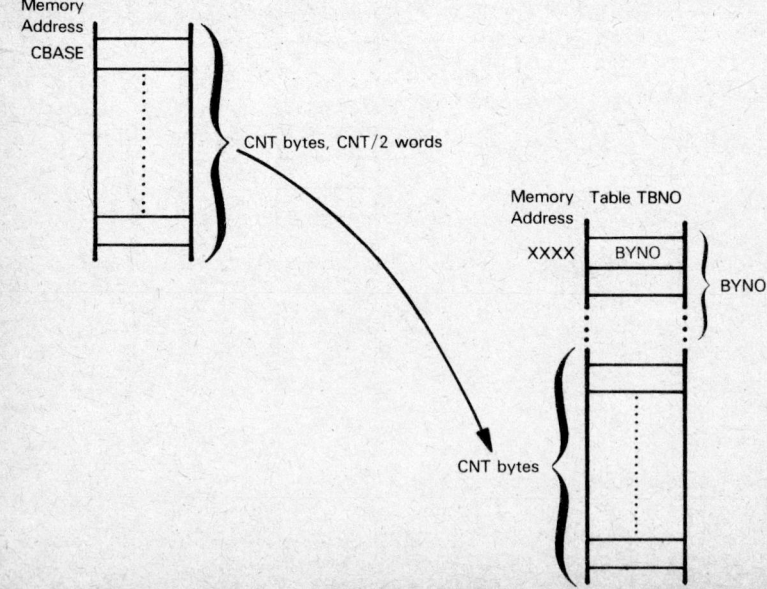

Here is the appropriate instruction sequence:

```
LD    HL,(TABX+TABNO)   ;LOAD TARGET TABLE ADDRESS INTO HL
LD    E,(HL)            ;LOAD DISPLACEMENT (BYNO) TO FIRST
                        ;FREE BYTE INTO DE
INC   HL
LD    D,(HL)
ADD   HL,DE             ;ADD TO HL, GIVING ADDRESS OF FIRST
                        ;FREE BYTE
EX    DE,HL             ;MOVE ADDRESS TO DE
LD    HL,CBASE          ;LOAD INPUT BUFFER BASE ADDRESS
                        ;(CBASE) INTO HL
LD    BC,(CNT)          ;LOAD BYTE COUNTER INTO BC
LDIR                    ;TRANSFER DATA
```

## SORTING DATA

Both of the programming examples we have described thus far simply move a block of data from one location to another. Reorganizing data is also very important; therefore, **we will illustrate a sort routine.**

The sort, as illustrated, takes a sequence of signed binary numbers stored in contiguous memory locations, and reorganizes them in ascending order so that the smallest number comes first and the largest number comes last.

**The sort routine we are going to program uses a bubble-up algorithm.** Consider a sequence of numbers where the label LIST identifies the address of the first number's storage location in memory. These are the necessary sort routine program steps:

1) Start a pass at the beginning of the LIST, and initialize a flag to indicate a "no swap" condition.
2) Compare a consecutive pair of numbers. If the first number is smaller than the second number do nothing; otherwise, exchange the two numbers and set the flag to indicate "swap made".
3) Compare the address of the second number to the end of list address, identified by the label ENDL. If not at the end, increment so that the second number of the current pair becomes the first number of the next pair, and return to step 2.
4) At the end of the list, check the "swap" flag. If any swap was made during the pass, return to step 1 to make another pass.
5) If a pass is made with no swaps, all numbers are in order. Exit.

As an example, consider the case where the numbers 1 through 10 are in reverse order. Nine exchanges will be made during the first pass, at the end of which the largest number will have been "bubbled up" to the top:

|      | START | AFTER 1 PASS |
|------|-------|--------------|
| LIST | 10    | 9            |
|      | 9     | 8            |
|      | 8     | 7            |
|      | 7     | 6            |
|      | 6     | 5            |
|      | 5     | 4            |
|      | 4     | 3            |
|      | 3     | 2            |
|      | 2     | 1            |
| ENDL | 1     | 10           |

Another eight passes will be needed to get all numbers in order, and then a tenth pass is needed to get a "no swap" exit condition.

SORT is implemented as a subroutine; prior to the subroutine call, HL is loaded with the beginning address (LIST) of the data to be sorted, and B is loaded with the length of the list.

```
            LD      HL,LIST
            CALL    SORT
            -
            -
            -
SORT:       LD      (SVAD),HL       ;SAVE LIST ADDRESS
LOOP1:      LD      HL,(SVAD)
            LD      B,ENDL-LIST
            RES     0,D             ;INITIALIZE SWAP INDICATOR
LOOP2:      LD      A,(HL)          ;LOAD 1ST BYTE INTO AC
            INC     HL              ;POINT AT NEXT BYTE
            CP      A,(HL)          ;COMPARE THE TWO BYTES
            JR      NC,SORT1
            LD      E,(HL)          ;NEXT 5 INSTRUCTIONS DO SWAP
            LD      (HL),A
            DEC     HL
            LD      (HL),E
            INC     HL
            SET     0,D             ;SET SWAP FLAG
SORT1       DJNZ    LOOP2           ;REPEAT LOOP IF LIST NOT TRAVERSED
            BIT     0,D             ;CHECK FOR SWAPS
            JR      NZ,LOOP1        ;RETURN IF NO SWAPS
            RET
```

# ARITHMETIC

**Addition, subtraction, multiplication and division will be described under this group.** Transcendental functions are complex enough to require entire textbooks devoted to the subject, so we will not even broach the subject.

Even within the simple bounds of addition, subtraction, multiplication and division, there is a degree of latitude that exceeds the scope of material we can cover. Significantly different algorithms are required, depending upon the magnitude of the number. Binary and decimal arithmetic also require different algorithms. Therefore, **for addition and subtraction we will consider large or small binary or decimal numbers. For multiplication and division we will consider small binary numbers only.**

## BINARY ADDITION
### First consider multibyte binary addition.

Two positive integer numbers, each CNT bytes long, are to be added. The number buffer starting addresses are given by BUF1 and BUF2. The answer is to be stored in a buffer starting at BUF3.

### The multibyte addition may be illustrated as follows:

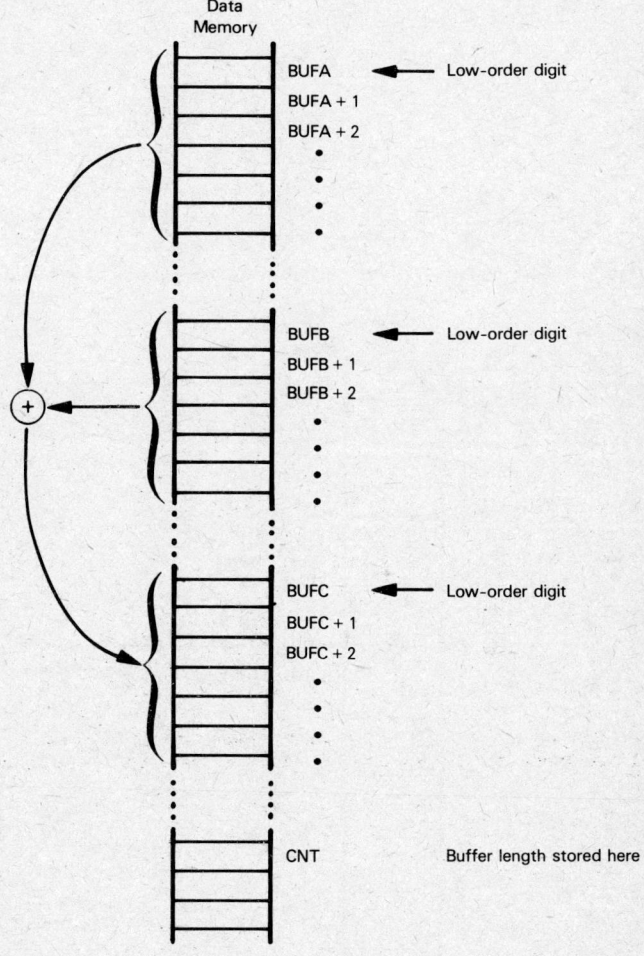

**This instruction sequence performs the illustrated addition:**

```
         LD    A,(CNT)        ;LOAD BUFFER LENGTH AND SAVE IN B
         LD    B,A
         LD    HL,BUFC        ;LOAD ANSWER BUFFER ADDRESS INTO HL
         PUSH  HL             ;SAVE ON THE STACK
         LD    DE,BUFA        ;LOAD FIRST BUFFER ADDRESS INTO DE
         LD    HL,BUFB        ;LOAD SECOND BUFFER ADDRESS INTO HL
         AND   A              ;CLEAR CARRY
LOOP     LD    A,(DE)         ;LOAD NEXT BUFA BYTE
         ADC   (HL)           ;ADD NEXT BUFB BYTE
         EX    (SP),HL        ;SAVE IN NEXT ANSWER BUFFER BYTE
         LD    (HL),A
         INC   HL             ;INCREMENT BUFC ADDRESS
         EX    (SP),HL
         INC   DE             ;INCREMENT BUFA ADDRESS
         INC   HL             ;INCREMENT BUFB ADDRESS
         DJNZ  LOOP           ;DECREMENT COUNTER AND RETURN FOR MORE
                              ;BYTES IF NOT ZERO
```

**Multibyte addition is simpler if you can store the sum in one of the source buffers:**

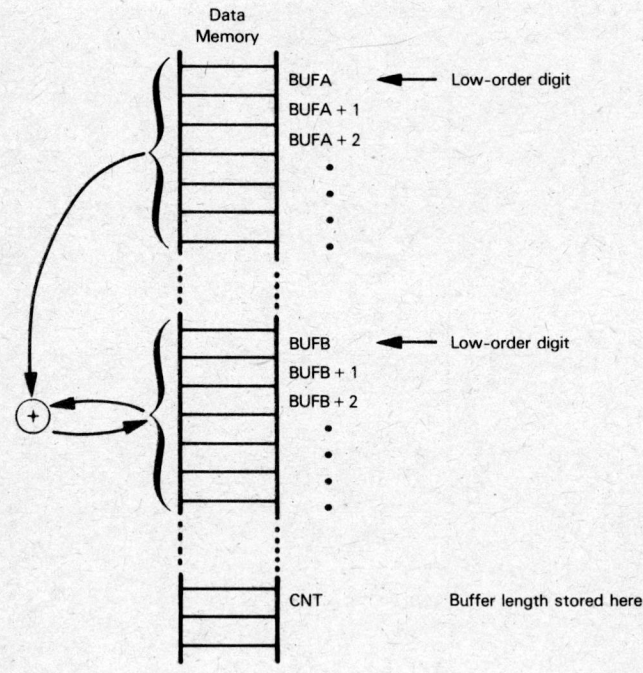

Here is the shorter instruction sequence:

```
         LD    A,(CNT)        ;LOAD BUFFER LENGTH AND SAVE IN B
         LD    B,A
         LD    DE,BUFA        ;LOAD FIRST BUFFER ADDRESS INTO DE
         LD    HL,BUFB        ;LOAD SECOND BUFFER ADDRESS INTO HL
         AND   A              ;CLEAR CARRY
LOOP     LD    A,(DE)         ;LOAD NEXT BUFA BYTE
         ADC   (HL)           ;ADD NEXT BUFB BYTE
         LD    (HL),A         ;STORE ANSWER
         INC   DE             ;INCREMENT BUFA ADDRESS
         INC   HL             ;INCREMENT BUFB ADDRESS
         DJNZ                 ;DECREMENT COUNTER AND RETURN IF NOT ZERO
```

## BINARY SUBTRACTION

**Because the Z80 has special subtraction instructions, binary subtraction is almost identical to binary addition.** In either subroutine, simply replace the ADC instruction with the SBC instruction and accurate binary subtraction will result.

## DECIMAL ADDITION

Decimal addition is also very easy using a Z80 microcomputer. **Simply insert a DAA instruction to follow the ADC** in either of the binary addition programs, and you have decimal addition.

```
         -
         -
         -
LOOP     LD    A,(DE)         ;LOAD NEXT BUFA BYTE
         ADC   (HL)           ;ADD NEXT BUFB BYTE
         DAA                  ;DECIMAL ADJUST RESULT
         LD    (HL),A         ;SAVE ANSWER
         -
         -
         -
```

**One caution, however: the decimal addition routine you create assumes that valid binary-coded decimal data is stored in your source buffers.** If, in error, you have invalid data in either of your source buffers, you will generate a meaningless answer — and not know it.

If your program is one which cannot guarantee that data in source buffers is valid binary-coded decimal, then you must write a routine to check buffer contents and ensure that no high or low 4-bit unit within any byte contains a binary code of A through F.

## DECIMAL SUBTRACTION

Because the Z80 has a special Subtract flag (N), **the Decimal Adjust Accumulator (DAA) instruction can also be used for decimal subtraction. Simply insert a DAA instruction following the SBC instruction, and you have decimal subtraction.** The same caution mentioned for decimal addition applies here: you must ensure that valid binary-coded decimal data is stored in your source buffers.

# MULTIPLICATION AND DIVISION

**Multiplication and division must be approached with an element of caution within microcomputer systems.** These are operations which are unsuited to the organization of a microcomputer; any nontrivial multiplication or division can take so long to execute that it will severely degrade overall performance. **If your microcomputer application is going to make extensive use of multiplication, division, or transcendental functions, you should seriously consider using one of the many calculator/arithmetic chips that are now commercially available.** Transferring complex arithmetic to such a chip can make the difference between a microcomputer system being viable or non-viable in your application.

You can implement simple multiplication and division in microcomputer systems that do not make extensive or time-consuming use of these routines; therefore, we will describe some simple program sequences.

## 8-BIT BINARY MULTIPLICATION

**Consider the multiplication of two unsigned 8-bit data values to generate a 16-bit product.** The simplest way to perform this multiplication is to add the multiplier to 0 the number of times given by the multiplicand. For example, **you can multiply 4 by 3 if you add 4 to 0 three times.**

Suppose that Register B contains the multiplicand and Register E contains the multiplier. The following routine performs the multiplication operation, returning the 16-bit result in Accumulator A (low order) and Register C (high order):

```
         LD    A,0            ;CLEAR A AND C TO
         LD    C,A            ;INITIALIZE ANSWER BUFFER
         CP    B              ;TEST FOR 0 IN B (MULTIPLICAND)
         RET   Z              ;IF 0, ANSWER IS 0 SO END
LOOP     ADD   E              ;ADD MULTIPLIER TO LOW-ORDER ANSWER BYTE
         JR    NC,NEXT        ;IF CARRY IS SET,
         INC   C              ;INCREMENT C (HIGH-ORDER BYTE)
NEXT     DJNZ  LOOP           ;DECREMENT MULTIPLICAND, IF NOT ZERO
                              ;JUMP TO ADD AGAIN
         RET                  ;RETURN WHEN MULTIPLICATION COMPLETE
```

This routine could be a very fast one (if the multiplicand is 0, then only four instructions will execute) or a very slow one (if the multiplicand is 255, then this routine could take up to 1025 instruction executions).

**In general, there is a faster way of executing multiplications. Using common decimal notation, consider the following multiplication:**

```
            1 4 2       Multiplicand
          x 3 1 7       Multiplier
          -------
            9 9 4  ⎫
            1 4 2  ⎬ partial
            4 2 6  ⎭ products
          -------
          4 5 0 1 4     Product
```

**This is the way we learned to do multiplication using a pencil and paper. Each partial product equals the multiplicand being multiplied by one digit of the multiplier.** We began by multiplying the multiplicand (142) by the rightmost digit (7) of the multiplier. Next we multiplied 142 by the second digit (1) of the multiplier. The partial result from this operation is shifted left one position. The leftmost digit of the multiplier was then used to multiply 142, and the partial result was shifted left one more position. After all multiplication operations have been performed, the partial products are then added together to obtain the final product. **This method is well-suited to pencil and paper operations; however, it is not the most efficient method for a computer to perform multiplication. Let us take a look at another method.**

**First, there is no need to wait until all multiplications have been completed before adding the partial products together;** we can generate a "running total" or intermediate result by immediately adding each partial product to the previous partial product. For example:

```
      1 4 2      Multiplicand
      3 1 7      Multiplier
      0 0 0      intermediate result (initial condition)
    + 9 9 4      partial product (7 x 142)
      9 9 4      intermediate result
    + 1 4 2      partial product (1 x 142)
    2 4 1 4      intermediate result
    + 4 2 6      partial product (3 x 142)
    4 5 0 1 4    Product
```

Although this method is more time-consuming when using pencil and paper, it is a much more efficient multiplication method for a computer.

Now, **we also must cause each partial product to be shifted to the left one digit before it is added to the intermediate result.** There are two ways of accomplishing this: we can actually shift the partial product to the left, or we can shift the intermediate result to the right — the effect will be the same. Let us defer our decision on this point for a moment, while we consider one more option.

**Although we have learned to perform multiplication by beginning with the least significant (rightmost) digit of the multiplier, there is nothing to prevent us from starting at the other end so long as we keep track of the significance of the multiplying digit being used.** For example,

```
      1 4 2      Multiplicand
      3 1 7      Multiplier
      0 0 0      intermediate result (initial condition)
     +4 2 6      partial product (3 x 142)
      4 2 6      intermediate result
    +  1 4 2     partial product (1 x 142)
      4 4 0 2    intermediate result
    +   9 9 4    partial product (7 x 142)
    4 5 0 1 4    Product
```

Notice in this example that, when we begin with the most significant digit of the multiplier, subsequent partial products must then be shifted to the right (instead of to the left) before being added. Once again, the shifting of the partial product could also be accomplished by shifting the intermediate result in the opposite direction.

**In summary, we can begin a multiplication operation using either the most significant digit or the least significant digit of the multiplier, and we can shift either partial products or intermediate results to obtain the proper alignment of significant digits.**

Which method should we use? Before deciding, let us look at what happens when multiplying binary numbers. **Since a binary digit is limited to having values of 0 or 1, this means that at the single-digit level multiplication degenerates to addition or no addition.** That is:

```
Multiplicand:              1 0 1 1      1 0 1 1
Multiplier digit:          x   1        x   0
Intermediate result:       0 0 0 0      0 0 0 0
Partial product:          +1 0 1 1     +0 0 0 0   (no add needed)
New intermediate result:   1 0 1 1      0 0 0 0
```

With this fact in mind, let us take another look at the multiplication methods we have discussed. First, we can see that **we no longer need separate steps for the multiplication operation and subsequent addition of the partial product to the intermediate result;** multiplying the multiplicand by 1 is the same as simply adding the multiplicand to the intermediate result.

Next, we see that, **since we are merely performing add operations instead of multiply-and-add operations, we do not need to handle a partial product** — we can simply add the multiplicand directly to the intermediate result. If we eliminate the partial product, then we will want to perform the shift operation on the intermediate result. **Let us now write two sets of multiplication rules for binary numbers.**

Method #1:
a) Shift intermediate result one place to the right.
b) If least significant digit of multiplier is zero, skip step c and go to step d.
c) Add multiplicand to intermediate result.
d) Repeat steps a, b and c for next digit (more significant) of multiplier until all digits have been used.

Method #2:
a) Shift intermediate result one place to the left.
b) If most significant digit of multiplier is zero, skip step c and go to step d.
c) Add multiplicand to intermediate result.
d) Repeat steps a, b and c for next digit (less significant) of the multiplier until all digits have been used.

Now that we have examined the mechanisms used in multiplying binary numbers and developed a few sets of rules, let us see how we can implement these algorithms using the Z80 CPU.

## AN 8-BIT BINARY MULTIPLICATION PROGRAM

**We will now write a program which will multiply two unsigned 8-bit values to generate a 16-bit product.**

Let us first consider register assignments: we need an 8-bit register for the multiplier, an 8-bit register for the multiplicand, a 16-bit register for the product, and a register to use as a bit counter during the multiplication operation.

**We will assign the registers as follows:**

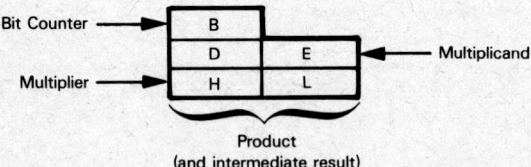

Now, some of the register assignments may seem a little strange, especially placing the multiplier in the H register while also assigning the H register as the most significant byte of the product. However, let us proceed to write our program, and then the reasons for assigning registers as shown above will make more sense.

**Here is our program:**

```
MULT:   LD    B,8        ;LOAD B WITH COUNT
        LD    D,0        ;CLEAR D REGISTER
        LD    L,D        ;CLEAR L REGISTER
LOOP:   ADD   HL,HL      ;SHIFT HL ONE PLACE LEFT
        JR    NC,DECB    ;IF NO CARRY, MULTIPLIER BIT=0, SKIP ADD
        ADD   HL,DE      ;ADD MULTIPLICAND TO INTERMEDIATE RESULT
DECB:   DJNZ  LOOP       ;DECREMENT COUNT IN B. IF NOT ZERO, REPEAT
                         ;LOOP
        RET              ;RETURN
```

**We have used "Method #2" from our preceding discussion for this program.** The program is written as a subroutine, and assumes that, upon entry, the E register will hold the 8-bit multiplicand and the H register will hold the 8-bit multiplier. **If you compare the program to "Method #2" it should seem quite straightforward, with the possible exception of the first ADD instruction.** Why add HL to itself? The non-obvious answer is that **we are** actually **using the ADD instruction to shift the H and L registers one bit to the left.** (Adding a binary number to itself results in the number being shifted one bit position to the left.) Now, it would seem to be more straightforward to simply use a shift instruction instead of the ADD instruction. However, the Z80 instruction set does not provide instructions for performing shift operations on a 16-bit register pair. Therefore, **we would have to use two shift instructions to accomplish the same thing as the single ADD instruction.**

Notice that when we shift the HL register pair to the left we are accomplishing two things. We perform the left shift of the intermediate result as required by our multiplication algorithm, and we also shift the most significant bit of the multiplier out into the Carry flag. The Jump instruction that follows the ADD then tests to see whether the multiplier bit that was shifted out was a 1 or a 0.

As we shift HL to the left we also shift the multiplier out of the way, so that the register pair can be used for the intermediate result. After we have gone through the loop the required eight times, the multiplier will have been shifted completely out of the H register, and HL will now contain the 16-bit product.

## 16-BIT BINARY MULTIPLICATION

**Now consider the multiplication of two 16-bit numbers, yielding a 32-bit result.** The algorithm we will use is the same as that used for the 8-bit multiply; however, a few additional instructions will be required to manipulate registers. **Here are the register assignments:**

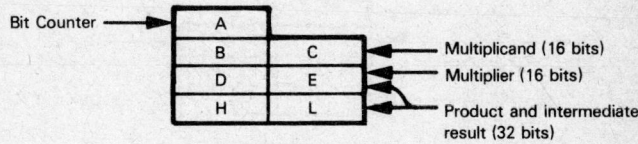

The DE register pair will contain the 16-bit multiplier at the beginning of the program, and will contain the most significant 16 bits of the 32-bit product when the multiplication has been completed.

```
MPY:    LD      HL,0000H        ;INITIALIZE PARTIAL PRODUCT IN
                                ;HL TO ZERO
        LD      A,16            ;INITIALIZE COUNT
LOOP:   ADD     HL,HL           ;SHIFT INTERMEDIATE RESULT
                                ;LEFT INTO CARRY
        EX      DE,HL           ;EXCHANGE DE AND HL
        ADC     HL,HL           ;SHIFT MULTIPLIER LEFT INTO CARRY
        EX      DE,HL           ;RETURN SHIFTED MULTIPLIER TO DE
        JR      NC,DECA         ;JUMP IF NO ADD (MULTIPLIER
                                ;BIT IN CARRY=0)
        ADD     HL,BC           ;ADD MULTIPLICAND IN BC TO
                                ;PARTIAL PRODUCT IN HL
        JR      NC,DECA         ;JUMP IF NO CARRY OUT OF ADDITION
        INC     DE              ;INCREMENT DE TO PROPAGATE
                                ;CARRY FROM ADD
DECA:   DEC     A               ;DECREMENT COUNT
        JP      NZ,LOOP         ;LOOP BACK IF NOT ZERO
        RET                     ;RETURN
```

## BINARY DIVISION

**The procedure used to perform binary division is quite similar to that used for multiplication. Here the process involves subtraction rather than addition.**

**Consider simple 8-bit division.** $B3_{16}$ divided by $15_{16}$ may be illustrated as follows:

```
                        1000        Quotient
    Divisor     10101)10110011     Dividend
                      10101
                        1011       Remainder
```

The result is $8_{16}$ with a remainder of $B_{16}$.

The division algorithm works by shifting the dividend into a register that is initially cleared. Whenever the dividend shift buffer contents equal or exceed the divisor, the divisor is subtracted from the shift buffer contents and a binary 1 digit is inserted into the appropriate quotient bit position.

**Consider the following register assignments:**

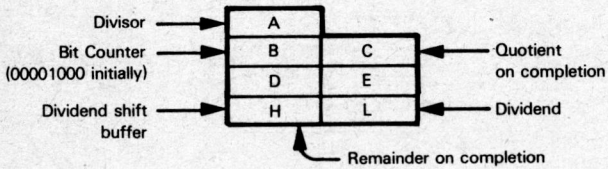

Initially, assume that the divisor is in Register A and the dividend is in Register L. The quotient will be generated in Register C. **Here is the division program which results:**

```
DIV:         LD      BC,0800H    ;LOAD BIT COUNTER AND CLEAR QUOTIENT
                                 ;REGISTER
             LD      H,C         ;CLEAR DIVIDEND SHIFT BUFFER (H)
             LD      E,H         ;LOAD ZERO IN REGISTER E
             LD      D,A         ;COPY DIVISOR INTO REGISTER D
LOOP         ADD     HL,HL       ;SHIFT DIVIDEND LEFT INTO REG H
             LD      A,H         ;COPY DIVIDEND SHIFT BUFFER INTO REG A
             CP      D           ;COMPARE DIVIDEND SHIFT BUFFER TO DIVISOR
             JR      C,NEXT      ;IF DIVIDEND SMALLER THAN DIVISOR DO NOT
                                 ;SUBTRACT
             SBC     HL,DE       ;SUBTRACT DIVISOR FROM DIVIDEND SHIFT
                                 ;BUFFER
NEXT         CCF                 ;COMPLEMENT CARRY FLAG
             RL      C           ;SHIFT 1 OR 0 (FROM CARRY) INTO QUOTIENT
             DJNZ    LOOP        ;DECREMENT COUNTER AND REPEAT LOOP TILL
                                 ;DONE
             RET                 ;RETURN TO CALLING PROGRAM
```

At the end, the quotient is in Register C and the remainder is in Register H.

Notice that we have once again used the ADD instruction to perform a left shift of the 16-bit register pair HL. We have also used the 16-bit subtract instruction (SBC); however, since we initially set the contents of Register E to zero, we are actually using the SBC instruction simply to subtract the contents of Register D from the contents of Register H — an 8-bit subtract operation. We used the 16-bit version of the subtract instruction here to reduce the number of register move instructions that would otherwise be required, since the 8-bit subtract instructions require the use of Register A, which is already in use.

## PROGRAM EXECUTION SEQUENCE LOGIC

### THE JUMP TABLE

**There is really only one program sequence that needs to be described under this heading; it is the jump table.**

**Remember that the Z80 instruction set is rich in conditional instructions;** Jump, Call and Return instructions all have eight conditional variations, which means that special routines are not required when your logic can only go one of two ways.

**When you have three or more options, the jump table becomes an effective programming tool.**

At the heart of a jump table there will be a sequence of 16-bit addresses:

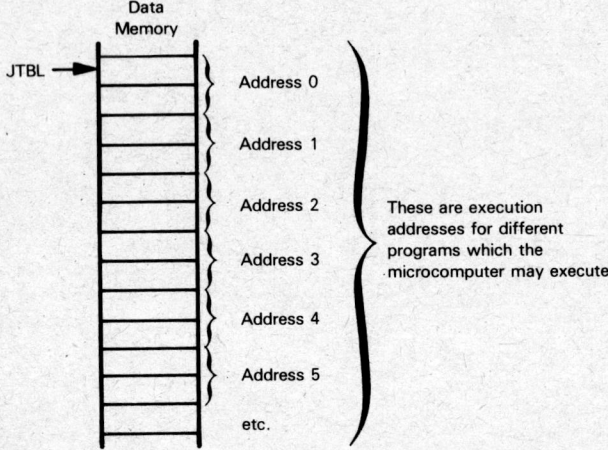

We will presume that these contiguous memory addresses represent the starting addresses for a number of different programs. Assuming that the required program is identified by a program number in the Accumulator, **the following instruction sequence causes execution to branch to the program whose number is stored in the Accumulator:**

```
;JUMP TABLE PROGRAM
        LD      HL,JTBL     ;LOAD JUMP TABLE BASE ADDRESS INTO HL
        ADD     A           ;MULTIPLY PROGRAM # BY TWO AND
        LD      E,A         ;MOVE RESULT TO REGISTER E
        LD      D,0         ;SET REGISTER D TO ZERO
        ADD     HL,DE       ;ADD PROGRAM # TIMES 2 TO JTBL
        LD      E,(HL)      ;LOAD E WITH LOW-ORDER ADDRESS BYTE
        INC     HL          ;INCREMENT THE POINTER IN HL
        LD      D,(HL)      ;LOAD D WITH HIGH-ORDER ADDRESS BYTE
        EX      DE,HL       ;PUT ADDRESS FOR START OF PROGRAM IN HL
        JP      (HL)        ;JUMP TO START OF PROGRAM
```

Randall Library - UNCW
TK7888.4 .Z18    NXWW
Z80 programming for logic design

3049002486632